COMPUTER-AIDED DESIGN,
ENGINEERING, AND MANUFACTURING
SYSTEMS TECHNIQUES AND APPLICATIONS

VOLUME
VII

ARTIFICIAL INTELLIGENCE AND ROBOTICS IN MANUFACTURING

COMPUTER-AIDED DESIGN,
ENGINEERING, AND MANUFACTURING
SYSTEMS TECHNIQUES AND APPLICATIONS

VOLUME
VII

ARTIFICIAL INTELLIGENCE AND ROBOTICS IN MANUFACTURING

EDITOR
CORNELIUS LEONDES

CRC Press
Boca Raton   London   New York   Washington, D.C.

Library of Congress Cataloging-in-Publication Data

Catalog record is available from the Library of Congress.

This book contains information obtained from authentic and highly regarded sources. Reprinted material is quoted with permission, and sources are indicated. A wide variety of references are listed. Reasonable efforts have been made to publish reliable data and information, but the author and the publisher cannot assume responsibility for the validity of all materials or for the consequences of their use.

No claim to original U.S. Government works
International Standard Book Number 0-8493-0999-9
Printed in the United States of America 1 2 3 4 5 6 7 8 9 0
Printed on acid-free paper

Preface

A strong trend today is toward the fullest feasible integration of all elements of manufacturing, including maintenance, reliability, supportability, the competitive environment, and other areas. This trend toward total integration is called concurrent engineering. Because of the central role information processing technology plays in this, the computer has also been identified and treated as a central and most essential issue. These are the issues that are at the core of the contents of this volume.

This set of volumes consists of seven distinctly titled and well-integrated volumes on the broadly significant subject of computer-aided design, engineering, and manufacturing: systems techniqes and applications. It is appropriate to mention that each of the seven volumes can be utilized individually. In any event, the great breadth of the field certainly suggests the requirement for seven distinctly titled and well-integrated volumes for an adequately comprehensive treatment. The seven volume titles are:

1. Systems Techniques and Computational Methods
2. Computer-Integrated Manufacturing
3. Operational Methods in Computer-Aided Design
4. Optimization Methods for Manufacturing
5. The Design of Manufacturing Systems
6. Manufacturing Systems Processes
7. Artificial Intelligence and Robotics in Manufacturing

The contributions to this volume clearly reveal the effectiveness and significance of the techniques available and with further development, the essential role they will play in the future. I hope that practitioners, research workers, students, computer scientists, and others on the international scene will find this set of volumes to be a unique and significant referance source for years to come.

Cornelius T. Leondes
Editor

Editor

Cornelius T. Leondes, B.S., M.S., Ph.D. Emeritus Professor, School of Engineering and Applied Science, University of California, Los Angeles has served as a member or consultant on numerous national technical and scientific advisory boards. He has served as a consultant for numerous Fortune 500 companies and international corporations. He has published over 200 technical journal articles and has edited and/or co-authored over 120 books. Dr. Leondes is a Guggenheim Fellow, Fulbright Research Scholar, and Fellow of IEEE as well as the recipient of the IEEE Baker Prize Award and the Barry Carlton Award of the IEEE.

Contributors

G.M. Acaccia
University of Genova
Genova, Italy

M. Callegari
University degli Studi di Ancona
Ancona, Italy

Rahul De′
Rider University
Lawrenceville, New Jersey

Feng Gao
Hebei University of Technology
Tianjin, China

G.S. Hong
National University of Singapore
Singapore

N.T. Hua
National Taiwan University of
    Science and Technology
Taipei, Taiwan

G.J. Huang
National Taiwan University of
    Science and Technology
Taipei, Taiwan

Sung Hoon Jung
Hansung University
Seoul, Korea

Tag Gon Kim
Korea Advanced Institute of
    Science and Technology
Taejon, Korea

Heungsoon Felix Lee
Southern Illinois University
Edwardsville, Illinois

R.C. Michelini
University of Genova
Genova, Italy

R.M. Molfino
University of Genova
Genova, Italy

Grantham K.H. Pang
The University of Hong Kong
Hong Kong, China

Kyu Ho Park
Korea Advanced Institute of
    Science and Technology
Taejon, Korea

Samuel Pierre
École Polytechnique Montréal
Montreal, Québec, Canada

Wilfried G. Probst
Université du Québec
Montreal, Québec, Canada

Monjy Rabemanantsoa
École Polytechnique Montréal
Montreal, Québec, Canada

M. Rahman
National University of Singapore
Singapore

R.P. Razzoli
University of Genova
Genova, Italy

Bijan Shirinzadeh
Monash University
Clayton, Victoria, Australia

Raymond Tang
Esso Petroleum Canada
Don Mills, Ontario, Canada

Y.S. Tarng
National Taiwan University of
    Science and Technology
Taipei, Taiwan

Y.S. Wong
National University of Singapore
Singapore

Stephen S. Woo
Esso Petroleum Canada
Don Mills, Ontario, Canada

Contents

1

Knowledge-Based System Techniques in the Design, Implementation, and Validation of Resource Scheduling on the Shop Floor of Manufacturing Systems

Rahul De'
Rider University

1.1 Introduction

In this chapter we will consider the design, implementation, and validation issues of building knowledge-based systems for scheduling resources on the shop floor. Such systems are typically defined as decision support tools that support the human operators whose task it is to plan, schedule, execute, and control shop floor operations. The issues of decision support are reflected in the design of these systems, either directly supporting the cognitive goals of the schedulers or indirectly providing task-specific information germane to the decision situation. In either case, there appear to be a common set of data and processing needs that can be delineated as essential requirements of such systems.

Knowledge-based systems are used in diverse scheduling applications that include scheduling space shuttle repair, scheduling space telescope observations, scheduling shipbuilding, assigning personnel for projects, scheduling operating theaters at hospitals, retail logistics scheduling, military operations planning, etc. A large number of applications deal with allocating resources on the shop floor. Much research has focused on this latter problem given the tremendous complexity of the domain and the relative lack of traditional OR techniques that can be effectively applied. This chapters explores the problems of shop floor scheduling.

Software engineers choose to sharply distinguish between the design and implementation of systems. The former implies the logical and physical specifications of data structures and process (or objects) for the system, and the latter the realization of the design specifications in coded modules that are installed at the facility. Most methodologies, proposing normative system building methods, insist on keeping the two activities in distinct phases. The construction of knowledge-based systems, as is evident from the literature, follows an approach of iterating the design-implementation phases, where each iteration involves the construction of another layer of the system. In describing some of the systems in this paper, the design rationale will be examined first, followed by the implementation choices considered.

Validation of a knowledge-based system entails measuring its performance within its working environment. This is usually a difficult task in the domain of scheduling because it is difficult to obtain standards against which to measure the performance. Given that it is virtually impossible to obtain optimal solutions for most real scheduling situations, researchers have to rely on measuring changes in certain process characteristics specific to the particular factory. For example, in steel making, the parameter observed by a team of researchers to measure the performance of the scheduling system was the amount of reduction in the wait time of the molten metal (before pouring into casts). Other approaches to validation include measuring the quality of the schedules against those proposed by a human scheduler or measuring the improvement in the manufacturing processes via simulations. This chapter explores the different approaches.

1.2   Design of Knowledge-Based Scheduling Systems

The design of knowledge-based scheduling systems has been approached from many perspectives. The complex nature of the scheduling problem, and that of the domain, have ensured that researchers have had to continually seek newer and more innovative designs. A core theory for the representation of such designs has emerged and is used as a basis for many systems. However, the fact remains that there is much diversity in the final systems designed. This diversity includes variations in the manner of searching for solutions, the manner in which the system modules are linked up the kind of support provided to the scheduler the manner in which knowledge is extracted for the system, and the adaptability of the system to different tasks.

The knowledge in knowledge-based scheduling systems pertains to the rules, procedures, and heuristics used by the schedulers or workers to enable smooth functioning of the factory. The knowledge is usually obtained or extracted from the scheduler and is then included within the design of the system. It is worth pointing out that knowledge-based systems in manufacturing do not strictly follow the approach of expert systems where the expert's rules are the sole basis of reasoning within the system. To construct such systems, considerable effort is spent in extracting a set of rules that is as near complete as the expert's knowledge of the domain. Once available, it is encoded in a declarative form and used by the reasoning mechanism (the pattern matcher) of the system. The declarative rules extracted from the expert are a part of the total reasoning capabilities of the system. The knowledge base also contains knowledge about the factory, the processes, and other details that may be represented in many ways, not only in the declarative manner.

A number of reasons are provided by researchers as to why just the rules obtained from an expert scheduler are inadequate for constructing effective scheduling systems. First, it is not always possible for schedulers to articulate, and for knowledge engineers to capture, all of the complexities of the work they do (Kerr, 1992; May and Vargas, 1996). Their work involves assessing the state of materials and machines on the shop floor, assigning or reassigning duties to work centers, locating the source of problems in work flow, responding to the directives of higher management regarding manpower or work-in-process levels, not all of which can be captured as rules. Secondly, the domain of the shop floor is complex and consists of a number of interconnected parts for which the scheduler is not directly responsible and consequently does not control. Third, owing to political and cultural reasons, schedulers were not always willing to provide complete information to the system builders (McKay et al., 1995).

A core idea that many scheduling systems use is that of a constraint-based representation. Constraints are the restrictions of process times, process routings, or machine availability that are imposed on the resources of the plant along with the restrictions that originate from management directives, engineering concerns, operator preferences, etc. (Fox and Smith, 1984). The scheduler has to respond to all these constraints. Consequently, the task of scheduling revolves centrally around satisfying the constraints. All the constraints that can exist in a shop floor situation have to be represented within the system in order to complete the scheduling task. A constraint-based representation provides a formal method by which constraints can be represented and manipulated to create schedules.

Constraint-based representation forms the basis of many scheduling systems; however, a number of other important design issues that are relevant are:

- Constructive versus repair methods. In the constructive methods of schedule or plan construction, schedules are constructed by incrementally extending partial schedules. In the repair methods, schedules are completed by beginning with a complete, but possibly flawed, set of assignments and then iteratively modifying or repairing them. Algorithms designed for either of these methods rely on the underlying constraint representation to construct or modify the schedule.

- Predictive versus reactive scheduling. Predictive scheduling implies creating resource assignments for future periods of time, whereas reactive scheduling refers to adjusting the schedules to respond to current situations on the shop floor. Most systems have facilities to respond to both types of scheduling situations.

- Distributed scheduling systems versus "monolithic" scheduling systems. In the former, scheduling solutions are provided by a number of relatively independent modules that cooperate in different ways to arrive at a solution. In the latter, the system is designed as a large, comprehensive system consisting of a number of integrated modules. The level of aggregation in the system determines both its implementation strategy as well as the problem solving process that it uses.

- Cooperative problem solving versus independent machine-generated solutions. Cooperative problem solving systems are designed to work actively with the scheduler in supporting his or her goals and tasks. These are distinguished from systems that generate independent solutions with constraint or data input from the scheduler. In the former case, the solution generation proceeds with active input from the scheduler, but not in the latter.

- System knowledge based on one (or two) schedulers versus that based on input from many supervisors/workers on the shop floor. There are many implemented systems based on knowledge obtained from a senior scheduler in a factory. These systems tend to be smaller and focus on scheduling important (bottleneck) resources. Input from a large number of potential users is collected to build a comprehensive scheduling system that includes all activities on the shop floor.

- Inclusion of established operations research algorithms versus creating special implementations of general search algorithms. Many designers have tried to build their systems algorithms from operations research that can be applied at a very fine level of detail. For the situations in which they can be applied, these algorithms ensure that an optimal or near-optimal solution is obtained. However, in most cases, it is difficult to reduce the problem to the level in which such algorithms will work so designers usually end up modifying search algorithms to suit the problem situation.

- Generalized knowledge-based scheduling systems versus those specialized for a domain. Some designers have chosen to create designs that are general in scope and can provide scheduling solutions for a number of different domains. These systems provide knowledge representation and reasoning mechanisms, as in expert system "shells," which can be specialized for a domain. Special systems are designed for a particular domain, to solve the scheduling problems in that domain alone.
- Learning of scheduling heuristics versus static implementations. Some systems have been designed to learn scheduling heuristics and improve on their scheduling behavior. These systems are able to use their experience of arriving at schedules as a basis on which to build further heuristics. Most systems, however, tend to be static in that they use the problem solving methods already built in.
- Use of simulations for providing support. Some systems rely on intelligently simulating the processes on the shop floor to support the scheduler's decision making. The system enables the scheduler to 'visualize' the effect of the decisions over a given time horizon.

The following sections examine the issues highlighted above in greater detail. Examples of various systems reported in the literature are used to discuss the issues.

Constraint Representation

Constraints as a means of representing scheduling knowledge was first explored by Fox and Smith (1983, 1984). Their goal was to create a system to schedule a job shop. Analysis of the scheduling tasks revealed that "...the crux of the scheduling problem is the determination and satisfaction of a large variety of constraints." They found that the scheduler spent 10–20% of his time scheduling or determining assignments, while the balance of his time was spent communicating with other employees to determine additional 'constraints' that would affect the schedule. Thus, they concluded that the constraint-based representation would be most appropriate for building the scheduling system.

Constraints are built into the basic modeling units (schemas), as meta-information for any slot, slot-value or the schema itself. Schemas are frame-based (Minsky, 1975) knowledge representation units consisting of slots and their values, which may relate different schemas in a hierarchical inheritance network. Schemas are used to model the factory details from physical machine descriptions to organizational structures and relations (Fox and Smith, 1984). Schemas are created and updated for order definitions, lot definitions, work area definitions, plant organization, etc. Within the hierarchical organization of schemas, the lowest level consists of primitive concepts of states, objects, and actions, on top of which domain specific concepts are defined and related. Constraints in this hierarchy of schemas may define the duration of some activity, the due dates for some job, the precondition for some state, etc.

Constraints themselves are represented as schemas with certain properties. One property that is defined for all constraints is the relaxation that is possible for that constraint in a case of conflict with another. Constraint boundaries are relaxed, by a specified mechanism, to resolve conflicts. Relaxations are specified by discrete or continuous choice sets that have utility functions associated with them to assign preference amongst the alternatives. Constraints are also characterized by their relative importance or priority order. This measure is used to determine which of conflicting constraints may be relaxed. Another property is the relevance of the constraint which specifies the conditions under which the constraint should be applied. Interactions is another property that specifies the effects of variations of one constraint's values on others that are related to it, and the direction and extent of these variations. The generation of constraints may be dynamic, created during schedule construction, or they may be created along with the instantiation of schemas.

Constructive Approach

The constraint-based representation was used to build a scheduling system called ISIS. ISIS used the constraints in a sophisticated search procedure to construct schedules. The system conducts a hierarchical, constraint-directed search in the space of all possible schedules (Fox and Smith, 1984). Control is passed from the top to the lower levels and communication between the levels occurs via constraints. At each level, the system does three kinds of processing: pre-analysis (where bounds of the current level's search space are determined); search (where the actual assignment solution is sought); and post-analysis (where the results of the search are assessed). If the post-analysis finds the results acceptable, these are coded as constraints to be passed on to the next lower level. In case the post-analysis rejects the results, the search space is altered at the current or previous level and the control is transferred there. The search itself is conducted at four levels where level one selects an order to be scheduled based on its priority and due date, level two does a capacity analysis to determine availability of machines for this order, level three schedules the resources to satisfy the order, and level four finalizes the schedule by reserving the machines for the order. The control is returned to the top and another order is picked for assignment.

Using the same representation scheme, several researchers have tried different search procedures and architectures with which to solve the scheduling problem. Some of these have resulted in fielded systems while others have helped find faster and more efficient ways of solving the scheduling problem.

Iterative Repair

Iterative repair methods of scheduling are based on modifying a complete but possibly flawed set of assignments that are iteratively adjusted to improve the overall schedule. Zweben et al. (1994) describe a system, called GERRY, that is used to schedule operations for space shuttle flights. The system uses iterative repair with a stochastic approach that enables the construction of very good schedules. GERRY uses a general purpose scheduling algorithm that could be deployed in any situation. Its architecture is domain independent.

GERRY uses two important constraints known as the milestone and temporal constraints. Milestones are fixed metric times beyond which a task cannot be moved, whereas temporal constraints show the start-end times of tasks and their relationship to each other. The system never violates temporal constants. Resource constraints are classified as classes or pools, where a class represents a type and a pool represents the entire collection of a type. Capacity constraints restrict overuse or over allocation of resources.

State constraints depict certain attribute values that have to hold for a task to be accomplished. These are domain specific attributes related to the tasks. A data structure called a *history* is used to track the changes in attribute values, what caused the changes, and for what are they required. As the system runs, this data structure is constantly updated or read to maintain control over the changes to the task schedules. Reading and maintaining the data structure constitutes a major activity for the system.

The system also uses preemptive schedules where each task is associated with a calendar of legal work periods when it can be performed. If a task cannot be completed in one period, then this requires breaking it up into a set of subtasks to be allocated to different periods.

The iterative repair method proceeds with a complete but flawed schedule, which is iteratively modified until its quality is found to be satisfactory or until it reaches a time bound (for computation). The quality is measured by a cost function which is a weighted sum of penalties for constraint violations. The weights and penalties are non-zero values identified independently. In each iteration, constraints are repaired. Each repair causes local and global perturbations in the network which are resolved in later iterations. After each iteration, the system recomputes the evaluation (cost) function to compare the new schedule with the old. If the new schedule is better, it is accepted for the next iteration. If the new schedule is more costly than the old, then it is accepted with a certain probability. This technique is called *simulated annealing* and is used to avoid local minima. If the new schedule is rejected, then the old one is used to continue the iterations.

The results from tests performed with the system showed that in most cases the system converged and a solution was found. Even in the cases where the system timed out before finding a solution, the cost of the solution at the end was much lower than the initial cost.

Predictive and Reactive Scheduling

Predictive scheduling was the norm in planning and scheduling systems when computation was slower and more expensive. With some given input conditions, a plan or schedule was generated and shop floor managers had to rely on this output alone. Godin (1978) expressed the need for interactive scheduling systems: "Scheduling problems change so rapidly that the systems are not flexible or sophisticated enough to keep up with them." Systems worked more in the batch than in the interaction mode of operation.

The situation changed rapidly when interactivity was made central to scheduling systems, known as *reactive scheduling systems*. Design of reactive scheduling systems was centered around recognizing conflicts in the schedules arising from changes in the environment and modifying the schedules to resolve the conflicts (Ow et al., 1987). Conflict resolution in this fashion tended to affect the entire schedule, as the changes introduced in one part would spread to other parts. Research on reactive systems focused on devising algorithms that would handle such "ripple effects" efficiently (Szelke and Kerr, 1994).

One idea that remains constant throughout the work on reactive scheduling systems is that the predictive and reactive parts of scheduling cannot really be viewed separately. They are viewed as complementary functions for generating and revising schedules. An example of an implemented scheduling system, where the predictive and reactive parts are fully integrated, is that of the system for scheduling semiconductor wafer fabrication at Intel Corporation (Kempf, 1994).

The semiconductor manufacturing process at Intel is considered to be a linear flow with loops for rework. The process consists of hundreds of steps that take many weeks to complete. The factory consists of hundreds of processing resources, including machines and tools, and the machines exhibit a wide variety of characteristics, including cycle times from minutes to tens of hours, load sizes varying from one wafer to many, and multiple setup switches. In this situation, scheduling consists of assigning lots of wafers to active resources. The predictive part of scheduling produces a set of assignments of lots to specific resources in future times and the reactive part of scheduling attempts to realize this set of assignments in the light of unforeseen events on the shop floor.

Kempf argues that the idea of the predictive-reactive system was unavoidable because the predictive scheduler by itself would have been useless (given the complexity of the scheduling task and the number of unexpected events that occurred) and a reactive scheduler by itself would only find grossly sub optimal solutions. Thus, the predictive scheduler was designed to produce a 'stake in the ground' schedule that would be used until a change occurred and then the reactor would be used to recompute the schedule, as 'tethered to the stake.' Jobs were ordered according to priority and the higher priority ones were assigned by the predictor as fully as possible, leaving the lower priority ones for later assignment. The reactor focused on the changes made to the scheduled jobs.

Distributed Scheduling Systems

Distributed scheduling system architectures are constituted of a number of autonomous or independent modules (also referred to as agents) that act upon different parts of the scheduling problem to provide a comprehensive solution. A central coordinating mechanism controls the behavior of the modules to a certain extent; however, within the overall problem solving process, the modules act autonomously and opportunistically. These architectures are distinguished from monolithic, mostly hierarchic structures whose modules are integrated and act in a defined, structured manner. A number of possible types of distributed architectures have been proposed, some of which are described below.

Ow and Smith (1987) describe the use of multiple knowledge sources in the OPIS 0 scheduling system. In this first version of the OPIS family of systems, the emphasis was on opportunistic problem solving where a set of knowledge sources seek out solution opportunities in a problem situation. Knowledge sources are collections of heuristics that can bear on a particular aspect of the problem. These sources continuously monitor the solution process and wherever an opportunity arises for any one of them to act, they do so and provide their part of the solution to the global solution. The activities of the knowledge sources are controlled through a "blackboard architecture" (Hayes-Roth, 1985).

In OPIS 0, the scheduling problem was broken into two types of sub-problems: order-based and resource-based. In the order-based approach, a schedule would be created for a particular order and then refined, whereas in the resource-based approach, each resource or machine would be assigned according to availability after which the complete schedule would be built up. To manage the application of the knowledge sources, a hierarchical control structure was devised that consisted of manager knowledge sources. These would be responsible for decomposing the problem into two types of sub-problems: applying knowledge sources to them and collecting the partial schedules thus obtained. Depending on the nature of the problem, an order-based, or a resource-based solution, would be tried first.

Later implementations (Ow et al., 1987; Smith, 1987; Ow and Smith, 1987; Smith, 1994) of the OPIS architecture relaxed the rigid control of OPIS 0 and allowed for greater flexibility of applying the knowledge sources. The OPIS systems were tested for various problems and returned reasonable schedules.

The ReDS architecture (Hadavi et al., 1992; Hadavi, 1994) consists of small, independent modules (agents) based on a generic design that are recursively deployed for different purposes. Each module, called a *planning agent*, consists of a "scheduling gene" that has a predictive and a reactive component. The scheduling gene changes its state over time and beginning as a reactive agent, it evolves its predictive abilities. The agents are cooperative and perform tasks given to them by other agents and other inputs to the agents are feedback about their actions, responses from "higher" agents in the hierarchy, and information about the environment. The asynchronous inputs are used by the agents to decide on a course of action: their outputs. Two forms of feedback are provided to the agents: information about their model of the state of the world and about their reasoning. Within the architecture, agents respond to broadcast messages and provide their own inputs for a specific time horizon. There is no central controlling mechanism.

Each planning agent in ReDS consists of various modules which are as follows: 1. the SEQ module decides on sequencing and dispatching of jobs and deals with the details of lot combining, setups, machine loading, etc. 2. The DS or detailed scheduling module first considers the feasibility of scheduling the jobs before making specific assignments. It makes periodic checks on the progress of the orders, and if any are lagging, it reschedules them (makes inputs to the SEQ module). 3. The FA or feasibility analysis module makes "quick and dirty" analysis of the orders to determine release times. Release times are carefully considered so that when orders are released, they do not spend much time in queues. This is particularly important for semiconductor manufacturing for which ReDS is designed (Hadavi, 1994).

O-Plan2 is also an agent-based, distributed architecture that uses three agents to accomplish the planning, scheduling, and execution tasks (Currie and Tate, 1991; Tate et al., 1994). The first agent provides an interface to the user for accepting requests for scheduling tasks. The second agent is a planner that generates a plan to perform the specified task. The third agent is an execution system that monitors the execution of the planned activities. The planner responds to problems or failures in the execution of the plan, as reported by the execution system, by either replacing activities or re-planning from the start.

The main components of each agent O-Plan2 are: domain information describing the application and the tasks in the domain to the agent; plan state—the emerging plan of activities; knowledge sources—or plan modification operators that are the processing capabilities of the agent; support modules that assist the agent in its functioning; and, controller — the module that decides the order in which decisions are made. The planning proceeds in a least commitment manner with successive refinement or repair of the plan or schedule.

Burke and Prosser (1994) make a strong case for distributed scheduling arguing that a distributed problem solving architecture most closely resembles the structure of the enterprise to be scheduled. Most enterprises are distributed in nature, whether physically in terms of the location of productive resources or logically reflecting the organization structure. They propose a distributed asynchronous scheduler (DAS) that decomposes the scheduling problem and distributes it across a hierarchy of intelligent autonomous agents. Each agent has a defined role and communication path depending on its position in the hierarchy. Three types of agents are defined: the strategic agents are responsible for assigning tasks to lower agents and resolving conflicts; the tactical agents are responsible for delegation of tasks to individual resources; and the operational agents are responsible for the execution of tasks on individual resources. The agents are loosely coupled, in terms of communication between each other, but since they operate in a tightly coupled environment where a small change in a situation at one resource is rippled to other areas, they work in a reactive mode.

The system functions in a hierarchical manner by the strategic agents accepting inputs for new orders and assigning them to tactical agents and on to operational agents. The latter then try to include the tasks in their local schedule. On failure, this is communicated upward to the tactical agents that tries to rearrange the load balance. Upon failure of this, it goes up to the strategic agents to re-assign the resources. Agents keep track of their backtracking and use this knowledge to further guide the solution process.

Agent-based scheduling is an active area of research in scheduling systems. Sikora and Shaw (1997) describe an agent-based system for coordinating scheduling where each agent, though acting autonomously, depends on others to solve parts of the problem it cannot handle. Agents coordinate their activities by communicating their partial results to each other by using a "tradeoff function" to arrive at a common objective function.

Integration of design, process planning and scheduling within the same agent-based system is also an active area of investigation. Gu et al. (1997) describe a system that uses a bidding-based approach to coordinate the functions of process planning and scheduling. Each product is an agent that carries its production and due-date information and upon arriving at the manufacturing facility, its requirements are broadcast to other agents bound to resources. The agents bid for different tasks and if there is a conflict, a negotiation process is initiated. A resolution defines the process routing and resource assignment for the product. In Interrante and Rochowiak, 1994, a multi-agent system is used for assisting in dynamic scheduling and rescheduling where the focus is on concurrent engineering. Each agent represents a sub-system of the factory and mechanisms are designed for collaboration between them. The AARIA project (Parunak et al., 1997; Baker et al., 1997) uses agents to manage the entire process from ascertaining customer demand to final production. In an implemented prototype, customers can directly state their demands to the system which responds by providing a set of cost and due date schedules. Customers may choose the best cost alternative after which agents, through negotiation, create a production schedule for the product.

Cooperative Problem Solving Systems

The MacMerl system was designed to "understand and support scheduling from the perspective of the human scheduler" as opposed to implementing the expert's methods which is the approach taken by traditional expert systems (Prietula et al., 1991; Hsu et al., 1993; Prietula et al., 1994). The rationale for MacMerl's design was based on the fact that problem solving, by any intelligent agent, can be characterized as search conducted in a problem space of alternatives. The space is a representation of aspects of the task. The interactive scheduler was seen as one permitting the human and machine to be operating in coincident problem spaces. The idea was to find those decisions in which the human could be supported by the machine. To achieve this, a "backbone" system was designed based on the human scheduler's methods and supporting the achievement of goals with strong computational support. The design also had a cooperative approach for generating and reviewing schedules because the entire set of parameters to generate acceptable solutions could not be specified.

MacMerl was designed and implemented to solve the scheduling problems at a particular factory—one making windshields for automobiles. The basic production activities here were those of cutting the glass to the right size and then bending or shaping it. After this, further processes were done to attach the glass to different parts. Bending was the most critical activity where the glass had to be heated in an oven, called a lehr, and then pressed or simply shaped by gravity. Scheduling jobs for the lehr was the bottleneck activity at the plant. The scheduling task had hard constraints defining the quantity and type of glass to bend and the time in the lehr. It also had softer constraints, called preferences, that were used for determining preferences over schedules.

Given MacMerl's design philosophy of having mechanistic support states that assist task-specific human expertise, complemented by the flexible and judgmental scheduling knowledge of the human, the system was realized in three steps: first, identifying the expert's view of the scheduling problem, the expert's scheduling behavior, and the reasoning behind the expert's actions; second, defining a fundamental data structure to implement the scheduling knowledge and defining the processes (preprocessing and schedule generation); and third, interactions with the expert to review and revise the operators. After the third step, a set of operators were available that corresponded to specific goals of the scheduler.

Numao (1994) proposes an architecture where the user, procedures, and rules are used cooperatively to solve the scheduling problem. The objective of the architecture is to collaborate with the user in finding a solution. The problem is studied in a steel producing OIC plant.

The steel-making process consists of three major steps: first is that the converter refines the pig iron into steel of the desired composition by blowing oxygen through the hot molten metal; second is that the ladle-refining adds alloy ingredients or removes impurities; and third is that a continuous caster casts the molten steel into slabs, blooms or billets. The objective of scheduling is to determine the sequence of operations from the converter to the caster for a charge. This requires determining the number of charges per day (based on the demand and due date information), the waiting time limitation given that the molten steel cannot wait for certain processes for a specified amount of time, and the order of the refinement processes for different qualities of steel. The scheduling tries to minimize the waiting time and maximize the number of charges per day.

Traditional combinatorial optimization techniques are shown to be NP-complete for this kind of problem so the authors explore cooperative scheduling. The emphasis here is on decision making and decision support rather than on constraint-satisfaction. The authors rely on the fact that it is easier for an expert to suggest what is wrong with a schedule and to suggest improvements than to "extract" his knowledge into a complete set of rules. Thus, the system criteria are that: it should produce a feasible solution, it should be interactive, it should provide on-line re-scheduling, and it should be modular and compatible with existing systems. The system essentially provides a feasible solution to the user who improves it interactively.

The architecture consists of three major components: a scheduling engine, a rule-base and an interface. The scheduling engine works as a general constraint satisfier to solve general primitive constraints. The rule-base then solves the domain dependent constraints. The user refines the solution via the interface. This process is iterated until a feasible solution is available.

The scheduling process has two steps: sub-scheduling and merging followed by interactive refinement. In the first step, a schedule is generated. Human experts determine the starting of each charge set, keeping in mind the final processes have to be completed together. They disregard machine conflicts. Using the constraints of waiting time limitations and continuous casting, the system resolves the conflicts of machine scheduling. Scheduling tasks are broken into subtasks and these are backward scheduled from the casting process. Overlaps are removed, loads are balanced, and the subschedules are merged. In the second step, the user interactively modifies the schedule. The user may create conflicts in the process which are then removed by the system.

The system described by Esquirol and others (Esquirol et al., 1997) uses a constraint-based approach to provide a set of solutions for scheduling problems, combining these with a cooperative human-machine problem solving framework. The domain is that of a flanged-element manufacturing workshop that includes: cutting-out metal sheet pieces, heat treatments; flanging on a hydraulic press, and manual finishing. Routings can differ for each job, or part cut out of the sheet, so the shop operates as a job shop.

Using a cognitive modeling approach, the authors wanted to integrate the humans in the decision process rather than simulate them. Cooperation in this case implies the sharing of goals by cooperative entities which provide complementary knowledge and skills. The system, called *Scoop*, enabled users to entirely construct the scheduling solution while it checked the consistency of the constraints used.

Constraint-based analysis is used to depict processing constraints and temporal constraints. This analysis shows infeasibilities to the users in the current set of decisions and a priority order of possible solutions is presented to them in different modes.

The architecture consists of two major modules: the Selection aid and the Placement aid. The Selection aid creates the subset of orders that have to be prioritized and presented to the user, while the Placement aid indicates where certain orders can be placed successfully. The users also see a graphical representation of the sheet that is being cut and flanged.

Knowledge Elicitation for Scheduling Systems

The manner in which knowledge was extracted from expert schedulers to implement in the different systems varied according to the design envisaged for the system. In some cases, it was sufficient to extract knowledge about the domain without finding out how the scheduler actually did the scheduling, while in other cases the latter was important also. For constructing the ISIS system, Fox (1994) interviewed experts to identify the scheduling knowledge required for building the system. They found that the most interesting and important knowledge related to constraints that bind the scheduling the problem. Fox identified five broad categories of constraints which included; organizational, physical, causal, availability, and preference constraints. Organizational constraints included: due date constraints restricting the lateness of orders, work-in-process limits, and physical constraints specifying characteristics of resources that limit functionality (such as the length of the milling machine's workbed). Causal constraints specified the preconditions for use of resources such as the precedence requirements (job has to be cleaned before it can be annealed) or resource requirements (requirements of tools, or trained operators for a job), while availability of resources constraint restricted the simultaneous use of any resource by multiple jobs. Preference constraints were "soft" constraints that indicated priorities or preferences for certain resources.

For extracting knowledge from the expert to construct MacMerl, Prietula and others went through several phases of analysis and study of the expert's problem solving behavior. Their objective was to not only understand the characteristics of the factory but also to have a trace of the expert's cognitive activities. To begin, they used protocol analysis as the expert developed schedules, direct questioning, post-task analysis, and extended apprenticeship to understand the problem solving methods, representation, and causal reasoning used by the scheduler. Detailed interviews were used to understand the scheduler's goals and heuristics. Combinations of parts, determined from the part characteristics that the scheduler used, were recorded to limit the search space. They also identified a general algorithm that the scheduler followed based on specific goals. These goals attempted to reduce setups, minimize stock-outs, and prioritized high-volume sales parts. A causal model of the expressed constraints and preferences was identified by continuously asking the expert to explain and justify all decisions.

A similar technique was used by May and Vargas (1996) and De' (to appear) to identify characteristics of the factory and also to elicit the process by which the scheduler constructed a mental simulation of the factory. This process was studied by extensive *in situ* protocol analysis, wherein the researchers observed and recorded the expert as he went about his daily tasks and talked about what he was doing. After spending several months acting as trainees, they were in a position to codify the problem solving behavior of the expert. May and Vargas also noted that the eventual model of behavior that emerged from their apprenticeship was different from the set of scheduling rules that the expert had mentioned to them at the beginning of the exercise (in the presence of management).

In Esquirol et al. (1997), the knowledge acquisition was done by using a modification of the personal construct theory where users were asked to identify their most important constraints on a grid. All current and potential users of the system were asked to participate and give their most important concerns. The analysis of the grid led to a design of the cooperative strategy including flexibility, support, and interactivity.

OR Algorithms in Systems

Shah et al. (1992) outline a system that provides meta-knowledge about the kinds of scheduling algorithms to use. They argue that researchers and practitioners waste effort searching the literature to seek out algorithms that will fit their particular scheduling situation. So they present a knowledge-based system, where the expert is a scheduling researcher. The expert takes input about the type of setup configuration of a particular shop floor situation, the type of constraints binding on the situation (such as due dates, delay costs, etc.), and the type of objective function to be used (such as minimize tardiness, or average completion time, etc.). The expert then gives, as output, a sequence of OR algorithms by which the problem can be solved. The system does not solve the problem itself.

Kempf (1994) and others designed their system to use published OR algorithms as part of their system. This was to be in the predictive component of their system which was designed to have an "importance-ordering" mechanism at a higher level to order tasks according to priority, not according to the time-order of tasks where priority is not regarded. Importance-ordering would enable a global view of the situation and not rely entirely on the local details. At a lower level, this component used a "multi-algorithm" approach where a number of algorithms from the published literature were used for given situations on the factory floor. (The reactive component was tied to the predictive part in that it made decisions that had been specified by the predictive component. In situations where the system could not react, the predictive component was restarted.)

Generalized Scheduling Systems

The constraint-based scheduling system designs discussed above are all attempts to realize a general representational framework within which any type of scheduling problem can be tackled. There are, however, attempts to take this idea one step further by designing entire architectures that can, with little modification, be adapted for various scheduling applications.

The ARPA-Rome knowledge-based planning and scheduling initiative (Fowler et al., 1995) was created based on the observation that although constraint-based frameworks have been successful in solving scheduling problems, they do not scale to larger problems very well. There are few that fully integrate the planning, scheduling and database components within a single architecture. At the system's (ARPI) core is a distributed network of graphics-based planning cells sharing a common reasoning infrastructure that enables continuous concurrent planning. Human users collaborate with the automated decision aids to rapidly generate large scale plans and schedules, with full use of all the background and relevant data. This initiative led to the creation of the common prototype environment (Burstein et al., 1995) consisting of a repository of software tools and a testbed for evaluation of planning and scheduling systems.

Another generic framework for building practical scheduling systems is described by Sauer and Bruns (1997). The system would enhance the problem solving capabilities of human domain experts. The generic framework is based on two design principles–the combination of standard components (such as user interfaces), databases with knowledge-based concepts, (such as heuristic scheduling algorithms), and declarative knowledge representation and the explicit representation of scheduling knowledge for flexible reuse and adaptation.

The first principle provides support for predictive, reactive, and interactive scheduling. The user interface provides a graphical depiction of the scheduling situation, allowing the user to change priorities, resources, etc., while simultaneously checking for consistency. The knowledge-base contains the production management knowledge of the application domain—knowledge about products, resources, and solution schedules—in a relational form. It also contains rules that represent hard or soft constraints.

Algorithms for scheduling are used to create predictive schedules from given input conditions. Problem specific knowledge is used to guide the search, including heuristics from an expert scheduler. Reactive scheduling algorithms enable appropriate reactions to unexpected events by adjusting the schedules. The reactions can range from simple manipulations to complete rescheduling.

The second principle enables the reusability of algorithms that have been designed for highly specific scheduling situations. The framework separates predictive and reactive scheduling algorithms in underlying scheduling strategies (order-based or resource-based) and the selection rules to be used in these strategies. Predictive and reactive "skeletons" are used to separate the two approaches, leading to separate deployment and adaptability. The rules and strategies were determined from experts or from existing systems.

The framework was tested by building schedulers for three different applications–two in manufacturing and one in medicine. All three knowledge-based systems greatly reduced the time to create and maintain production schedules. (All were implemented in Prolog on Sun SparcStations.)

Learning in Scheduling Systems

The learning of scheduling heuristics has concerned researchers for many reasons (Aytug et al. 1994). It enables the learning of scheduling heuristics automatically, overcoming the problem of knowledge acquisition from experts. Learning methods also enable the systems to learn about the environment and improve on their performance through experience.

Piramuthu and others (Piramuthu et al., 1994; Piramuthu et al., 1993) describe a system that learns to apply scheduling heuristics in response to certain patterns in the environment. The system consists of two major components–a simulation module that generates training examples of various shop floor conditions, and an inductive learning module that learns the best scheduling rule (such as SPT or EDD) to apply to the given conditions. Over time, the system learns more decision rules. The system has a bi-level model, where the first level deals with part-release and the second level deals with dispatching at individual machines.

The system is designed to simulate a surface mount technology process consisting of various stages through which fourteen different part types are processed. The objective of the inductive learner is to learn rules of the form: "If $(b_{i_1} \geq a_{i_1} \geq c_{i_1})$ And $\cdots(b_{im} \geq a_{im} \geq c_{im})$ Then τ," where a_{ij} represents the ith level of the jth attribute and b_{ij}, c_{ij} define the range for a_{ij}. τ denotes the class (Piramuthu et al., 1994). The a_{ij}'s could represent parameters such as buffer content, machine utilization, or coefficient of variation of processing times. τ would be a part-release or dispatching rule.

A training example generator creates examples of shop floor patterns for the learning module of the system. The learning module passes the examples through a feature extraction preprocess and then through an inductive learning algorithm, which is a refinement of the ID3 algorithm (Quinlan, 1986). The decision rules obtained are used by a pattern-directed scheduling module. This module uses a smoothing function to remove the "chattering" that results from too many changes in the scheduling heuristic with changes in the input patterns. The authors use two options for smoothing–one has a constant threshold value and another that weighs current patterns more heavily. A critic module is used to evaluate the performance of the scheduler, in cases where performance tends to deteriorate (due to over-generalization, for instance). Results of running multiple simulations for many sets of patterns over the system showed its performance to be better by using the pattern-directed learning approach rather than by using the part-release or dispatch heuristics alone.

Simulations are also used to create training examples for a multi-dimensional classification algorithm in Chaturvedi and Nazareth's system (1994). The authors state the theory underlying conditional classification as an extension of simple classification where, in the former, the classification is performed once along one dimension and the output from this is used as priors in the second classification. This method is able to include more complexity of the shop floor scheduling phenomena than the simple classification.

In their experiments, the authors generate decision rules for a FMS shop floor using two output classification dimensions of production rate and utilization for machines. The shop floor is simulated and training and testing examples are obtained. The decision rules are able to accurately classify all the test examples. The training and test sets used are small, and the authors caution against over generalizing the results of their research.

Chen and Yih (1996) trained a neural net to identify the attributes required for developing a knowledge base. Their neural net could identify a set of relevant attributes, for a given problem situation, from a general pool of attributes. Their test case showed that the system could successfully select important attributes from a pool.

The CABINS system (Miyashita and Sycara, 1994; Miyashita et al., 1996) implements a methodology for learning a control level model for selection of heuristic repair actions based on experience. The scheduling of resources is based on constraint-directed scheduling methods. Case-based reasoning is used for the acquisition and flexible reuse of scheduling preferences and selection of repair actions. The case-base includes examples that collectively capture performance trade-offs under diverse repair situations. The case description captures the dependencies among the scheduling features, the repair context, and a suitable repair action. The dependencies in the case-base are used to dynamically adjust the search procedure. Users provide evaluations of the outcome of using cases to select search heuristics and these are later reused to select heuristics in similar situations. User preferences are reflected in the case-base in two ways: as preferences for selecting a repair tactic depending on the features of the repair context and as evaluation preferences for the repair outcome.

Experimental results with the system showed that the approach outperformed dispatch heuristics for similar problems and constraint-based scheduling using only static search procedures. It was useful in capturing user preferences not present in the scheduling model.

Zweben et al. (1992) describe a system that learns search control knowledge for a constraint-based scheduling system. The authors maintain that the efficiency of scheduling systems, based on constraints, is affected by resource contention. Their system learns the conditions under which chronic resource contention occurs and modifies its search control to avoid repeating those conditions. They modify the existing method of explanation-based learning to learn from multiple plausible explanations. Zhang and Dietterich (1995) use a neural net to learn a heuristic evaluation function (to control search), for the same problem of scheduling studied by Zweben. Their system uses the evaluation function for a one-step look-a head procedure. The results indicate that the performance of the system improves on the performance of the iterative repair method used by Zweben.

Simulation-Based Scheduling Support

Simulations allow users to see the effects of their decisions on the conditions on the shop floor. Systems that use heuristics to guide the simulation allow the users' expert heuristics to be encoded within simulation. The system by Jain et al. (1990) used deterministic, backward simulation to provide real-time support to shop floor personnel. Backward simulation enabled the system to simulate backward, from a given event (say the completion of a job), to the starting point. Heuristics obtained from an experienced scheduler were used to make decisions at crucial choice points which included part dispatching, machine selection, interval selection, and secondary resource constraints.

The SIMPSON system (May and Vargas 1996), also simulated the factory floor based on heuristics obtained from an experienced scheduler. The focus was on bottleneck resources where heuristics were used to 'feed the bottleneck' that would also ensure the desired utilization of other resources. In the same domain, the PLANOPTICON system, (De' 1996), also used a knowledge-based simulation approach to suggest setup changes for bottleneck machines. In both cases, the user could input hypothetical data to arrive at what-if scenarios.

1.3 Implementation Issues

In this section, we consider the issues related to the implementation and fielding of shop floor scheduling systems. Although the specific factory and organization determines in a large part the eventual implementation, there are a number of issues that are common to the implementation process. These issues include the access to transaction databases, the implementation of user interfaces, the use of scheduling horizons, and cultural issues.

The implementation issues are studied in reported instances of fielded systems. There are a fairly large number of fielded scheduling systems that have been reported in the literature. For this study, we focus on a few of these that have been implemented specifically in manufacturing settings and have been used for a significant portion of time for regular production scheduling. The systems studied are: LMS (Fordyce and Sullivan, 1994), ReDS (Hadavi, 1994; Hadavi et al., 1992), the systems at Intel (Kempf, 1994), the systems developed at Texas Instruments (Fargher and Smith, 1994), the DAS system (Lee et al., 1995), the MicroBOSS scheduler (Fowler et al., 1995), the high-grade steel-making scheduler (Dorn and Shams, 1996; Dorn and Slany, 1994), the cooperative scheduler for steel-making (Numao, 1994), the MacMerl scheduler (Prietula et al., 1994; Prietula et al., 1991; Hsu et al., 1993), the SIMPSON scheduler (May and Vargas, 1996), and the system at General Motors (Jain et al., 1990). Two other fielded systems that were not used to schedule the entire factory, but only specific resources, are those by Burke and Prosser (1994 and De', 1996).

Transaction Databases

Transaction databases are crucial for the implementation of any realistic scheduling system. These databases log the transactions occurring on the shop floor related to starting of processing of jobs, status of machines, completion of jobs, operator assignment, setup changes, etc. Scheduling systems rely on these to obtain current status reports from the shop floor.

In all the reported instances of scheduling system implementations, the developers have first had to create or deploy interface software to retrieve the transaction data from the management information systems. Depending on the currency of the data, or the frequency at which shop floor status reports were available, the scheduling system could update its records and respond to the situation. All the developers note that culturally this data access was the first and possibly the most important hurdle to overcome to have a viable implemented system.

The common issues related to transaction databases are:

- Building interfaces with the MIS databases to obtain the relevant data. This involves building a working relationship with the MIS department to identify the nature of the data, its manner of capture, and its representation.

- Ensuring that the data is entered correctly and in a timely manner. In most cases, this requires re-setting the data entry patterns and re-training employees to do this.

- Collecting distributed data stored in diverse forms (i.e., on different types of application software), and translating it into a form understandable to the scheduling system.

- Building additional data collection mechanisms if the existing system is not collecting data on all the steps in the manufacturing process.

User Interfaces

Constructing attractive and user-friendly interfaces was a deliberate implementation tactic employed by some developers. The motive was to get the user, in some cases, the expert involved in the development process as a provider of scheduling heuristics and system requirements. Some important issues with regard to user interfaces are:

- The interfaces have to show timely and useful information to the users. To build the LMS system, Fordyce and Sullivan built interfaces that showed the shop floor in a graphical manner and also had signals to indicate problem situations or warnings. These interfaces helped to build support for their scheduling system.

- The interfaces should be as close as possible to the current forms or screens used by the users. May and Vargas designed input screens and output reports for the user of SIMPSON that were identical to what he was used to on paper and preferred. For the MacMerl system, the developers made the interface used by the scheduler as user-friendly as possible.

- The schedule items depicted on the screen should be easily manipulable (such as being mouse-sensitive) or modifiable by the user. Numao used such an approach to allow the user to make on-fly-changes to input conditions and see the resulting effects.

Scheduling Horizons

Implementations of scheduling systems, particularly with hierarchical architectures, necessitate the explicit use of horizons. The horizons are time boundaries within which decisions are located. They are usually pre-determined and the system is designed to limit the schedules assignments up till those points.

In the LMS system, the lowest level of the hierarchy, called the dispatch scheduling tier, has a horizon of a few hours to a few weeks. Here are where the pertainings to decisions running test lots, prioritizing late lots, determining maintenance downtime, determining downstream needs, etc., are made. A short horizon was needed for most decisions at this level because of the dynamic nature of the shop floor where the validity of decisions degenerated quickly.

Kempf distinguishes the predictive and reactive modules of the system (Kempf 1994) on the basis of the scheduling horizons they address. The predictive part would work on a horizon of a few shifts to allocate capacity between production, maintenance, etc., whereas the reactive part would work on a horizon of a few minutes to respond to occurrences on the shop floor.

In the ReDS system (Hadavi), the authors had to build a system corresponding to a particular scheduling horizon, primarily because it was so desired by the management. This was a predictive system which would refresh the schedule after every time horizon with new data from the shop floor. The horizon was set by the frequency by which the new data was available.

For setting explicit scheduling horizons, De' (1993) argues that horizon parameters have to be established whose values are determined from shop floor parameters. Long or short horizon values are determined from parameters such as the number of products. For different and persisting shop-floor conditions, different horizon lengths may be required.

Cultural Issues

Cultural and political issues play as much of an important role in the deployment and eventual success of a scheduling system as do the technological issues. These issues arise from the community of people who are concerned with the manufacturing process and whose cooperation and support is needed for any system implementation. The incidents below show the cultural reasons why systems may be supported or opposed, regardless of the merit of the systems themselves.

For the LMS project, Fordyce and Sullivan (Fordyce and Sullivan, 1994) recalled two incidents that facilitated the system's implementation. The first was a call by a supervisor to be provided a real-time monitoring system that would warn of any impending deviations from the plan and the second was a problem observed in the simulation systems. A review of the manufacturing systems revealed several shortcomings, chief among them was insufficient integration between real-time data, knowledge bases, and tactical guidelines. To build support for their eventual scheduling systems, the authors first built a real-time monitoring system that would enable shop supervisors and engineers to observe events on the shop floor through desktop terminals. Users were weaned on the systems, after which they demanded more features such as alerts and rule bases, that further propelled the development of the system.

In the second phase, the system design shifted from a reactive mode to a more proactive mode. This was achieved in a time of booming business, when greater throughput and planning was required from the existing systems. Using a mix of knowledge engineering and operational analysis, the authors were able to implement a set of dispatch heuristics that improved throughput and profitability. Following this success, the value of knowledge-based systems was firmly established and, along with a renewed interest by IBM in participating in the manufacturing marketplace, further development and implementation of such systems was assured.

At Intel, (Kempf, 1994), the scheduling research group had to market their ideas to two different sets of people–the managers and the users on the shop floor. The first group had to be persuaded to support the project, given that the further any senior executive was from the shop floor, the less concerned he or she was with scheduling problems. This group was persuaded through promises and assurances of greater responsiveness, better resource utilization, and integration with automated tools. The second group was cognizant of the scope of the problems but remained skeptical of any systemic solution for various reasons–the dynamic and changing nature of the shop floor, the complexity of the problem, the adherence to manual procedures, and the history of failed attempts to solve the problems. This latter group was persuaded to support the development effort by involving them with the development process. Resistance to change remained, however, because changing work processes for even one person meant changing the grading, reporting, and training of possibly an entire group working with the individual.

The first modules of the system implemented on the shop floor were the user interface and the data interface. Users were asked to comment on the design of the interface, and in a situation where personnel had a very wide degree of familiarity with graphical interfaces, the suggestions were diverse and many. The system builders could not cope with them. The database interface suffered from inconsistencies and inaccuracies in data being logged in on the shop floor. The scheduling system could not proceed without an accurate snapshot of the shop floor; thus, the developers had to persuade those in charge of collecting data to improve their system and practices.

In the ReDS system (Hadavi, 1994), the developers first implemented a research prototype that they used to show management the possibilities of such a system and gain support for it. Users were aware of such a prototype but when they saw it, they were disappointed with its limited capabilities. The developers tried to deploy the prototype itself, which led to further annoyance and friction. The biggest drawback was that there were no means of collecting shop floor data for input to the system. The developers had to start another implementation from scratch. This time, they collaborated with the MIS personnel to obtain useful and timely data from the shop floor system. Eventually there were no objections to the introduction of the system; however, some users expressed doubts as to its usefulness over the older simulation-based system.

The eventual introduction of the system brought many changes in the practices on the shop floor. Management now had a means by which to directly track the progress of jobs on the shop floor. Any delays would be immediately visible and liable to questioning. This introduced a form of *panopticism* (Zuboff, 1988) that forced shop floor personnel to be alert at all times.

1.4   Validation Issues

Testing implemented systems involves the twin tasks of verification and validation. Simply stated "verification is building the system right, validation is building the right system" (O'Keefe et al., 1987). Validation evaluates the system in the context for which it is designed. Knowledge-based systems are often validated by seeking acceptable performance levels for specified criteria (O'Keefe and Preece, 1996).

Validation of implemented scheduling systems is very difficult and few developers do so in any detailed manner. The reasons for this difficulty are the lack of standards by which to measure the performance of the system (Gary et al., 1995). Real-life shop floors are so complex that generating a theoretical model and solving one to arrive at an optimal standard is impossible. Scheduling researchers try to validate the system by various means. Measuring its performance on specific goal parameters is the one most widely used. There are, however, many ways by which knowledge-based systems in general can be validated (O'Keefe and O'Leary, 1993). Internal functioning of the systems are verified by unit and system testing.

Validation Against System Objectives or Goals

In most cases, support for the development and implementation of large-scale and expensive factory scheduling systems is sought from management on conditions of meeting specified goals. These goals usually pertain to improvements in productivity, throughput, tardiness, utilization, cost savings, etc. Subsequent validation of the system then relies on showing that the goals have been met.

- Sadeh's MicroBoss scheduler (Fowler et al., 1995) showed a 55 to 60% improvement in lead times, for actual load conditions, for scheduling over 1000 part types at a Raytheon manufacturing facility. The system reduced average tardiness by 14 to 16% and inventories by 20 to 30%.

- IBM's LMS was able to improve productivity by about 35% and avoided a capital outlay of $10 million.

- The ReDS system improved productivity and cycle times by a "considerable amount." The developers contend that, due to the many changes on the shop floor, it is hard to assign any particular amount of productivity increase to the system alone.

- Numao (Numao, 1994) shows that their scheduling system, used for scheduling steel-making, enabled a reduction in scheduling time from 3 hours to 30 minutes. This made real-time scheduling and re-scheduling possible. It also improved the quality of the schedule by reducing the average waiting time for the charge, molten iron, from 16 minutes to 8 minutes. This resulted in a saving in energy costs totalling about $1 million a year.

- Dorn and Shams (1996) report no direct impact on the production process of their expert system implemented for supporting steel-making; however, they do mention a number of qualitative benefits. The system eliminated a lot of paperwork and reduced weekly planning times. It enabled the schedulers to seek better sequencing through what-if analysis. It supported a deterministic decision framework that improved overall production quality.

- For the DAS scheduler, Lee et al., (1995) estimate an annual benefit of about $4 million to Daewoo Shipbuilding Company. This estimate is arrived at by adding a production productivity improvement of 30% and a planning productivity improvement of 50%.

System developers warn against ascribing the improvements in certain criteria to the systems performance alone. Many changes take place on the shop floor, as well as the tasks performed by personnel, when a large factory-wide scheduling system is installed. Simply the access to control information by management affects the performance of shop floor personnel. The requirement for entering the data accurately and punctually, the ability to see the status of upstream and downstream machines, and the facility of having tasks prepared by an automation device, adds to the changes in the shop floor practices. Thus, when changes in productivity occur, whether positive or negative, these cannot be simply credited or blamed on the new system.

Validation Against Manual Scheduler's Criteria

Often the scheduling expert, on whose knowledge the expert system has been based, becomes the judge for the performance of the system. The scheduler verifies that the schedules produced for particular situations are appropriate and in accordance with what he would have done. Though this approach has certain merits, it is often criticized on grounds that there is no assurance that the scheduler's solution is near the optimal one. Schedulers simply do not rely on the shop-floor situation information available to the system to make their schedules but actively influence the various parameters (such as demand and machine capacity) (McKay et al., 1995). Schedulers may not be able to critique large scale schedules produced by the system, and the problems identified by the schedulers may not be entirely significant (Kempf et al., 1991a).

For situations where only a few resources are being scheduled by the system, the method of using the scheduler's judgement is adequate. De', (1996) used a historical record to compare the system's output against what was actually scheduled on the shop floor. The historical record reflected a combination of the schedulers decisions and contingency moves made by shift supervisors and workers. The system performed reasonably well by this measure. While implementing the MacMerl system, Prietula et al. (1994) constantly asked the scheduler about the quality of the schedules and the user interface. The system was fine tuned until the schedules it produced resembled those of the scheduler, only they were more consistent and exhaustive.

Validation With Simulations

A simulation model of the factory processes makes available a fairly large set of test cases that can be used to validate the scheduling system. Such a method is deployed by Fargher and Smith, (1994) where plan validation was performed by comparing the output from the system with simulation results, both using data from a type of factory modeled by the system. Close agreement between the system and the simulation suggested that the performance of the system was adequate.

1.5   Conclusions

In 1991, Kempf and others (Kempf et al., 1991b), while reviewing the state of knowledge-based scheduling systems, stated that despite there being a large amount of research targeted at this problem, there were few fielded systems being used in daily manufacturing practice. Ten years later, the current situation is different since there are a fair number of fielded systems that have survived the uncertainties of the shop floor and are being used daily.

The research in designs of scheduling systems continues unabated. Of the design issues discussed in this chapter, the two that are receiving the most attention, and are likely to continue to do so, are those of distributed scheduling and learning in scheduling systems. With the widespread integration of information technology within all aspects of the manufacturing processes and the organization as a whole (through enterprise-wide integration systems), distributed agent-based systems have emerged as the most appropriate technology with which scheduling can be performed. Agent programs are designed to be autonomous, relatively independent of the users and other modules for their functioning, and can reside on various machines from which data can be directly obtained for scheduling purposes. Agents can also act as monitors, informing about the performance of certain resources and as negotiators, using utility-based or game-theoretic models to resolve assignment conflicts.

Learning in scheduling systems addresses the single bottleneck issue facing developing knowledge-based systems–that of extracting knowledge. Learning systems try to obtain, from a set of training examples, a set of generalized rules for scheduling. This task is especially difficult because of the tremendous complexity of any shop floor system for which generalizations are being sought. A few systems have obtained promising results but it is clear that much more research needs to be done in this area.

Another design area that is gaining importance is that of generalized scheduling systems. These systems will, in some measure, obviate the need for extensive knowledge elicitation from schedulers and users. They are being designed with the aim of having entire production systems built around them in order to simplify the product design, production planning, and scheduling tasks. In such situations, users would learn to use the scheduling system while they are learning to use the production system. Also, the scheduling system will be composed of distributed, autonomous agents that have adaptive capabilities, which will further reduce the need for knowledge gathering from shop floor personnel.

It was evident from the review of the implemented systems that the cultural and political issues were very important for the final acceptance of the scheduling system. Once the users were actively involved with the activities of providing specifications and testing the software, a certain measure of commitment and acceptance was obtained from them. The lesson to be learned here has been described in detail before, in the discipline of software engineering, where developers are urged to include the users and the sponsors of the system within the development life cycle.

References

Baker, A., Parunak, H., and Erol, K. (1997). Manufacturing over the internet and into your living room: Perspectives from the aaria project. Technical report, Dept. of Electrical and Computer Engineering and Computer Science, University of Cincinnati.

Burke, P. and Prosser, P. (1994). The distributed asynchronous scheduler. In Zweben, M. and Fox, M. S., editors, *Intelligent Scheduling,* chapter 11, pages 309–340. Morgan Kaufman Publishers.

Burstein, M., Schantz, R., Bienkowski, M., desJardins, M. E., and Smith, S. (1995). The common prototyping environment: A framework for software technology integration, evaluation, and transition. *IEEE Expert,* 10(1):17–26.

Chaturvedi, A. and Nazareth, D. (1994). Investigating the effectiveness of conditional classification: An application to manufacturing scheduling. *IEEE Transactions on Engineering Management,* 41(2): 183–193.

Chen, C. and Yih, Y. (1996). Identifying attributes for knowledge-based development in dynamic scheduling environment. *International Journal of Production Research,* 34(6):1739–1755.

Currie, K. and Tate, A. (1991). o-plan: the open planning architecture. *Artificial Intelligence,* 52:49–86.

De', R. (1993). *Empirical Estimation of Operational Planning Horizons: A Study in a Manufacturing Domain.* Ph.D. thesis, University of Pittsburgh.

De', R. (1996). A knowledge-based system for scheduling setup changes: An implementation and validation. *Expert Systems With Applications,* 10(1):63–74.

De', R. (To appear). An implementation of a system using heuristics to support decisions about shop floor setup changes. In Yu, G., editor, *Industrial Applications of Combinatorial Optimization.* Kluwer Academic Publishers.

Dorn, J. and Shams, R. (1996). Scheduling high-grade steelmaking. *IEEE Expert,* 11(1):28–35.

Dorn, J. and Slany, W. (1994). A flow shop with compatibility constraints in a steelmaking plant. In Zweben, M. and Fox, M. S., editors, *Intelligent Scheduling,* chapter 22, pages 629–654. Morgan Kaufman Publishers.

Esquirol, P., Lopez, P., Haudot, L., and Sicard, M. (1997). Constraint-oriented cooperative scheduling for aircraft manufacturing. *IEEE Expert,* 12(1):32–39.

Fargher, H. E. and Smith, R. A. (1994). Planning in a flexible semiconductor manufacturing environment. In Zweben, M. and Fox, M. S., editors, *Intelligent Scheduling,* chapter 19, pages 545–580. Morgan Kaufman Publishers.

Fordyce, K. and Sullivan, G. (1994). Logistics management system (Ims): Integrating decision technologies for dispatch, scheduling in semiconductor manufacturing. In Zweben, M. and Fox, M. S., editors, *Intelligent Scheduling,* chapter 17, pages 473–516. Morgan Kaufmann Publishers.

Fowler, N., Cross, S., and Owens, C. (1995). Guest editor's introduction: The arpa-rome knowledge-based planning and scheduling initiative. *IEEE Expert,* 10(1):4–9.

Fox, M. and Smith, S. F. (1984). ISIS — a knowledge-based system for factory scheduling. *Expert Systems,* 1(1):25–49.

Fox, M. S. (1983). *Constraint-Directed Search: A Case Study of Job Shop Scheduling.* Ph.D. thesis, Carnegie-Mellon University.

Gary, K., Uzsoy, R., Smith, S., and Kempf, K. (1995). Measuring the quality of manufacturing schedules. In Brown, D. and Scherer, W., editors, *Intelligent Scheduling Systems.* Kluwer, Boston.

Godin, V. (1978). Interactive scheduling: Historical survey and state of the art. *AIIE Transactions,* 10(3):331–337.

Gu, P., Balasubramanian, S., and Norrie, D. (1997). Bidding-based process planning and scheduling in a multi-agent system. *Computers and Industrial Engineering,* 32(2):477–496.

Hadavi, K., Hsu, W., Chen, T., and Lee, C. (1992). An architecture for real-time distributed scheduling. *AI Magazine,* 13(3):46–56.

Hadavi, K. C. (1994). A real time production scheduling system from conception to practice. In Zweben, M. and Fox, M. S., editors, *Intelligent Scheduling,* chapter 20, pages 581–604. Morgan Kaufman Publishers.

Hayes-Roth, B. (1985). A blackboard architecture for control. *Artificial Intelligence,* 26:251–321.

Hsu, W. L., Prietula, M. J., and Ow, P. S. (1993). A mixed-initiative scheduling workbench: Integrating ai, or, and hci. *Decision Support Systems,* 9(3):245–257.

Interrante, L. D. and Rochowiak, D. M. (1994). Active rescheduling and collaboration in dynamic manufacturing systems. *Concurrent Engineering: Research and Applications,* 2(2):97–105.

Jain, S., Barber, K., and Osterfeld, D. (1990). Expert simulation for on-line scheduling. *Communications of the ACM,* 33(10):55–60.

Kempf, K., Pape, C. L., Smith, S., and Fox, B. (1991a). Issues in the design of ai-based schedulers: A workshop report. *AI Magazine,* 11(5):37–46.

Kempf, K., Russell, B., Sidhu, S., and Barrett, S. (1991b). Ai-based schedulers in manufacturing practice. *AI Magazine,* 11(5):46–55.

Kempf, K. G. (1994). Intelligently scheduling semiconductor wafer fabrication. In Zweben, M. and Fox, M. S., editors, *Intelligent Scheduling,* chapter 18, pages 517–544. Morgan Kaufmann Publishers.

Kerr, R. (1992). Expert systems in production scheduling: Lessons from a failed implementation. *Journal of Systems Software,* 19:123–130.

Lee, J., Lee, K., Hong, J., Kim, W., Kim, E., Choi, S., Kim, H., Yang, O., and Choi, H. (1995). Das: Intelligent scheduling systems for shipbuilding. *AI Magazine,* 16(4):94.

May, J. H. and Vargas, L. G. (1996). Simpson: An intelligent assistant for short-term manufacturing scheduling. *European Journal of Operational Research,* 88:269–286.

McKay, K., Buzacott, J., and Safayeni, F. (1995). 'Common sense' realities of planning and scheduling in printed circuit board production. *International Journal of Production Research,* 33:1587–1603.

Minsky, M. (1975). A framework for representing knowledge. In Winston, P., editor, *The Psychology of Computer Vision,* pages 211–277. McGraw-Hill, New York.

Miyashita, K. and Sycara, K. (1994). Adaptive case-based control of schedule revision. In Zweben, M. and Fox, M. S., editors, *Intelligent Scheduling,* chapter 10, pages 291–308. Morgan Kaufman Publishers.

Miyashita, K., Sycara, K., and Mizoguchi, R. (1996). Modeling ill-structured optimization tasks through cases. *Decision Support Systems,* 17:345–364.

Numao, M. (1994). Development of a cooperative scheduling system for the steel-making process. In Zweben, M. and Fox, M. S., editors, *Intelligent Scheduling*, chapter 21, pages 607–628. Morgan Kaufman Publishers.

O'Keefe, R., Balci, O., and Smith, E. (1987). Validating expert system performace. *IEEE Expert*, 2(4):81–90.

O'Keefe, R. and O'Leary, D. (1993). A review and survey of expert system verification and validation. *Artificial Intelligence Review*, 7(1):3–42.

O'Keefe, R. and Preece, A. (1996). The development, validation, and implementation of knowledge-based systems. *European Journal of Operations Research*, 92:458–473.

Ow, P., Smith, S. F., and Thiriez, A. (1987). Reactive plan revision. In *Proceedings of the Seventh National Conference on Artificial Intelligence*, pages 77–82, San Mateo, CA. Morgan Kaufmann.

Ow, P. S. and Smith, S. F. (1987). Two design principles for knowledge-based systems. *Decision Sciences*, 18(3):430–447.

Parunak, H., Baker, A., and Clark, S. (1997). The aaria agent architecture: An example of requirements driven agent-based system design. In *Proceedings of the First International Conference on Autonomous Agents (ICAA '97)*, Marina del Rey, CA.

Piramuthu, S., Raman, N., and Shaw, M. (1994). Learning-based scheduling in a flexible manufacturing flow line. *IEEE Transactions on Engineering Management*, 41(2):172–182.

Piramuthu, S., Raman, N., Shaw, M., and Park, S. (1993). Integration of simulation modeling and inductive learning in an adaptive decision support system. *Decision Support Systems*, 9:127–142.

Prietula, M., Hsu, W.-L., Ow, P., and Thompson, G. (1994). Macmerl: Mixed-initiative scheduling with coincident problem spaces. In Zweben, M. and Fox, M. S., editors, *Intelligent Scheduling*, chapter 23, pages 655–682. Morgan Kaufman Publishers.

Prietula, M. J., Hsu, W. L., and Ow, P. S. (1991). A coincident problem space perspective to scheduling support. In *Proceedings of the Fourth International Symposium on Artificial Intelligence*, Cancun, Mexico.

Quinlan, J. (1986). Induction of decision trees. *Machine Learning*, 1:81–106.

Sauer, J. and Bruns, R. (1997). Knowledge-based scheduling systems in industry and medicine. *IEEE Expert*, 12(1):24–31.

Shah, V., Madey, G., and Mehrez, A. (1992). A methodology for knowledge based scheduling decision support. *Omega, International Journal of Management Science*, 20(5/6):679–703.

Sikora, R. and Shaw, M. J. (1997). Coordination mechanisms for multi-agent manufacturing systems: Applications to integrated manufacturing scheduling. *IEEE Transactions on Engineering Management*, 44(2):175–187.

Smith, S. (1987). A constraint-based framework for reactive management of factory schedules. In *Proceedings of the International Congress on Expert Systems and Leading Edge in Production Planning and Control*, Charleston, SC.

Smith, S. F. (1994). Opis: A methodology and architecture for reactive scheduling. In Zweben, M. and Fox, M. S., editors, *Intelligent Scheduling*, chapter 2, pages 29–66. Morgan Kaufmann Publishers.

Szelke, E. and Kerr, R. M. (1994). Knowledge-based reactive scheduling. *Production Planning & Control*, 5(2):124–145.

Tate, A., Drabble, B., and Kirby, R. (1994). O-plan2: An open architecture for command, planning, and control. In Zweben, M. and Fox, M. S., editors, *Intelligent Scheduling*, chapter 7, pages 213–239. Morgan Kaufmann Publishers.

Zhang, W. and Dietterich, T. (1995). A reinforcement learning approach to job-shop scheduling. In *Proceedings of the 14th International Joint Conference on Artificial Intelligence*.

Zuboff, S. (1988). *In the Age of the Smart Machine: The Future of Work and Power*. Basic Books, Inc.

Zweben, M., Daunn, B., Davis, E., and Deale, M. (1994). Scheduling and rescheduling with iterative repair. In Zweben, M. and Fox, M. S., editors, *Intelligent Scheduling*, chapter 8, pages 241–255. Morgan Kaufmann Publishers.

Zweben, M., Davis, E., Daun, B., Drascher, E., Deale, M., and Eskey, M. (1992). Learning to improve constraint-based scheduling. *Artificial Intelligence*, 58:271–296.

2

Neural Network Systems Techniques in the Intelligent Control of Chemical Manufacturing Plants

Sung Hoon Jung
Hansung University

Tag Gon Kim
Korea Advanced Institute of Science and Technology

Kyu Ho Park
Korea Advanced Institute of Science and Technology

2.1 Introduction

Neural networks have been widely used in many control areas [1, 2, 3, 4, 5]. However, as controlled systems have been more and more complex, no one control paradigm is enough to control especially where the controlled systems are complex hybrid ones composed of discrete event systems and continuous systems. In order to control such hybrid systems, intelligent control methodologies must be embedded into an integrated intelligent control system [6, 7, 8, 9, 10, 11]. This integration provides an intelligent system with some capabilities such as self-learning, self-planning, and self-decision making [7, 9, 10]. The structure of the integrated system must be well defined and constructed for getting synergy effect from all modules. Thus, the basic framework for constructing an integrated system is very important.

Recently, Zeigler [12, 13] introduced an *event-based intelligent control* paradigm based on the simulation theory for discrete event systems [14, 15]. This control paradigm is devised using a simulation formalism called *DEVS (discrete event system specification)*. The DEVS provides mathematically-sound semantics to specify operations of discrete event systems in a hierarchical, modular manner [14, 15]. Therefore, this control paradigm can be a good framework especially where the controlled processes are highly complex, such as chemical plants. With this framework, high-level modules, such as planners and schedulers, can be easily constructed and other modules, such as neural networks and fuzzy logic control, can be easily incorporated with hierarchy and modularity [16].

In event-based control, a internal controller has an event-based model of a controlled plant that characterizes the operational properties of the plant [12, 13]. Using this model, the event-based controller can simulate the state of transitions in the plant and can diagnose the operations of the plant using time constraints. We used a neural network model as the event-based model. This neural network model has the dynamics of the controlled plant with discrete levels for simulating operations of the controlled plant, as well as for decision outputs to control. This event-based control system, with neural network models, can control any type of plant such as continuous plants, discrete event plants, and hybrid plants composed of continuous and discrete event systems.

The neural network model maps the dynamics of a controlled plant so that it can automatically generate dynamics for control and diagnosis. However, the mapped dynamics of a neural network model can not be exactly the same as those of the original plant, owing to incomplete learning. When a plant has a saturation property, this incomplete learning results in serious problems. This is because of the output of a nonlinear dynamic plant, with saturation property, is very sensitive to small changes of its input values.

In [17], the authors proposed a neural model predictive control strategy combining a neural network and a nonlinear programming algorithm. The control performance may be considerably degraded when the predicted value differs greatly from the actual value. This is because the control input depends greatly on the predictive value and there are no methods to compensate the predictive error. The application of the method to highly nonlinear plants may be very difficult or even impossible.

To cope with this problem, our method partitions a block, between a current state and a target one, into several intermediate blocks. Then control inputs are repeatedly applied to the plant until the target state is reached. This provides the event-based controller with a state feedback mechanism of the conventional control. The plant output may be slightly erroneous owing to incomplete learning. For this problem, a state window is employed that provides a state tolerance about a steady state error. These two windows, namely the time window and the state window, make a cross-check area to check the state and time constraints. This scheme may be viewed as a combination of a time-based diagnosis mechanism in an event-based control system [12] and a state-based control mechanism in a neural network control system [18].

We experimented with a continuously stirred tank reactor (CSTR) plant using our event-based control system with a neural network mapping model. The CSTR plant is a chemical process that has strong non-linearity and complicated dynamics. Experimental results show relatively good control performance in spite of the strong nonlinearity and complicated dynamics.

2.2 Neural Network Construction for Event-Based Intelligent Control

This section describes the neural network construction method of a continuous system for event-based intelligent control. First, we briefly review the event-based control paradigm and state the neural network construction method next.

Brief Review of Event-Based Intelligent Control Paradigm

In an event-based control, the event-based model of a plant is specified by an event-based control DEVS [19]. This event-based control DEVS is a modified version of the discrete event system specification (DEVS) formalism [14, 15, 13]. An event-based control DEVS is defined as a 7-tuples [19]:

$$M = \langle X, S, Y, \delta_{int}\delta_{ext}, \lambda, ta \rangle \qquad (2.1)$$

where

- X is the external input events set; and
- $S = B \times X$, where B is the finite set of elements, each called a boundary; and

- Y is the finite output events set; and
- δ_{int} is the internal transition function; and
- δ_{ext} is the external transition function; and
- $\lambda: S \rightarrow Y$ is the output function; and
- $ta: S \rightarrow R^+_{0,\infty} \times R^+_{0,\infty}$, i. e., $ta(s) = [r, r']$, where $r, r' \in R^+_{0,\infty}$ and $r \leq r'$.

In an event-based control, an event-based model specified by the event-based-control DEVS is composed of an input set, a state set, and an output set. These three sets correspond to control inputs, control points, and threshold outputs of the plant in event-based control, respectively. The internal and external transition functions for the event-based model provide the behavioral characteristics of the plant. Each function in the tuple represents constraints on the system dynamics. The output function maps the discrete event states to threshold-like outputs. The *time advanced function* differs slightly from the original DEVS formalism [15] owing to its need for a time interval. The minimum and maximum times of this time interval, called a *time window*, provide time tolerance that acts as the conventional controller's counterpart to state tolerance. That is, the event-based controller regards the control operation as correct if the target state arrives within the time window.

After the event-based controller has output the control input to the real plant, the controller waits for a sensor signal from the threshold sensors. If the signal arrives prior to the minimum time of a time window, then the event-based controller issues a *"too-early"* error. If the signal arrives after the maximum time of the time window, the controller issues a *"too-late"* error. Even if the signal arrives within the time window, the controller issues an *"unexpected-state"* error if the signal is different from the target state. In summary, the event-based controller regards the operation of the plant as correct only if the signal arrives within the time window and is the same as the expected planned target state. The event-based controller sends diagnostic information—*too-early, too-late,* and *unexpected-state*—to diagnostic personnel to diagnose errors. This event-based paradigm has two advantages over the traditional ones; the error informations produced by the event-based controller could be used for diagnostic purpose and the use of a time window provides the controller with robustness against expected values from sensors.

Figure 2.1 shows the architecture of an event-based controller. An event-based controller is composed of a control part and a diagnosis part. The control part is composed of a *goal-driven-planner* (GDP) and a *control-output-generator* (COG). The diagnosis part is made up of a *simulator* and an *event-based model*. These two parts are managed by the central controller which provides a main control algorithm.

The logic of the control is as follows:

- The event-based controller receives an external input X_c from threshold sensors, and sends control signals Y_c to a plant if no error is detected.
- If errors are detected, the event-based controller sends diagnostic information Y_d to the diagnoser.
- The central controller internally sends a *current state* signal to the GDP to get the next target state, then sends the received target state with the current state to COG to get control output.
- The controller then sends the received control output to the plant. At the same time, it sends this output to the simulator to simulate the plant behavior.
- The simulator simulates the plant using an event-based model and generates two outputs, i.e., expected outputs Y_s of the plant and time windows TW.
- With the reference of the outputs and time windows, the controller diagnoses the state of the plant.
- If the expected state of a plant is sensed within the time window, then the event-based controller regards the current state of this plant as being correct.
- Otherwise, an error is assumed to have occurred, and the controller invokes diagnoser functions to find where it has occurred.
- The event-based controller repeatedly performs the described control logic in accordance with the sensor readings.

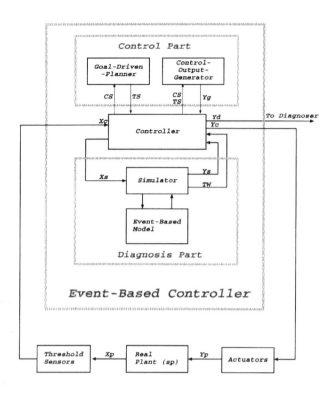

FIGURE 2.1   Event-Based Intelligent Control Environment.

Neural Network Construction Method

With mapping capability of neural networks, nonlinear plants could be easily modeled and these models have applied control in nonlinear plants in many different ways [1, 2, 4, 20]. Also, the on-line learning capability of neural networks makes it possible for control systems to have dynamic modelling capability. We use the mapping and on-line capability of neural networks.

In order to use a neural network model as an event-based model of an event-based intelligent controller, the controlled continuous system must be first abstracted into an event-based model and then the abstracted event-based model is mapped to a neural network. By doing this, the neural network model can be isomorphic to the continuous plant at the discrete input-output level [19].

Abstraction Process

Figure 2.2 shows the abstraction processes. The abstraction processes are as follows:

- When a continuous system is given, a system designer must first decide the operational objectives of the continuous system.
- The operational objectives can be derived from functional requirements and constraints of the system.
- To get the discrete inputs and outputs of the continuous system, the designer must do output partitioning first and input sampling second.
- The output partitioning is to divide the outputs of the continuous system into several mutually exclusive blocks to quantize the output levels considering operational objectives of the system.
- The input sampling is to select some inputs for state transitions specified by the operational objectives.

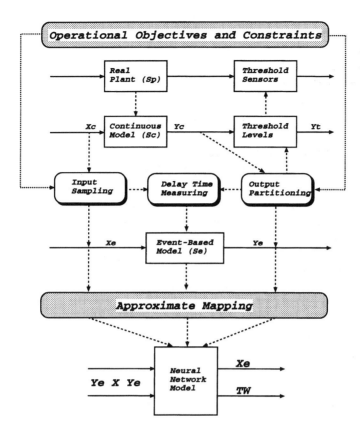

FIGURE 2.2　Abstraction Process and Neural Network Mapping.

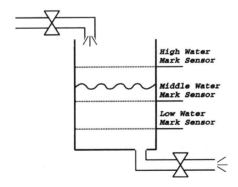

FIGURE 2.3　Water Tank Continuous System.

- Third delay times for the state transitions under given inputs must be measured using a real continuous system or a model such as a set of differential equations.
- With this information, we can construct an event-based model for the continuous system.

With a simple water-tank example, we illustrate the abstraction process. Figure 2.3 shows a water-tank continuous system. Let the constraints of the water tank system and operational objectives of this example be given as follows:

- Constraints
  - The water tank must be able to supply the water to another device with two rates (not zeros).
  - The water can be supplied from another device with two rates (not zeros).
- Operational Objectives
  - The water level must be kept in the vicinity of the middle of the water-tank not to go over the top of the water-tank and not to go under the bottom of the water-tank.
  - Time to rise the water level from the bottom level to the middle level must not be over 2 minutes.
  - Time to fall the water level from the top level to the middle level must not be over 3 minutes.

From the first operational objective, we can do output partitioning with three levels: *High – Mark*, *Midd – Mark*, and *Low – Mark*. Of course, three threshold sensors for detecting the three levels must be equipped on the water-tank for real control. Using the second and third operational objectives and two constraints, we can next do input sampling. For simplicity, four values are selected for the input valve and output valve: *High – Input*, *Low – Input*, *High – Output*, and *Low – Output*. Of course, the symbolic elements for each set is assigned for real value: *High – Mark* = 3 m, *High – Input* = 10 lb/min and so on. Finally, delay times for state transitions, under given inputs, should be measured using a real continuous system or a differential equations model.

The water-tank operation is simply modeled using a first-order differential equation as follows.

$$C\frac{dh}{dt} = q_i - q_o \tag{2.2}$$

where C is the capacity of water-tank, h is the height of water-tank, q_i is the input flow rate of water, and q_o is the output flow rate of water. The height of water-tank is obtained by solving the equation:

$$h(t) = h_0 + \frac{q_i - q_o}{C}t \tag{2.3}$$

From this differential equation, the delay time can be calculated by the following equation

$$dt = \frac{h_c - h_0}{\frac{q_i - q_o}{C}} \tag{2.4}$$

where dt is delay time, h_c is a target state, and h_0 is a current state. Of course, the elements of h_c, h_0, q_i, q_o must be in the predefined set.

The dynamics of a real water tank will not be exactly the same as those of the differential equation model. This is caused not only by modeling errors but also by parameter changes of the real water tank and its environment. Let the minimum and maximum parameter variation be P_{min}, P_{max} respectively, and the minimum and maximum times to reach a specific height h_c are as follows.

$$t_{min} = \frac{h_c - h_0}{\left(\frac{q_i - q_o}{C}\right)P_{max}}$$

$$t_{max} = \frac{h_c - h_0}{\left(\frac{q_i - q_o}{C}\right)P_{min}} \tag{2.5}$$

TABLE 2.1   Real Values of States and Inputs of Water Tank

Water Mark			Two Input			
HIGH_TANK	MID_TANK	LOW_TANK	HIGH_IN	LOW_ IN	HIGH_ OUT	LOW_OUT
22.5	15.0	3.5	12.5	2.5	8.5	4.5

TABLE 2.2   Gathered Data from Differential Equation Model of Water Tank

Water Mark		Two Input		Time Window	
Current State	Target State	Input Rate	Output Rate	Minimum Time	Maximum Time
LOW_TANK	MID_TANK	HIGH_IN	HIGH_OUT	31.1	38.8
LOW_TANK	MID_TANK	HIGH_IN	LOW_OUT	15.5	19.4
LOW_TANK	MID_TANK	LOW_IN	HIGH_OUT	−20.7	−25.9
LOW_TANK	MID_TANK	LOW_IN	LOW_OUT	−62.1	−77.6
LOW_TANK	HIGH_TANK	HIGH_IN	HIGH_OUT	51.3	64.1
LOW_TANK	HIGH_TANK	HIGH_IN	LOW_OUT	25.7	32.1
LOW_TANK	HIGH_TANK	LOW_IN	HIGH_OUT	−34.2	−42.8
LOW_TANK	HIGH_TANK	LOW_IN	LOW_OUT	−102.6	−128.2
MID_TANK	LOW_TANK	HIGH_IN	HIGH_OUT	−31.1	−38.8
MID_TANK	LOW_TANK	HIGH_IN	LOW_OUT	−15.5	−19.4
MID_TANK	LOW_TANK	LOW_IN	HIGH_OUT	20.7	25.9
MID_TANK	LOW_TANK	LOW_IN	LOW_OUT	62.1	77.6
MID_TANK	HIGH_TANK	HIGH_IN	HIGH_OUT	20.2	25.3
MID_TANK	HIGH_TANK	HIGH_IN	LOW_OUT	10.1	12.7
MID_TANK	HIGH_TANK	LOW_IN	HIGH_OUT	−13.5	−16.9
MID_TANK	HIGH_TANK	LOW_IN	LOW_OUT	−40.5	−50.6
HIGH_TANK	LOW_TANK	HIGH_IN	HIGH_OUT	−51.3	−64.1
HIGH_TANK	LOW_TANK	HIGH_IN	LOW_OUT	−25.7	−32.1
HIGH_TANK	LOW_TANK	LOW_IN	HIGH_OUT	34.2	42.8
HIGH_TANK	LOW_TANK	LOW_IN	LOW_OUT	102.6	128.2
HIGH_TANK	MID_TANK	HIGH_IN	HIGH_OUT	−20.2	−25.3
HIGH_TANK	MID_TANK	HIGH_IN	LOW_OUT	−10.1	−12.7
HIGH_TANK	MID_TANK	LOW_IN	HIGH_OUT	13.5	16.9
HIGH_TANK	MID_TANK	LOW_IN	LOW_OUT	40.5	50.6

Consequently, we can get a time window by taking these minimum and maximum times:

$$t_{min} \leq t_{win} \leq t_{max} = \frac{C(h_c - h_0)}{(q_i - q_o)P_{max}} \leq t_{win} \leq \frac{C(h_c - h_0)}{(q_i - q_o)P_{min}}$$

Let the C, P_{min}, and P_{max} be 12.2, 0.9037, and 1.12963 respectively, then the time window is given as:

$$\frac{10.8(h_c - h_0)}{(q_i - q_o)} \leq t_{win} \leq \frac{13.5(h_c - h_0)}{(q_i - q_o)} \tag{2.6}$$

These gathered data is mapped to the neural network with a back-propagation algorithm.

Let the output states (level of the water) and inputs (flow rates of input and output valves) be partitioned and sampled as shown in Table 2.1. To satisfy the first operational objective and two constraints, the high input rate of water to fill the water tank must be greater than the high output rate not to be emptied the water in the tank. Also, the low input rate of water to sink the water must be lower than the low output rate not to be overflowed the water. We can measure the minimum and maximum times for each state transition under given inputs. Table 2.2 shows the gathered data from Eq. 2.6. In this table, the minus time means that the delay time is infinity. That is, the target state will never be reached under given inputs.

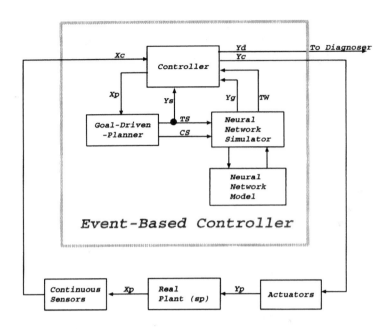

FIGURE 2.4   Event-Based Intelligent Control Environment with a neural network Model.

Neural Network Mapping

Figure 2.2 shows how the event-based plant model is mapped into a neural network. For mapping an event-based model to a neural network, the inputs and outputs of the neural network should be defined. The input and output events sets cannot be assigned for outputs of the neural network. This is because the neural network will generate outputs different from those of the training data owing to incomplete training. This makes it impossible to decide which event of the input and output events set are matched to the outputs of the neural networks. However, time windows of state transitions can be assigned for the output of the neural network because the time windows are able to have any real values in $R_{0,\infty}^{+}$ as shown in Eq. 2.1. Also control inputs can be assigned for the output of the neural network because the control inputs are not exactly the same as the sampled inputs. From this observation, we can decide the inputs and outputs of neural network. The inputs of neural network are composed of a current state value and a target state value. With these two inputs, the neural network generates the time window for the state transition and control inputs. Of course, all values of states must be in predefined state sets as elements. Note that the values of the elements for the event-based model are approximately mapped into a neural network model. Data not learned is generated by the neural network model in an approximate form. The structure of an event-based controller with an neural network model is shown in Fig. 2.4.

Model-Plant Mismatch Caused by Incomplete Learning

A back propagation algorithm, which is a typical supervised learning algorithm, is used for learning in our neural network model. The dynamics of an neural network model may not be exactly the same as those of the original continuous system owing to incomplete learning. This mismatch may cause the neural network model to generate inaccurate values. In the case of monotonically increasing or decreasing systems, the time window offers a tolerance against inaccurate time of state changes in the plant and uncertainty in the environmental changes. Initial state and plant parameter variations are examples of such inaccuracies and uncertainties [19].

However, this incomplete learning causes serious problems in the case of a saturated plant. This is because the output of a nonlinear saturated plant is very sensitive to changes of input values. Thus, the saturated value of the controlled plant may be greatly different from that of the target state. That is, the output of a saturated plant may not reach the target state under given inputs. This is because the control inputs generated by the neural network model no longer make the output of the plant to be saturated with the exact target state. This problem can happen, not only because of the model-plant mismatch, but also because of some other factors (such as external disturbances and environment changes). We call these three elements *perturbation factors*. These error factors will not result in a fatal problem in monotonically increasing/decreasing systems because the output of the controlled plant will eventually reach the target state.

Recently, Song and Park [17, 21] proposed a predictive control scheme based on a neural network model for control of highly nonlinear chemical plants. The proposed scheme combines a neural network for plant identification with a nonlinear programming algorithm for solving nonlinear control problems. The method first generates a one-step-ahead predictive value of a controlled plant using the predictive neural model and then calculates a control input using an optimization algorithm, SQP (*Successive Quadratic Programming*) module, with a predictive value and a desired output value. The method shows good performance when the predictive value is approximately equal to the actual value. However, the performance may be considerably degraded when the predicted value differs greatly from the actual value. This is because the control input depends greatly on the predictive value and there are no methods to compensate for error between the two. Finally, the plant may have a large steady state error due to the predictive error.

To cope with this problem, we partition the block between a current state and a target one into several intermediate blocks. Afterwards, we apply control inputs repeatedly to the plant until set points are reached. This provides the event-based controller with a state feedback mechanism used in conventional control. In fact, a control system that controls a nonlinear continuous plant does not work well without a state feedback mechanism. Thus, direct application of an event-based control system to control a nonlinear continuous plant may cause problems, such as those that occur in saturated plants. In this scheme, time windows are used to evaluate the saturated state of the plant, not to diagnose its state. That is, even though the state of the plant does not reach a target state within a time window, the controller does not generate the "*LATE*" error signal. Instead, it applies new inputs to the neural network model with a new current state.

This operation is applied to the plant repeatedly until the state of the plant reaches its target state. In spite of this operation, the plant output may be slightly erroneous, due to incomplete learning. That is, the neural network model may always generate the same control output Y_c for a specific current and target state.

This error can be seen as a steady state error which can be tapered by the continuous learning of the dynamics of a plant during control. To solve this problem, we adopt a state window—an interval of values of a state—which provides a state tolerance.

The width of the state tolerance accounts for the plant dynamics and the plant environment. These two windows—the time window and a state window—create a cross-check area that checks the state and time constraints with a tolerance. The event-based controller issues an error message only when the plant state is not within the cross-check area although n control operations are applied. The two dimensional error constraints of time and states are to provide the event-based controller with a 2-D maximum error tolerance in each direction. Consider the two situations shown in Fig. 2.5. In case 1, the output of a plant is saturated with a value within the error constraints although only one input u_1 is applied to the plant. This situation occurs when the current state and target state of the GDP are the same or slightly different from those of collected data. Even if the current state and target state are exactly the same as those of collected data, the output of a plant may not be exactly the same as the target state because the perturbation factors make the neural network model different from the collected data.

In case 2, the output of a plant is saturated with a value within the error constraints after receiving three inputs. This situation occurs when the planned current and target pair are not in the collected data, but in the transitive closure for the state transition such as $A \rightarrow B \rightarrow C \rightarrow D$ for $A \rightarrow D$ as shown in Fig. 2.5. This situation is also regarded as correct. Thus, the saturation of a plant output provides the event-based controller with the state feedback mechanism to avoid the saturation. In the two situations, the time constraints can be used to check the time requirements. However, if the target

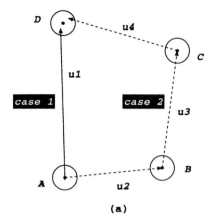

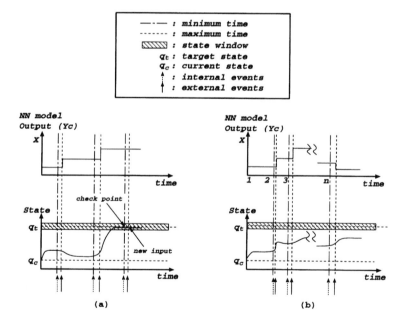

FIGURE 2.5   Two cases of Control (a) state space (b) a data table (c) another data table.

FIGURE 2.6   Two Control Situations of a Saturated Plant: (a) success case (b) error case.

state is not reached after n trials, then the event-based controller regards the state of plant itself or of environment as an error. The number n should be decided carefully in considering plant characteristics and operational constraints. This scheme may be viewed as a combined method of a time-based diagnosis mechanism of an event-based control system [12] and a state-based control mechanism of a neural network control system [18]. Figure 2.6 shows the success case and the error case of the control operation.

Neural Network Model for Supervisory Control

Recently, supervisory control methods have been extensively researched [22, 23, 24, 25, 26]. A neural network model, represented by the above mentioned neural network modelling strategy, can also be used for supervisory control with small modification. In this application, the input/output flow rates of the water tank are assigned to inputs of a neural network model. This is because the supervisory control just allows some events in controllable events to occur for a state transition. The control scheme does not calculate the control inputs directly. Thus, the allowable control inputs (in other words, enabled controllable events) should be exactly defined and in a predefined set.

This is the reason why the control inputs of an event-based model cannot be assigned to outputs of a neural network model. With the simple water-tank example, we illustrate the application of a neural network model to the supervisory control. In supervisory control, the water tank model abstracted can be defined as a 5-tuple

$$M = (Q, \Sigma, \delta, q_0, Q_m) \tag{2.7}$$

where

- Q is the state set, $Q = \{Empty, LOW\text{-}TANK, MID\text{-}TANK, HIGH\text{-}TANK, Over\ flow\}$
- Σ is the events set, $\Sigma = \{HIGH\text{-}IN, LOW\text{-}IN, HIGH\text{-}OUT, LOW\text{-}OUT, LOW\text{-}TANK\text{-}IND, MID\text{-}TANK\text{-}IND, HIGH\text{-}TANK\text{-}IND\}$
- $\delta: Q \times \Sigma \rightarrow Q$ is the transition functions, for examples,
  - $\{LOW\text{-}TANK, (HIGH\text{-}IN, HIGH\text{-}OUT)\} \rightarrow MID\text{-}TANK$
  - $\{LOW\text{-}TANK, (LOW\text{-}IN, HIGH\text{-}OUT)\} \rightarrow Empty$
  - $\{HIGH\text{-}TANK, (LOW\text{-}IN, HIGH\text{-}OUT)\} \rightarrow MID\text{-}TANK$
  - $\{HIGH\text{-}TANK, (HIGH\text{-}IN, HIGH\text{-}OUT)\} \rightarrow Over\ flow$
  - $\{HIGH\text{-}TANK, (HIGH\text{-}IN, LOW\text{-}OUT)\} \rightarrow Over\ flow$
- $q_0 \in Q$ is the initial state, $q_0: = LOW\text{-}TANK$
- $Q_m \subset Q$ is the maker states, $Q_M: = MID\text{-}TANK$

The events, *LOW-TANK-IND*, *MID-TANK-IND*, and *HIGH-TANK-IND*, are indicative events of the three states, respectively. In this model, let the controllable events and uncontrollable events be given as:

$$\begin{aligned} \Sigma &= \Sigma_c \cup \Sigma_{uc} \\ &= \{HIGH\text{-}IN, LOW\text{-}IN, HIGH\text{-}OUT, LOW\text{-}OUT\} \cup \\ &\quad \{LOW\text{-}TANK\text{-}IND, MID\text{-}TANK\text{-}IND, HIGH\text{-}TANK\text{-}IND\} \end{aligned} \tag{1.8}$$

where Σ_c and Σ_{uc} are controllable and uncontrollable events, respectively. Then, a supervisor can drive from any states to the maker state because we select that any composition of flow rates of inputs and outputs can fill or sink the water.

In timed supervisory control, the abstracted neural network model can also be used for diagnosis of controlled systems. For example, when a supervisory controller enables two events, *LOW-IN, HIGH-OUT* and disables two events, *HIGH-OUT, LOW-OUT*, for state transition from *HIGH-TANK* to *MID-TANK*, the event *MID-TANK-IND* should occur within a time window generated from the neural network model. In all applications of abstracted neural network model, learning capability of neural network makes it possible for dynamic modelling of controlled system by on-line learning. This scheme is very similar to human control strategy in that human can learn more and more information about the controlled system as the control actions proceed.

2.3 Simulation Environment

This section describes the controlled plant and the neural network learning strategy.

Continuously Stirred Tank Reactor (CSTR)

A CSTR is a chemical process that produces chemical products. The CSTR model is a part of a larger test system introduced by Williams and Otto [27] and McFarlane et al. [28]. The CSTR system, shown in Fig. 2.7, supports the following multiple reactions:

$$A + B \xrightarrow{k1} C$$
$$C + B \xrightarrow{k2} P + E \qquad\qquad (2.9)$$
$$P + C \xrightarrow{k3} G$$

The desired product is P, while G, C, and E are byproducts subject to quality and environmental constraints. Reactants A and B enter as pure components in separate streams with flow rates F_{ai} and F_{bi}, respectively.

The flow rate F_{ai} and cooling water temperature T are variables in this chemical process. The input stream F_{bi} is considered to be a disturbance variable. The equations describing the kinetic behavior of the above reactions and the dynamic mass balance for the CSTR are a coupled set of nonlinear algebraic

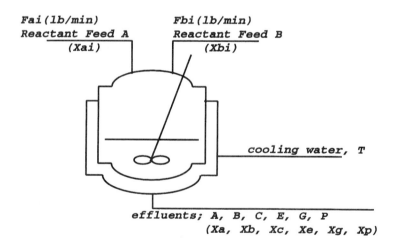

FIGURE 2.7 CSTR Plant.

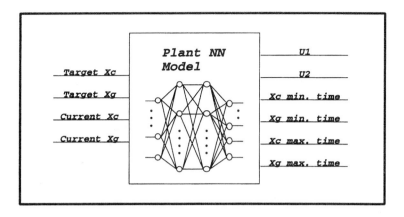

FIGURE 2.8   Neural Network Structure for CSTR.

and ordinary differential equations. A description of these equations has been provided by Williams and Otto [27]:

$$
\begin{aligned}
dXa/dt &= F_{ai}/Fr - rx1 - Xa \\
dXb/dt &= F_{bi}/Fr - rx1 - rx2 - Xb \\
dXc/dt &= 2rx1 - 2rx2 - rx3 - Xc \\
dXe/dt &= 2rx2 - Xe \\
dXg/dt &= 1.5rx3 - Xg \\
dXp/dt &= rx2 - 0.5rx3 - Xp \\
rx1 &= 5.9755E9 \exp(-12000/T)\, XaXb\rho\, V/(60Fr) \\
rx2 &= 2.5962E12 \exp(-15000/T) XbXc\rho\, V/(60Fr) \\
rx3 &= 9.6283E15 \exp(-20000/T) XcXp\rho\, V/(60Fr) \\
Fr &= F_{ai} + F_{bi} \\
\rho &= 50\,\text{lb/ft}^3,\ V = 60\,\text{ft}^3,\ 580°R < T < 680°R \\
X_i &= mass\ fraction
\end{aligned}
\tag{2.10}
$$

The reaction constants, initial conditions, and constant parameters are as follows:

$$
\begin{aligned}
Xa &= 0.075 & Xe &= 0.208 & F_{ai} &= 170\ \text{lb/min} \\
Xb &= 0.57 & Xg &= 0.0398 & F_{bi} &= 679\ \text{lb/min} \\
Xc &= 0.015 & Xp &= 0.019 & T &= 645°R
\end{aligned}
$$

The control objective of this system is to maximize the yield of the desired product P by regulating the two related state variables Xc and Xg. In this example, our neural network has 4-10-10-6 morphology as shown in Fig. 2.8.

Neural Network Learning Strategy

As shown in Eq. (2.1) of the event-based DEVS definition, the states of a continuous system are represented by the cross product of boundaries and inputs. Some initial control variable values can also be employed as initial states. If a system is represented by n boundaries, m inputs, and k initial states, then

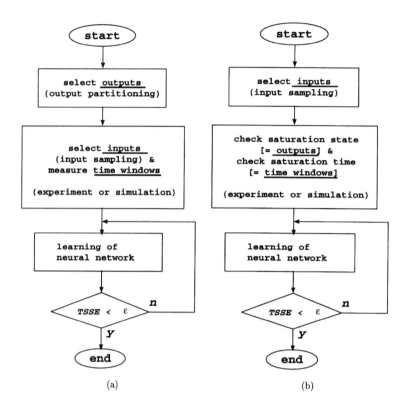

FIGURE 2.9   Flow Chart of State Space Mapping (a) original scheme (= in case of monotonic systems) (b) modified scheme (= in case of saturated systems).

$n \times m \times k$ number of state spaces are necessary to represent the system completely at discrete levels. For completeness, the states in the state space should be mapped to the neural network. For simplicity, we took only the input variables except for the initial states as independent variables.

Figure 2.9 shows the two flow charts of state space mapping algorithm. Figure 2.9(a) is the original state mapping scheme for monotonically increasing/decreasing systems. However, this scheme should be modified so as to fit with saturated plants as shown in Figure 2.9(b). First, in case of monotonic systems, control outputs should be determined using output partitioning. Then a sequence of control inputs should be selected such that the inputs move the current state of the plant to the target one. Finally, the control inputs are applied to the plant, and time windows measured. On the other hand, inputs within an operating range are first selected with random distribution for saturated plants. This is because it is difficult for the designer to find the inputs that make the output of the controlled plant saturated with a specific desired value. Thus initially, randomly selected inputs are applied to the plant and then the output of the plant is determined to be saturated or not.

The saturated state becomes an output state which can be used as training data. The saturation time is used to decide time windows. After collecting data, it is mapped to a neural network. To determine the mapping rate of the neural network model, a total sum square error (TSSE) is employed. The learning procedure is stopped only when the predetermined error bound, ϵ, is satisfied. As adequacy of the neural network model is dependent on ϵ, the value of ϵ should be carefully selected.

Algorithm 2.1 shows a detailed description of the collecting method of the mapping data.

> **Algorithm 2.1** Mapping-data-Collecting()
> // N: the number of learning patterns //
> 1 **for** $i = 0$ to $N - 1$ // getting of N learning patterns //
> 2   select uniformly distributed random inputs within the operating range

```
 3    record initial states
 4    determine saturation time (Algorithm 2.2)
 5    set collected data (Algorithm 2.3)
 6 end for
```

The original mapping process from a plant to a neural network is shown in Fig. 2.2. However, this method cannot be applied to a plant that has a saturation property because the output partitioning does not affect the mapping process. For plants with a saturation property, inputs to the plant should be selected first; then the plant is determined to be saturated or not. If saturated, the saturation time is stored and used as a time window. That is, the target states are decided not from output partitioning in the original scheme but from the saturation property of the plants. Thus, the initial state and saturation state are taken to be the current state (*CS*) and target state (*TS*) respectively. In brief, the original method first determines the current state and target state with the output partitioning and then determines inputs with the input sampling. Afterwards, the length of time it takes for current state to reach it's target state is determined. Our method, on the other hand, first determines the inputs with the random sampling and then checks the saturated state with time. The saturated state and saturation time are regarded as a target state and a time window respectively.

The detailed description of the procedure for the determination of saturation time (time windows) is shown Algorithm 2.2.

Algorithm 2.2 Saturation-time-determination()
// P: the number of plant output variables //
// plant(): a simulation plant with discrete time-based simulation //
// st(j): a saturation time of j'th output variable //
// ct: a simulation time //
// dt: a delta time //

```
 1 do // determination of the saturation times //
 2    for k = 0 to P − 1 // saving the previous plant outputs //
 3       p_output_old(k) ← p_output(k)
 4    for k = 0 to P − 1 // checking saturation //
 5       plant (ct, initial(P), p_input(M), p_output(P) )
 6       if p_output(k) = p_output_old(k) then
 7          st(k) ← ct
 8       end if
 9    end for
10    ct ← ct + dt
11 until all p_output(P) = all p_output_old(P)
```

Although the neural network model represents the dynamics in a broader range, it cannot be exactly the same as the real plant itself. This is because the learning of the neural network is not complete. Thus, we add a Gaussian noise to the saturation time to provide a robustness against the external disturbances. Algorithm 2.3 shows details of the data setting procedure.

Algorithm 2.3 Data-setting()
// N: the number of learning patterns //
// M: the number of plant input variables /
// P: the number of plant output variables //

```
 1 for i = 0 to N − 1
 2    for j = 0 to P − 1
 3       nn_inp_pat(i, j) ← initial(j) // setting initial states //
 4       nn_inp_pat(i, j + P) ← p_output(j) // setting target states //
 5    end for
```

```
 6    for j = 0 to P − 1 // adding Gaussian noise //
 7        time1 ← st(j) * gasdev()
 8        time2 ← st(j) * gasdev()
 9        min_time(j) ← min(timel, time2) // determining time windows//
10        max_time(j) ← max(timel, time2)
11    end for
12    for j = 0 to P − 1 // setting min. and max. times //
13        nn_out_dat(i, j) ← min_time(j)
14        nn_out_dat(i, j + P) ← max_time(j)
15    end for
16    for j = 0 to M − 1 // setting inputs //
17        nn_out_dat(i, j + 2P) ← p_Input (j)
18 end for
```

The whole algorithm is outlined as follows. First, inputs within operating ranges are randomly selected with uniform distribution and are stored for later use. Second, current plant outputs are stored as initial states. Third, the selected inputs are applied to the plant, which is simulated with discrete time-based simulation. Finally, the time at which the output of the plant is saturated is measured and stored. Using this information, the state space of the plant is mapped into a neural network as an event-based model of the plant.

2.4   Simulation Results

Based on our event-based control scheme, we realized an event-based control system, HICON (High-level Intelligent CONtroller), with a neural network model of a plant on an expert system ART-IM/windows [29]. In this experiment, the CSTR plant was simulated using Eq. (2.10) with a discrete time based simulation method at 50 ms time intervals. The neural network plant model must complete learning the elements of the table model before the control operation begins.

Using the neural network model, a neural network manager generated the time windows and control outputs for the controller. New data obtained by the control operation can also be learned to dynamically adapt to the control environment. This dynamic learning provides flexible modelling capability to the event-based control system. Figure. 2.10 shows the overall controlled situation. The input/output states of the controlled plant are respectively depicted in the middle/bottom boxes of "CSTR Plant" window. In the output state, the desired set points are represented by bold lines and the controlled output by thin lines. As previously mentioned, the steady state error due to the incomplete learning is observed. This steady state error is generated only when the controlled plant has a saturation property. That is, when the plant output is saturated with respect to an input, model-plant mismatch causes the steady state error.

2.5   Conclusions

This chapter discussed a neural network application method for event-based intelligent control and supervisory control. We showed the usefulness of neural network modeling through an experiment of a chemical plant with a saturation property. This control method can be applied to any type of systems even hybrid systems composed of discrete event and continuous systems. In application to supervisory control, its diagnosis capability will enhance the performance of the supervisory control. The scheme may be viewed as a combined method of time-based diagnosis and state-based control. Experimental results showed that this scheme could be used to control a complex nonlinear plant and could be associated with the higher knowledge-based task management modules to construct more autonomous control system in further works [30, 31, 16].

Acknowledgments

The authors would like to thank Dr. J. J. Song for his helpful suggestions.

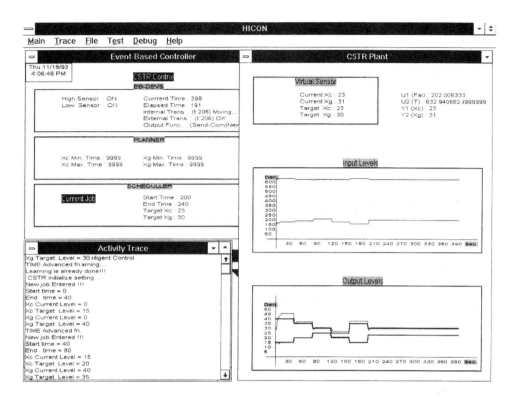

FIGURE 2.10   Result of CSTR Plant Control.

References

1. K. S. Narendra and K. Parthasarathy, "Identification and Control of Dynamical Systems Using Neural Networks," *IEEE Trans. on Neural Networks,* vol. 1, pp. 4–27, Mar. 1990.

2. Y. Ichikawa and T. Sawa, "Neural Network Application for Direct Feedback Controllers," *IEEE Trans. on Neural Networks,* vol. 3, pp. 224–231, Mar. 1992.

3. C.-C. Lee, "Intelligent Control Based on Fuzzy Logic and Neural Net Theory," *Proceedings of the International Conference on Fuzzy Logic,* pp. 759–764, July 1990.

4. C.-T. Lin and C. G. Lee, "Neural-Network-Based Fuzzy Logic Control and Decision System," *IEEE Trans. on Computers,* vol. 40, pp. 1320–1336, Dec. 1991.

5. S. ichi Horikawa, T. Furuhashi, and Y. Uchikawa, "On Fuzzy Modeling Using Fuzzy Neural Networks with the Back-Propagation Algorithm," *IEEE Trans. on Neural Networks,* vol. 3, pp. 801–806, Sept. 1992.

6. F. Highland, "Embedded AI," *IEEE Expert,* pp. 18–20, June 1994.

7. P. Antsaklis, "Defining Intelligent Control," *IEEE Control Systems,* pp. 4–5,58–66, June 1994.

8. R. Shoureshi, "Intelligent Control Systems: Are They for Real ?," *Journal of Dynamic Systems, Measurement, and Control,* vol. 115, pp. 392–401, June 1993.

9. B. P. Zeigler, "High Autonomy Systems: Concepts and Models," *Proceedings of AI, Simulation, and Planning in High Autonomy Systems,* pp. 2–7, Mar. 1990.

10. S. Chi, *Modelling and Simulation for High Autonomy Systems.* PhD thesis, University of Arizona, 1991.

11. S. H. Jung, *Multilevel, Hybrid Intelligent Control System: Its Framework and Realization.* PhD thesis, KAIST, Feb. 1995.

12. B. P. Zeigler, "DEVS Representation of Dynamical Systems: Event-Based Intelligent Control," *Proceedings of the IEEE,* vol. 77, pp. 72–80, Jan. 1989.

13. B. P. Zeigler, *Object Oriented Simulation with Hierarchical, Modular Models: Intelligent Agents and Endomorphic Systems*. Academic Press, 1990.

14. B. P. Zeigler, *Theory of Modelling and Simulation*. John Wiley & sons, 1976.

15. B. P. Zeigler, *Multi-Faceted Modelling and Discrete Event Simulation*. Academic Press, 1984.

16. S. H. Jung, T. G. Kim, and K. H. Park, "HICON: A Multi-Level, Hybrid Intelligent Control System," *IEEE Trans. on Systems, Man and Cybernetics*. submitted to IEEE Trans. on Systems, Man and Cybernetics.

17. J. J. Song and S. Park, "Neural Model Predictive Control For Nonlinear Chemical Processes," *Journal of Chemical Engineering of Japan*, vol. 26, no. 4, pp. 347–354, 1993.

18. A. Benveniste and P. L. Guernic, "Hybrid Dynamical Systems Theory and the SIGNAL Language," *IEEE Trans. on Automatic Control*, vol. 35, pp. 535–546, May 1990.

19. C.-J. Luh and B. P. Zeigler, "Abstracting Event-Based Control Models for High Autonomy Systems," *IEEE Trans. on Systems, Man and Cybernetics*, vol. 23, pp. 42–54, JANUARY/FEBRUARY 1993.

20. C. L. Giles, G. M. Kuhn, and R. J. Williams, "Dynamic Recurrent Neural Networks: Theory and Applications," *IEEE Trans. on Neural Networks*, vol. 5, pp. 153–155, Mar. 1994.

21. J. J. Song, *Intelligent Control of Chemical Processes Using Neural Networks and Fuzzy Systems*. Ph.D. thesis, KAIST, 1993.

22. P. J. Ramadge and W. M. Wonham, "Supervisory control of a class of discrete event processes," *SIAM J. Control and Optimization*, vol. 25, pp. 206–23O, Jan. 1987.

23. P. J. G. Ramadge, "Some tractable supervisory control problems for discrete-event systems modeled by buchi automata," *IEEE Trans. on Automatic Control*, vol. 34, pp. 10–19, Jan. 1989.

24. R. Kumar, V. Garg, and S. I. Marcus, "Predicates and prediate transformers for supervisory control of discrete event dynamical systems," *IEEE Trans. on Automatic Control*, vol. 38, pp. 232–247, Feb. 1993.

25. T. Ushio, "A necessary and sufficient condition for the existence of finite state supervisors in discrete-event systems," *IEEE Trans. on Automatic Control*, vol. 38, pp. 135–138, Jan. 1993.

26. J. S. Ostroff and W. M. Wonham, "A framework for real-time discrete event control," *IEEE Trans. on Automatic Control*, vol. 35, pp. 386–397, Apr. 1990.

27. W. T. I. and R. Otto, "A Generalized Chemical Processing Model for the Investigation of Computer Control," *Trans. Am. Inst. Elect. Engr.*, vol. 79, pp. 458–465, 1960.

28. C. McFarlane and D. Bacon, "Adaptive Optimizing Control of Multivariable Constrained Chemical Processes. 1. Theoretical Development, 2. Application Studies," *Ind. Eng. Chem. Res.*, vol. 28, pp. 1828–1835, 1989.

29. "ART-IM/Windows Programming Language Reference." Inference Corporation, 1991.

30. T. G. Kim and B. P. Zeigler, "AIDECS: An AI-Based, Distributed Environmental Control System for Self-Sustaining Habits," *Artificial Intelligence in Engineering*, vol. 5, no. 1, pp. 33–42, 1990.

31. T. G. Kim, "Hierarchical Scheduling in an Intelligent Environmental Control System," *Journal of Intelligent and Robotic Systems*, vol. 3, pp. 183–193, 1990.

3

A Rule-Based Expert System for Designing Flexible Manufacturing Systems

Heungsoon Felix Lee
Southern Illinois University

Designing advanced manufacturing systems like flexible manufacturing systems (FMSs) involves the solution of a complex series of interrelated problems. In this paper, we present an expert system to aid this design process for FMSs. The proposed system combines a rule-based expert system with computer simulation in order to capture dynamics of FMSs, evaluate design alternatives of FMSs, and seek effective ones with user friendly interface.

3.1 Introduction

Advanced manufacturing systems like flexible manufacturing systems (FMSs) are capital-intensive. Designing functional, yet cost-effective FMSs is a challenging task because it involves the solution of a complex series of interrelated problems. The importance of early design activities is emphasized for highly automated manufacturing systems. About 80% of the total budget is committed at the design stage (Vollbracht 1986) and 55% of the engineering cost is spent by the project authorization point (Harter and Mueller 1988).

A typical flexible manufacturing system (FMS) consists of groups of versatile numerically-controlled (NC) machines that are linked by a material handling system (MHS). Machines within each group are tooled identically and are capable of performing a certain set of operations. Operations and material movements are all under a central computer control. Since FMSs were introduced in the early 1960s, broader applications of FMSs have been developed in the areas of injection molding, metal forming and fabricating, and assembly. In 1989, roughly 1200 FMSs existed worldwide. According to forecasts, between 2500-3000 FMSs will be operating in the year 2000 (Tempelmeier and Kuhn 1993). FMSs, however, are highly capital-intensive and FMS designers are interested in seeking minimal-cost or minimal resource-usage design alternatives that satisfy performance and technical requirements such as throughput capacity and flexibility capacity.

Given selection of part types to be produced, we study a design problem of FMSs that consist of multiple types of NC machines. This problem seeks minimal cost design subject to meeting throughput requirements.

The decisions to be made include the number of machine groups, the number of machines at each group, the number of pallets, the number of transporters, and batch transfer size. When parts are small, a batch of parts can be mounted on a pallet and transferred together between machine groups in order to reduce material handling operations. At the early design stage of complex manufacturing systems like FMSs, different design issues are highly related and should not be treated independently (Heavey and Browne 1996). Highly significant interactions between the design factors may invalidate simple one-factor-at-a-time procedures for finding a minimum-cost system design. In FMSs, machines are flexible and versatile and there is a large latitude in allocating workload among machine groups. Clearly, there are strong interactions between the workload allocation and the optimal system configuration and between the batch size and the MHS capacity.

Research works on FMS design problems can be divided into three groups based upon the modeling techniques employed. These are queueing networks, integer programming, and simulation. Many researchers have used closed queueing network (CQN) models to solve design problems for flexible manufacturing systems (FMSs). Machines, at each machine group, are modeled as a multiserver station and pallets carrying work-in-process inventories modeled as the fixed job population circulating in CQN. A Markovian closed queueing network model is used by Vinod and Solberg (1985), Shanthikumar and Yao (1988), Dallery and Stecke (1990), Kouvelis and Lee (1995), and Tetzlaff (1995). All of these works deal with specific decisions, assuming that many other design decisions are already known. Also, they make several assumptions for ease of analysis such as exponential service times and large buffer spaces, which are often unrealistic.

Researchers have also used integer programming (Whitney and Suri 1985, Graves and Redfield 1988, Afentakis 1989). These integer programming models do not take into account the aspects of material handling issues and product flows, of resource contention and machine idle time, and of random events occurring on the assembly floor such as machine breakdowns or machine tool jams.

Simulation has been used by several researchers (Thompson et al. 1989, Nandkeolyar and Christy 1992, Winters and Burstein 1992). Simulation is the process of designing a model of a real system and conducting experiments with this model for the purpose of understanding the behavior of the system and/or evaluating various strategies for the operation of the system (Zeigler 1984). Simulation is flexible to represent an FMS at any level of detail realistically. However, it can be also costly and time consuming to develop, validate and run simulations for many design alternatives before one good alternative is chosen. Furthermore, one cannot tell how good the chosen alternative is because simulation does not usually provide an optimal solution or benchmark with which the chosen alternative can be compared. Simulation design processing is a numerical technique without the functions of reasoning and symbolic processing. A relatively high level of training is necessary to perform useful simulation studies. Managers, as unskilled users, may have difficulty building a simulation model, validating the simulation, and interpreting the results of simulations.

An expert system is a computer system that can solve problems using expertise and knowledge of the system environment in ways that mimic a human expert in a specialized problem area (Kusiak and Chen 1987, Rao and Lingaraj 1988). Thus, expert system technology can speed problem solving and address problems in complex and difficult problem domains (Tolar and Platt 1992). Expert systems were first applied in production and operations management. Today, expert systems and easy-to-use expert system shells are used in many business fields.

The relationship between expert systems and simulation is that expert knowledge often reflects time-dependent phenomena, even though that knowledge is usually in a rough form such as natural language or rules. By bridging gaps between qualitative and quantitative approaches, expert systems and simulation can greatly benefit each other (Fishwick 1991). This combined system is called a knowledge-based simulation system or a hybrid expert simulation system (HESS) in the literature. Its theoretical background is presented by Elzas etc. (1989) and Fishwick and Modjeski (1991) and examples of successful applications are presented by Stirling and Sevinc (1991), Eisenberg (1991), and Lee et al. (1996). In this paper, we present a new HESS application to FMS design.

The proposed HESS has advantages over other systems using queuing networks, integer programming, or simulation. Since the HESS uses simulation as a component, it obviates restrictions and assumptions required by queueing networks and integer programming. The proposed HESS has a nice user interface and obviates drawbacks of simulation as follows.

1. With user input for FMS design parameters, it automatically generates correct simulation programs so that users do not need to write simulation programs and verify them.
2. It interprets the simulation results and provides expert suggestions for improvement for FMS design.
3. It allows reiterations with changes of some FMS design parameters until a particular desired design or a small number of potential design alternatives are found. All these activities are seamless and users are not required to have knowledge on simulation or expert systems.

The HESS can be further improved by combining queuing network or integer programming approaches (Lee and Stecke 1996). The latter approaches can provide more effective initial FMS design to the HESS than the user input which usually relies on guess. This will help to reduce the number of iterations the HESS needs to undergo.

The remainder of this paper is organized as follows. In Section 3.2, we give a brief description of a typical FMS, in Section 3.3, we present the structure and user interface of the proposed HESS, and in Section 3.4, we illustrate the proposed HESS with an example.

3.2   Flexible Manufacturing Systems

An example of an FMS appears in Fig 3.1. This FMS produces different sizes of housings for automatic transmissions. It consists of four large 5-axis machining centers (called Omnimills), three 4-axis machining centers (Omnidrills), two vertical turret lathes (VTLs) and an inspection machine. Each machine has a limited-capacity tool magazine that hold tools assigned to it. The 16-station load/unload (L/UL) area provides a queuing area for parts entering the system, finished parts leaving the system, and in-process inventories. Three manual workers work in the L/UL area for loading/unloading and fixturing parts. Two transporters run on a straight track and carry 15 pallets among the machines and the L/UL stations.

Another example of an FMS appears in Fig. 3.2. This FMS has several CNCs and a centralized tool supply system where the tools of the cassettes are preadjusted and prepared for operation at a tool setup area. Afterwards they are either stored at a central tool magazine or transported to a local tool magazine at a machine. Two L/UL stations are located on the left hand side and represent the interface between the FMS and its production environment. Central buffer areas are placed along the transportation track. These buffer areas are temporary waiting spaces for parts that are waiting to be processed by the next machine after the current processing on another machine. These spaces are limited due to a limited floor space or storage facility. With use of the waiting spaces that can temporarily hold parts, a machine can continue operating while the following machines are busy or stopped and under repair.

Parts are loaded on pallets and enter an FMS at the L/UL station and then are routed through different processing resources (stations or machine groups) for various operations. Transport resources (automated guided vehicles (AGVs) or transporters) may be required to route parts from one station (machine group) to another with some attendant travel time. If conveyors move parts between machine groups, there is no or very

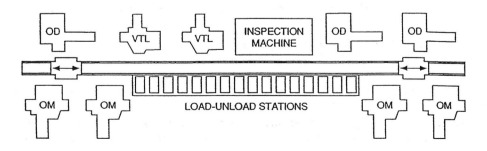

FIGURE 3.1   Sundstrand/Caterpillar FMS.

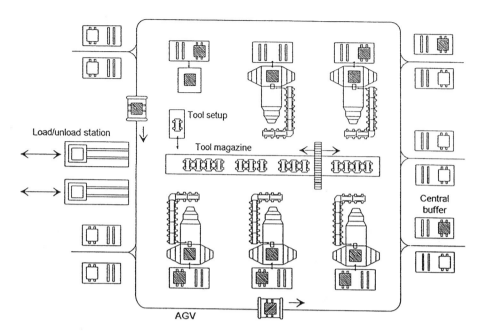

FIGURE 3.2 FMS with two AGVs and a central tool magazine. (From Tempelmeier, H. and Kuhn, H., *Flexible Manufacturing Systems: Decision Support for Design and Operation,* 1993, copyright ©John Wiley & Sons, Inc. Reprinted by permission of John Wiley & Sons, Inc.)

little waiting time for transport resources, and only a travel time between processing resources. A processing resource may go through an up and down cycle. For example, a machine breaks down and needs to be repaired.

The number of pallets circulating in the system remains more or less constant during a production period. This is the case when a base part fixed on a pallet enters a system at a loading station and travels through various machine groups for different operations. Upon completion, it travels to an unloading station where a finished part is taken off the pallet and is shipped and another base part is fixed on the pallet and the process repeats. This policy is common in both advanced and traditional manufacturing systems (Lee 1997, Spearman et al. 1990). The number of pallets circulating in the system affects the production rate and is an important decision variable.

Since FMSs simultaneously produce different products, an operating policy has to be specified concerning a process rule and an input rule. A process rule determines which part type is to be processed next on a machine while an input rule determines which part type is to be released into the system at the L/UL station. We use FCFS for the process rule. The input rule we use to choose a part type, such that ratios among completed plus work-in-process parts, are maintained throughout the entire production as close to ratios among their production requirements as possible. These rules help to achieve balancing workloads in machines since different part types require different processing times and at the same time to meet all production requirements. These rules are simple and easy to control yet effective in FMS operation (Lee and Stecke 1996).

3.3 A Hybrid Expert Simulation System

O'Keefe (1986) recognized the important roles of expert systems and simulation in support of decision making and the relative strengths of these two tools and proposed a taxonomy for combining simulation and expert systems into a HESS. The concept of HESS is to integrate existing simulation and expert system tools and exploit the knowledge of the expert system programmer as well as that of the simulation modeling expert (Shannon and Adelsberger 1985). One of the classes in O'Keefe's taxonomy is the use of an expert system as an intelligent user interface or front end to a simulation tool. When the skilled designer builds a hybrid expert simulation system of this sort, the relatively unskilled user is spared the problems of building,

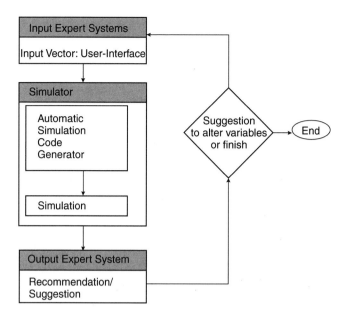

FIGURE 3.3   The conceptual structure of the proposed HESS.

validating, and interpreting the results of a simulation model and the time and effort necessary to get simulation results is considerably reduced. The structure of such a HESS consists of three sub-systems (Input expert system, Simulator, and Output expert system) and is illustrated in Fig. 3.3 (Lavary and Lin 1988).

We developed and implemented a HESS on a PC. The proposed HESS uses Siman V (Pegden et al. 1995) for the simulation tool and VP-Expert (Moose et al. 1989) for the expert system shell. Neither of these tools has provision for interfacing with the other, so we constructed interfaces in Turbo Pascal. Figure 3.4 shows the detail structure of the HESS.

The Input Expert Systems (IES)

The input expert system verifies the compatibility of the components of the input vector through reference to its knowledge base which contains realistic ranges of variables in the input vector. It eliminates unnecessary runs of the simulator by excluding erroneous input vectors and enhances the functionality of the input system with a user friendly interface.

The Input Expert System (IES) is composed of a VP-Expert program and IO.EXE program in Turbo Pascal. These programs provide an easy-to-use interface for entry of basic system variables by users, conveniently capturing the sequence of data necessary for the simulation. The input variables are summarized in Fig. 3.5. With the user input, the IES generates *file*.DAT, a user generated file name, which is the data file to be used by the Simulator for automatic simulation code generation. In addition to the input variables, the user can control the simulation by entering simulation variables such as a run length and a warm-up period to avoid the effect of transient behavior.

The Simulator

The Simulator captures the simulation model of FMSs. It is based on a discrete-state process-interaction modeling approach in which the system state changes at events on discrete-points in time and events are updated as entities (parts) arrive and flow through the system. The core of the simulator is the Automatic Siman Code Generator (ASCG). ASCG automatically generates a complete Siman program based on the system and simulation input data the user enters via IES. Since a Siman simulation program consists of a model frame and an experiment frame, the generated Siman Code leads to two files "File.MOD" and "File.EXP".

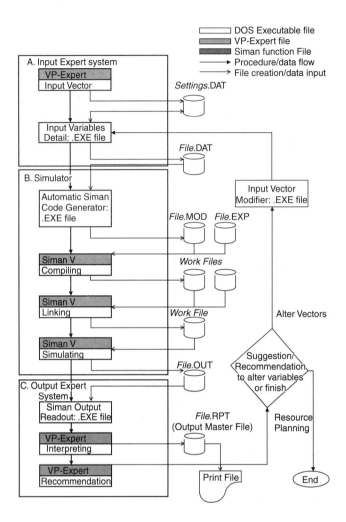

FIGURE 3.4   The detail structure of the proposed HESS.

The model frame describes the logical flow of entities within the system while the experimental frame provides the experimental conditions for executing the model frame. ASCG is written in Turbo Pascal. ASCG has been designed to be flexible and consider a wide range of variables, as the user is given choices and suggested changes on input variables that would lead to more effective FMS design. After the simulation program is generated, the HESS compiles and runs the simulation by use of Siman V.

The Output Expert Systems (OES)

The output expert system interprets and analyzes the simulation results and makes recommendations to the user on changes in the input vector in order to find a more effective FMS design. Thus, the user is shielded from the actual simulation output and is relieved of the task of its interpretation. The Output Expert System (OES) is composed of a VP-Expert and a Readout.EXE program. Readout.EXE reads *file*.OUT, which is the output of the simulator, makes numeric calculations, and returns the information to VP-Expert. The VP Expert program applies expert rules to make recommendations for an improved FMS design. Users then decide whether to accept the recommendations or not. As in the case of the Input Expert System, the Output Expert System is designed with a user-friendly interface.

The rules are based on the empirical experience and heuristic knowledge of experts that allow identification and elimination of bottlenecked and underutilized resources while, at the same time, meeting the system

<u>Objective</u>	<u>Input Variables</u>
Produce the desired number of finished parts with minimal use of resources	Number of stations (m/c groups)
	Number of machines at each station
	Number of product types
	Product mix (ratio of product types)
	Distribution type and parameter values for process times at each station[#]
	Sizes of waiting spaces at stations*
	Machine failure and repair time*
	Number of transporters (dispatchers)*
	Transfer times between stations - velocity and distances
	Route for each customer type among stations
	Demand for each product
	Number of pallets
	Batch size (parts per pallet)
	Machine jam probability and jam clear time*

[#] This is obtained from the initial operation assignment to stations and product mix.
* User-option variables

FIGURE 3.5   Objective and input variables of FMS design.

objective or design requirements. The rules use threshold values for machine and transporter utilizations to determine if each resource type is over-utilized or under-utilized. Both threshold values are tentatively set at 75%. In practice, the threshold values would depend on various factors used in the industries and be supplied by users who find these values through their knowledge of and experimentation with their systems.

The user can control the input variables to meet all product requirements at minimal use of resources. The rules in Fig. 3.6 show the decision tree for recommendations for an improved FMS design. The recommendations are suggested changes on one or more variables of the followings: the number of machines, the number of transporters, the number of pallets, and the batch size. When parts are small, a batch of parts can be mounted on a pallet using a special fixture. For example, a tombstone fixture allows up to four parts to be loaded on a pallet (Luggen 1991). A batch of parts are processed consecutively at a machine and then moved together to the next machine by a transporter. Batch sizing can be effective in increasing throughput when material handling resource causes a bottleneck. However, an unnecessarily large batch size increases the work-in-process inventories and system congestion without increasing throughput.

The OES also generates *File*.RPT which contains summary results of every simulation run as the user explores various FMS designs, following recommendations provided by the OSE. Results are written in the file, including a vector of input data, simulation summary result, and recommendations (Levary and Lin 1988). Currently the results are used to avoid redundant simulation runs, but in a more elaborated HESS, the knowledge base of past runs could be used for more sophisticated rules.

3.4   An Example on FMS design with HESS

An FMS produces three different products, each of which has a different route among four machine groups. These three types take 25%, 30%, and 45% of total parts produced. Travel time between machine groups is always 3 minutes. Each operation time is exponentially distributed. Each machine group fails every 4 hours on average and the average repair time is 15 minutes. Both follow exponential distributions. Thirty pallets are available and 2 transporters are used to move parts fixtured on pallets between machine groups. The batch size is fixed to one in this example. Machine type and the number of machines for each group are summarized in Table 3.1.

TABLE 3.1 Number of Machines in Each Machine Group

Machine Group	Machine Type	Number of machines
1	CNC-mill	7
2	VTL	3
3	Drill Presses	6
4	Shapers	4

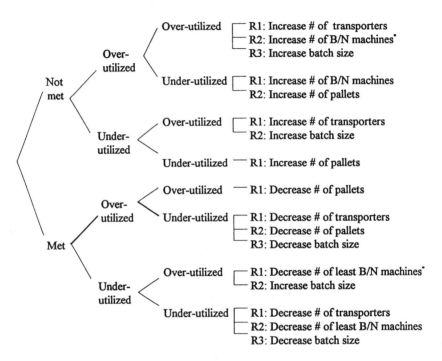

#Average machine utilization is an aggregate average of machine utilization's over all machine groups.

&Disregard this column and the related rules when a system does not use transporters.

* B/N: A bottleneck machine group is one in which the largest number of parts is waiting to be processed. The recommendation is to increase the number of machines in this machine group or to decrease the number of machines in the least bottleneck machine group.

FIGURE 3.6 Expert system decision tree for FMS design.

As each part moves through the FMS, it is processed at each machine group according to its visitation sequence. (See Table 3.2). At each group, a part waits in the queue, seizes the first available machine, is delayed by the processing time, is released by the machine, and then continues to its next group. On completion, a finished part leaves the system and at this time another base part enters the system.

The FMS is designed to meet weekly demands of three products, (125, 225, 150) with minimal use of resources. It operates in 5 days per week, 2 shifts per day, and 8 hours per shift. The simulation length is set to 5800 minutes, which is the warm-up period of 1000 minutes followed by one week, (i.e., 4800 minutes

TABLE 3.2   Processing Times for Three Product Types

Jobs Operation Type	Percent of Jobs	Sequence Number	Machine Type (group number)	Average Processing Time
1	25	1	CNC-mill (1)	35
		2	Drill Presses (3)	45
		3	Shapers (4)	18
2	45	1	CNC-mill (1)	35
		2	VTL (2)	25
		3	Shaper (4)	45
		4	Drill Press (3)	30
3	30	1	CNC-mill (1)	30
		2	Shaper (4)	20
		3	VTL (2)	15
		4	Drill Press (3)	45

[5 days/week * 2 shifts/day * 8 hours/shift * 60 minutes/hour]). The FMS designer initiates the proposed HESS with the number of machines 7, 3, 6, 4, for the four machine groups. In the first run, the following report is produced:

Summary Report 1

Project title: FMS design
Number of machines in each machine group: 7   3   6   4
Average machine utilization: 72.3%
Number of pallets: 30
Number of transporters: 2
Transportation utilization: 65.1%
Batch size: 1
Average time in system: 278.180
B/N machine group: #4 Queue Size = 7.787
Least B/N machine group: #1 Queue Size = 0.214
Total number of parts produced: 518
Type 1: 131, Type 2: 232, Type 3: 155

All demands are met and the least bottlenecked machine group is group 1. After the first run, the HESS gives two recommendations–R1: Decrease the number of transporters and R2: Decrease the machine number of machine group. The manager selects the second recommendation and decreases the number of CNC1s from 7 to 6. The simulation is run again. Table 3.3 lists a summary of 9 simulation runs and recommendations/actions taken for the 9 successive runs.

The recommended FMS design is 5 CNC-mills, 3 VTLs, 6 drill presses, 4 shapers with 2 transporters and 21 pallets, since this plan requires the minimal level of resources among those that meet demands (see Run number 7 in Table 3.3).   The summary report for this FMS design is provided below. The entire session with the proposed HESS for this case study took less than 30 minutes on a PC.

Summary Report 7

Project title: FMS design
Number of machines in each machine group: 5   3   6   4
Average machine utilization: 78%
Number of pallets: 21
Number of transporters: 2
Transportation utilization: 69%
Batch size: 1

TABLE 3.3   Summary of Simulation Runs and Actions Taken for the Successive Runs[&]

Run Number	Resource Plan: (m/c groups), (transporters, pallets)	Demand Met	Average m/c Utilization*	Transporter Utilization	Recommendation/ Action Taken
1	(7,3,6,4), (2,30)	met	72% (under)	65% (under)	decrease m/c group 1 by 1
2	(6,3,6,4), (2,30)	met	78% (over)	71% (under)	decrease transporters by 1
3	(6,3,6,4), (1,30)	not met	74% (under)	97% (over)	increase transporters by 1 and decrease pallets by 3
4	(6,3,6,4), (2,27)	met	79% (over)	73% (under)	decrease pallets by 3
5	(6,3,6,4), (2,24)	met	76% (over)	70% (under)	decrease pallets by 3
6	(6,3,6,4), (2,21)	met	73% (under)	69% (under)	decrease m/c group 1 by 1
7	(5,3,6,4), (2,21)#	met	78% (over)	69% (under)	decrease pallets by 3
8	(5,3,6,4), (2,18)	not met	71% (under)	66% (under)	increase pallets by 2
9	(5,3,6,4), (2,20)	not met	76% (over)	69% (under)	terminate HESS

[&] Batch size in this example is fixed to one.

* Average m/c utilization is an aggregate average of machine utilizations over all machine groups. 75% is used as the threshold value to determine if machines or transporters are over-utilized or under-utilized.

This resource plan is recommended since it requires the minimal level of resources among ones that meet demands.

Average time in system: 198.6
B/N machine group: #4 Queue Size = 2.655
Least B/N machine group: #1 Queue Size = .507
Total number of parts produced: 511
Type 1: 129, Type 2: 229, Type 3: 153

Acknowledgment

Dr. Lee's research is supported in part by grants from the National Science Foundation (Grant No. DDM-9201954) and from Southern Illinois University at Edwardsville.

References

Afentakis, P. (1989). A loop layout design problem for flexible manufacturing systems. *International Journal of Flexible Manufacturing Systems*, 1, p. 175.

Dallery, Y. and Stecke, K. E. (1990). On the optimal allocation of servers and workloads in closed queuing networks. *Operations Research*, 38, p. 694.

Eisenberg, M. (1991). The Kineticist's workbench: qualitative/quantitative simulation of chemical reaction mechanisms. *Expert Systems with Applications*, 3(3), p. 367.

Elzas, M. S., Oren, T. I. and Zeigler, B. P. (1989). *Modeling and simulation methodology: Knowledge systems' paradigms.* Amsterdam: North Holland.

Fishwick, P. A. (1991). Knowledge-based simulation. *Expert Systems with Applications*, 3(3), p. 301.

Fishwick, P. A. and Modjeski, R. B. (1991). *Knowledge based simulation: Methodology and application.* New York: Springer Verlag.

Graves, S. C. and Redfield, C. H. (1988). Equipment selection and task assignment for multiproduct assembly system design. *International Journal of Flexible Manufacturing Systems*, 1, p. 31.

Haddock, J. (1987). An expert system framework based on a simulation generator. *Simulation*, 48, p. 46.

Harter, J. A. and Mueller, C. J. (1988). FMS at Remington. *Manufacturing Engineering*, 100, p. 91.

Heavey, C. and Browne, J. (1996). A model management systems approach to manufacturing systems design. *International Journal of Flexible Manufacturing Systems*, 8, p. 103.

Kouvelis, P. and Lee, H. L. (1995). An improved algorithm for optimizing a closed queuing network model of a flexible manufacturing system. *IIE Transactions*, 27, p. 1.

Kusiak, A. and Chen, M. (1987). Expert systems for planning and scheduling manufacturing systems. *European Journal of Operational Research,* 56, p. 113.

Lee, H. F. (1997). Production planning for flexible manufacturing systems with multiple machine types: a practical method. To appear in *International Journal of Production Research.*

Lee, H. F., Cho., H. J. and Klepper, R. W. (1996). A HESS for resource planning in service and manufacturing industries, *Expert Systems with Applications,* 10, p. 147.

Lee, H. F. and Stecke, K. E. (1996). An integrated design support system for flexible assembly systems. *Journal of Manufacturing Systems,* 15, p. 13.

Levary, R. R. and Lin, C. Y. (1988). Hybrid expert simulation system (HESS). *Expert Systems,* 5, p. 120.

Li, Z., Tang, H. and Tu, H. (1992). An expert simulation system for the master production schedule. *Computers in Industry,* 19, p. 127.

Moose, A., Schussler, T., and Shafer, D. (1989). *VP-Expert.* Paperback Software International.

Moser, J. G. (1986). Integration of artificial intelligence and simulation in a comprehensive decision-support system. *Simulation,* 47, p. 223.

Nandkeolyar, U. and Christy, D. (1992). Evaluating the design of flexible manufacturing systems. *International Journal of Flexible Manufacturing Systems,* 4, p. 267.

O'Keefe, R. (1986). Simulation and expert systems — a taxonomy and some examples. *Simulation,* 47, p. 10.

Palaiswami, S. and Jenicke, L. (1992). A knowledge-based simulation system for manufacturing scheduling. *International Journal of Operations and Production Management,* 12, p. 4.

Pegden, C. D., Shannon, R. E., and Sadowski, R. P. (1995). *Introduction to simulation using Siman.* McGraw Hill, New Jersey.

Rao, H. R. and Lingaraj, B. P. (1988). Expert systems in production and operations management: classification and prospects. *Interfaces,* 18, p. 80.

Shannon, R. E. and Adelsberger, H. H. (1985). Expert systems and simulation. *Simulation,* 46, p. 275.

Shanthikumar, J. G. and Yao, D. D. (1988). On server allocation in multiple center manufacturing systems. *Operations Research,* 36, p. 333.

Spearman, M. L., Woodruff, D. L., and Hopp, W. J. (1990). CONWIP: a pull alternative to Kanban. *International Journal of Production Research,* 28, p. 879.

Stirling, D. and Sevinc, S. (1991). Combined simulation and knowledge-based control of a stainless steel rolling mill. *Expert Systems with Applications,* 3(3), p. 353.

Tempelmeier, H. and Kuhn, H. (1993). *Flexible Manufacturing Systems: Decision Support for Design and Operation.* John Wiley and Sons, Inc., New York, NY.

Tetzlaff, U. (1995). A model for the minimum cost configuration problem in flexible manufacturing systems. *International Journal of Flexible Manufacturing Systems,* 7, p. 127.

Thompson, G., Lafond, N., and Kekre, S. (1989). Managing operational costs in flexible assembly with asynchronous flows. *Proceedings of the 3rd ORSA/TIMS Conference on Flexible Manufacturing Systems,* Cambridge, MA, K. Stecke and R. Suri (Eds.), Elsevier Science Publishers B. V., Amsterdam, p. 199.

Tolar, K. and Platt, R. G. (1992). MAG-EX: a magnetic fabrication expert system. *Computers and Industrial Engineering,* 16, p. 165.

Vinod, B. and Solberg, J. J. (1985). The optimal design of flexible manufacturing systems. *International Journal of Production Research,* 23, p. 1141.

Vollbracht, G. R. (1986). The time for CAEDM is now. *Proceedings of the Fourth National Conference on University Programs in Computer-Aided Engineering, Design and Manufacturing,* Purdue University, W. Lafayette, IN, p. 86.

Whitney, C. K. and Suri, R. (1985). Algorithms for part and machine selection in flexible manufacturing systems. *Annals of Operations Research,* 3, p. 239.

Winters, I. J. and Burstein, M. C. (1992). A concurrent development tool for flexible assembly systems. *International Journal of Flexible Manufacturing Systems,* 4, p. 293.

Zeigler, B. P. (1984). Theory of modeling and simulation. Krieger.

4

Tool Condition Monitoring in Manufacturing Systems Using Neural Networks

G.S. Hong
National University of Singapore

M. Rahman
National University of Singapore

Y.S. Wong
National University of Singapore

4.1   Introduction

Tool condition monitoring is crucial to the efficient operation of any machining processes where the cutting tool is in constant or intermittent contact with the workpiece material and is subject to continuous wear. It presently acquires greater importance than ever as manufacturing systems are increasingly required to provide greater automation and flexibility while maintaining a high level of productivity [10]. The more recent computer numerical controllers can be programmed to monitor the time spent by a cutting tool in machining and automatically changes the tool when the total machining time spent by the tool reaches its tool life. This tool life is determined experimentally by conducting a controlled set of machining tests. The useful life of the tools tends to be conservatively taken and may be wasted, resulting in frequent tool changes and longer machine downtime, thereby decreasing the system productivity. On the other hand, there may be tools that fail prematurely compared with the average tool. Tools can also fail earlier when used in conditions not similar to those employed in the experimental determination of the tool life. A premature tool failure can result in damage to the workpiece and disrupt the automated machining operation. Hence, a suitably developed tool wear monitoring technique is needed to utilize the tool more efficiently while preventing premature tool failure. Several tool wear monitoring techniques have been developed and reported [7, 26, 6, 39, 8, 35, 43, 42, 44]. A few commercial monitoring systems have also been developed for use with CNC machines [31]. The monitoring approaches are typically based on acoustic emission, motor current, vibration, and force sensing, or their combination, and are primarily developed for application in roughing operations. Further research aims to improve the reliability of the monitoring system to minimize false alarms and allow the system to be used over a wider range of machining conditions. More recent attempts adopt sensor fusion techniques [33, 42] so as to rely on more than one type of sensor inputs for more robust deduction of the state of the cutting tool. A general scheme of such an approach is shown in Fig. 4.1. It consists of two stages: a sensing and preprocessing stage and a sensor fusion stage. In the sensing and preprocessing stage, the signal from a sensor is conditioned and processed to derive information pertinent to the machining process. This preprocessing is commonly referred to as *feature extraction*. Important features obtained from a sensor signal or a set of sensor signals are then integrated or "fused" together using an appropriate sensor fusion technique to determine the tool condition. The objective is to associate the different extracted feature pattern with a corresponding wear condition. One of the techniques used is to utilize the associative capability of neural network to arrive at a more reliable conclusion on the tool condition.

This chapter discusses some neural network applications to tool condition monitoring for turning process. Different types of tool conditions are first introduced in Section 4.2. Section 4.3 then presents various common sensor used for tool condition monitoring and sensing parameters that are sensitive to tool wear conditions. However, the information embedded in the sensors is convoluted such that the association of the tool condition with the measured sensor signal is not apparent. Some preprocessing techniques to extract the information from the relevant sensor signals are described. Thus, Section 4.4 discusses the importance of feature extraction process to enhance the sensitivity and robustness of tool condition predictions. Due to the variance in the cutting condition and the chaotic nature of the cutting process, a single feature is normally not sufficient for reliable tool condition prediction. However, an increased number of feature components inherently imply more complex heuristic rules to be employed to associate each feature pattern with a corresponding tool condition. A popular approach to such a problem is to utilize the learning capability of neural network [16, 8, 32, 33] to combine these feature components to produce some indices. Such reduced indices are usually more manageable such that simple heuristic law can be applied to associate these indices with a corresponding tool conditions. Section 4.5 introduces some common neural network architectures used in tool condition monitoring processes. In Section 4.6, various case studies are discussed to illustrate the application of these neural network examples to tool condition monitoring.

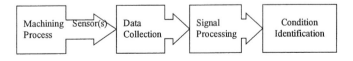

FIGURE 4.1   A general tool condition monitoring scheme.

4.2   Machining Tool Conditions

Machining can be performed efficiently if the tool geometries are very precise. Unfortunately, as the cutting process continues, wear changes the shape of the cutting edge and ultimately the tool life gets terminated when such changes are detected. In the following section, the wear mechanisms and the forms of tool wear are discussed.

Tool Wear Mechanism

The main mechanisms by which the cutting tools wear are attrition wear, abrasion wear, diffusion wear, edge chipping, and plastic deformation of cutting edge [25].

- Attrition wear: attrition wear is caused by the plucking out of microscopic fragments from the tool surface. In this wear mechanism, the work material seized to the tool is subsequently carried away by the moving chip and this imposes local tensile stresses on the cutting edge and many tiny particles may be torn out from the tool [41].

- Abrasive wear: abrasive wear is caused by the penetration and ploughing out of the hard particles from a softer surface. The inclusions in materials, such as carbides, oxides, and nitrides, that are harder than the tool material cause abrasive wear of the tool [36]. Since the built up edge (BUE) possesses high hardness, their fragmented torn out parts may also contribute to the abrasive wear of tools.

- Diffusion wear: diffusion wear takes place when the soluble atoms of the tool material diffuse into the work material across tool-work interface and are swept away with the chip [36]. This may also be caused by the diffusion of the work material into the tool and thus weakening the structure. Machining temperature, solubility, and cutting speed are the main controlling factors for diffusion rate.

- Edge chipping: at the start of a cut, when cutting with an uneven depth of cut or during interrupted cutting, sudden loads may be imposed on the cutting edge. Under such conditions, brittle carbide and ceramic tools may crack or fracture [40]. Cracks may also be formed due to thermal and mechanical fatigue arising from interrupted cutting [37]. If these cracks propagate it may cause small fragments of the tool to break away.

- Plastic deformation of cutting edge: during machining, the cutting edge of a tool is subjected to very high normal loads. At high cutting speeds, very high temperature is generated at the cutting edge of the tool and that may cause the tool material to soften and deform under high compressive stresses. This causes the tool tip to become rounded and blunt as a result of plastic deformation. The blunt cutting edge may become an additional source for further heat generation as it rubs against the freshly-machined work surface, causing further softening of the tool material and resulting in plastic collapse [40].

Different forms of tool wear and deformation processes, together with the regions on a tool in which these are likely to take place are shown in Fig. 4.2.

Forms of Tool Wear

Tool wear is a common cause of, and inevitable precursor, to tool failure in machining processes. The extent of the tool wear has a strong influence on the surface finish and dimensional integrity of the workpiece. Main forms of tool wear are flank war, crater wear, nose wear and groove (or notch) wear. Figure 4.3 shows the various wear and fracture surfaces that may be present on a worn cutting tool [38]. The measurement criteria for different types of wear are shown in the Fig. 4.4 [38].

Flank Wear

The most commonly observed phenomenon is the flank wear. The width of the wear land, VB (Fig. 4.4), is often used as a quantitative parameter of tool wear and it can easily be measured quite accurately. With

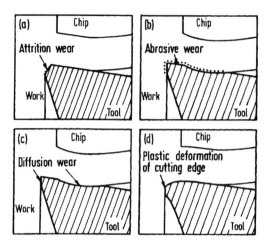

FIGURE 4.2   Wear mechanisms on cutting tools and their locations: (a) attrition wear, (b) abrasion wear, (c) diffusion wear, and (d) plastic deformation of cutting.

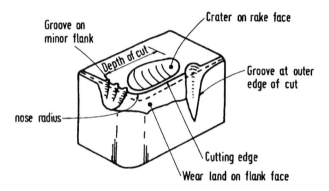

FIGURE 4.3   Various wear and fracture surfaces that may be present on a worn cutting tool.

the increase of flank wear, friction between the tool flank and the newly machined work surface increases and that leads to higher cutting forces and temperatures. Excessive flank wear may cause severe vibration leading to inefficient machining.

Crater Wear

Another form of common tool wear that appears is a crater on the rake face where the chips move over the tool surface. The formation of the crater usually starts at a distance from the cutting edge. With the elapse of time, it gradually becomes deeper and may lead to the weakening and breakage of the cutting edge. The crater wear is most common when cutting is done at relatively cutting speed on high-melting-point metals. The depth of crater, KT (Fig. 4.4), is often used as a quantitative parameter of tool wear.

Groove Wear

Groove wear may be observed at the flanks (both major and minor) at the position where the chip crosses the edge of the tool. If the groove wear becomes very deep, it may cause the tool to fracture. In practice, the tool life, which is a measure of the useful time span of the tool, is determined either by the flank wear or crater wear. Tool wear is attributed to several factors, such as the properties of the tool and workpiece materials, cutting conditions, tool geometry, and chip formation. When a tool wears, it affects

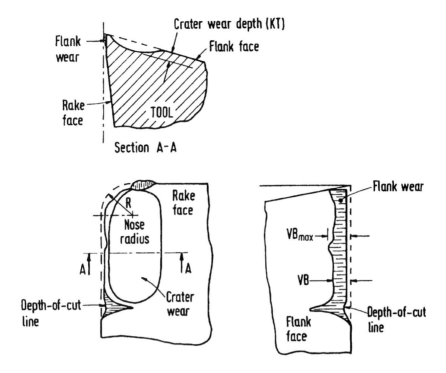

FIGURE 4.4 The measurement criteria for different types of wear.

the cutting force and the stress waves generated at the tool-workpiece interface due to friction and deformation. At times, there is increased vibration in the machine.

4.3 Sensors and Signal Processing

Tool conditions that have adverse effects on the machining process and need to be monitored include tool wear, various forms of tool damage, chatter, and chip breakage [30]. When the condition of a tool deteriorates, it affects the cutting force and the stress waves generated at the tool-workpiece interface due to friction and deformation. Sometimes, there is increased vibration in the machine. Thus, sensing methods measuring force, directly or indirectly (e.g., torque, current, or power of the spindle motor), acoustic emission, and vibration can be used to determine tool wear. The original force, (AE) and vibration signals require signal conditioning and processing to extract useful information. Commonly used tool wear sensing methods are based on cutting force and acoustic emission sensing. The following sub sections present examples of preprocesses force, AE and vibration signals and their correlation with tool conditions.

Dynamic Force

In the frequency domain, the dynamic tangential force has been found to exhibit a characteristic trend that can be used to indicate the extent of wear of the cutting tool [21]. Figure 4.5 shows the magnitude of the dynamic tangential force at the natural frequency of the tool overhang for a turning tool [27]. It shows a monotonic increase with the flank wear. As the tool approaches failure, it displays a relatively sharp decline. When the cutting tool is new, the sharp edge of the tool minimizes contact between the tool and the workpiece. Hence, the dynamic force is initially small. As the tool wears, the contact surface area between the tool and the workpiece increases, resulting in rubbing and increases in the dynamic tangential force. For an uncoated tool without grooves, as the tool approaches failure, crater wear becomes

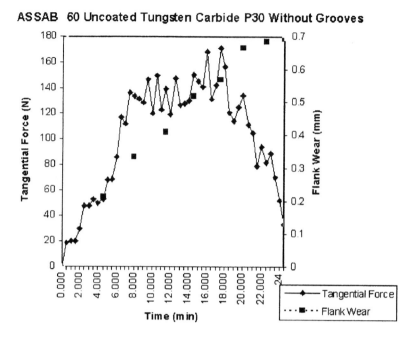

FIGURE 4.5   Dynamic tangential force at tool overhang frequency (turning).

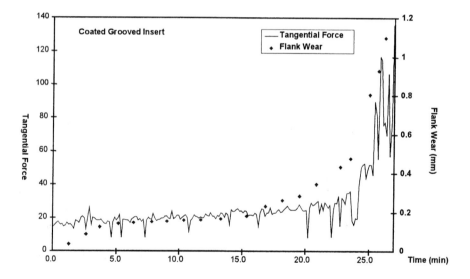

FIGURE 4.6   Increase in dynamic force component near tool failure.

significant. As the depth and width of the crater increases, it has the effect of increasing the effective normal rake angle (i.e., making the cutting edge relatively sharper) and correspondingly reduces the dynamic tangential force.

The decrease in the dynamic force component at the natural frequency of the tool overhang does not always occur. It depends on the manner at which the crater wear progresses and the geometry of the crater so formed. In the case of coated and grooved inserts, there is a sharp increase in the dynamic force component near tool failure. This continues until tool failure as can be seen in Fig. 4.6. During the cutting operation, the tool insert shows little crater wear because the chip-breaker groove does not allow chips to flow continuously across the rake face of the insert. Hence, crater wear is not the cause of tool failure for this type of inserts. In the case of a coated insert, flank wear increases at a slow rate at the beginning of the tool life because of the titanium nitride (TiN) coating. When the coating is worn off, the flank wear proceeds at a faster rate until tool failure.

Generally, the use of the amplitude change in the dynamic force component is dependent on the insert characteristics and the machining conditions. This is one of the problems of relying on only a single feature information, such as that obtained in either a time or frequency domain, which in the aforementioned case, is the peak frequency amplitude in the frequency domain.

Acoustic Emission (AE)

AE refers to the emission of elastic stress waves due to rapid changes in strain energy as a result of structural change in the material, such as during plastic deformation, fracture, or phase change. The frequency range of the AE signal is much higher than that of the vibrations in the machine tool and environmental noise. Therefore, a relatively uncontaminated signal can be obtained by using a high-pass filter. Figure 4.7 shows the flank wear of a turning insert, the corresponding resonant force at the natural frequency of the tool overhang and the root mean square (RMS) and band power in the frequency band of 300 to 600KHz of the AE signal. Both the resonant force and AE-RMS show a surge in amplitude as the tool approaches failure. The band power in the frequency band of 300 to 600 kHz indicates increased bursts of AE in the higher frequency range, due most probably to higher incidents of cracking and chipping of the insert.

Wavelet Packet Analysis of AE and Force Signals

Signal processing approaches of the AE and force signals for the purpose of feature extraction typically utilize the time-domain or frequency-domain analytical method. These approaches include the use of FFT, statistical analysis, and stochastic modeling (such as AR, ARMA). They provide only the time or frequency domain information that is generally more suitable for the analysis of stationary process. The machining process, on the other hand, may not necessarily be stationary. A global signal processing method which incorporates both the time and the frequency domains is more suitable for the analysis of a non-stationary process. The short-time Fourier transform (STFT) is currently the standard method for the analysis of a non-stationary process. The short-time Fourier transform (STAFT) is currently the standard method for the time-frequency analysis of signals [4]. Another commonly used method is the Wigner distribution, which originates from the classical works of Wigner in 1932. Du et al. [9] have used the exponential time-frequency distribution of acoustic emission for tool wear study in turning. In the last few years, a new method called the wavelet analysis has been developed with much progress both in the theoretical and applied areas. Like the Wigner distribution and short-time Fourier transform (STFT), the wavelet transform provides time-frequency analysis of signals but the wavelet transform adapts the window. A short window is needed to achieve the required fine resolution in the high-frequency range while a long window is needed to encompass the low-frequency range. The Wigner distribution and windowed Fourier transform use a uniform window size which can result in an uneven, and low frequency resolutions. Meanwhile, the Wigner distribution contains interference terms which are undesirable in practical applications. The wavelet analysis is a more advanced signal processing method. It offers the significant advantage of multi-resolution analysis of signals. Hence, both long and short windows can be used to capture the desired signal features. As a generalization of the wavelet transform, the wavelet packet (WP) decomposition together with the best basis algorithm have been developed by Coifman et al. [5], which can represent the signal in a most compact way. The wavelet packet analysis allows for the representation of the signal in both the time and frequency domains. Another appealing characteristic is that the computation time of this algorithm is as fast as the FFT and on-line implementation is possible. The wavelet packets contain modulated waveforms which have good time-frequency localization properties. Each waveform is called a time-frequency atom and can be assigned three parameters: frequency (f), scale(s), and position (p). The definition of the wavelet packets $\{W_{f,s,p}\}$ and wavelet packet decomposition are given in [11].

Materials	Workpiece	ASSAB760
	Tool Insert	Coated-Grooved Tungsten Carbide
Conditions	Cutting Speed (m/min)	230
	Feed Rate (mm/rev)	0.4
	Depth of Cut (mm)	2.0

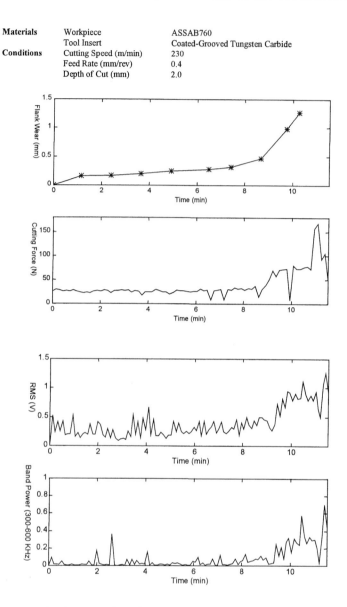

FIGURE 4.7 Flank wear, cutting force at tool overhang frequency and RMS and band power of acoustic emission (turning).

Given a signal x, the wavelet packet transform can be represented as a linear mapping:

$$WP: \begin{cases} X^n \Rightarrow W^n \\ \mathbf{x} \to \mathbf{w}(f, s, p) = WP(\mathbf{x}) = \langle \mathbf{x}, W_{f,s,p} \rangle \end{cases} \tag{4.1}$$

The above mapping represents the orthonormal wavelet transform. Hence, there exists a unique inverse transform to reconstruct the signal x:

$$WP^{-1}: \begin{cases} W^n \Rightarrow X^n \\ \mathbf{w}(f, s, p) \to \mathbf{x} = WP^{-1}(\mathbf{x}) = \langle \mathbf{w}(f, s, p), W_{f,s,p} \rangle \end{cases} \tag{4.2}$$

In the above wavelet packet transform, any basis sets from the orthonormal bases can be used to represent the analyzed signal. This provides more freedom in deciding which basis is to be used to represent the given signal. There must be a *best-basis* among them which can represent the signal in the most compact way (i.e., using the least number of coefficients). In order to find this best-basis, Coifman and Wicker-hauser [5] introduced the concept of the information cost function which is defined to be real-valued. Then the cost function is used as a measure to search for its minimum over all bases in the wavelet packet library to obtain such a *best basis* for the signal. It should be larger when the coefficients are roughly the same size and smaller when most coefficients are negligible, but only a few coefficients need to be retained. One of the cost functions proposed is the entropy cost function defined as follows. The Shannon-Weaver entropy of a sequence $x = \{x_j\}$ is:

$$H(x) = -\sum_j p_j \log p_j \tag{4.3}$$

where $p_j = \frac{|x|^2}{\|x\|}$. For this entropy, $\exp(H(x))$ is related to the number of coefficients needed to represent the signal to a fixed mean square error. The search for the best-basis requires $O(N \log N)$ operations.

Through the above process, a compact set of wavelet packet coefficients is obtained. A phase-plane plot is employed to graphically represent the time-frequency properties of the analyzed signal. In the phase-plane plot, the wavelet packet coefficients are displayed on a 2-D time-frequency plane. In this representation, each wavelet packet coefficient is associated with a time t and frequency f, with its time and frequency uncertainty amount Δt and Δf, respectively. The result is interpreted as a rectangular patch of dimensions Δt by Δf, located around (t, f) on the phase plane. The smallest area (Δt by Δf) of the rectangular patch is limited by the *Heisenberg uncertainty principle*. The patch is assigned a color or gray scale in proportion to the amplitude of the corresponding coefficient. Figure 4.8 gives a phase plane of a wavelet packet. It is obvious that the phase plane representation of the signal can give us the global view of its time-frequency feature patterns.

The best-basis wavelet packet decomposition provides an efficient and flexible scheme for time-frequency analysis of non-stationary signals. The major advantages of this method lie in the following:

- It globally optimizes the signal representation to provide a compact and sparsity representation of the signal pattern, i.e., the one with the fewest significant coefficients; and

- It possesses the computational speed of FFT.

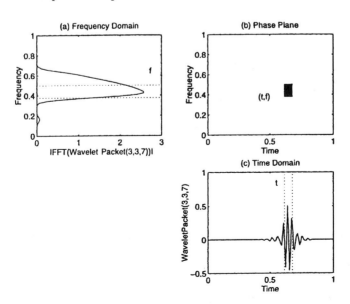

FIGURE 4.8 Phase plane of a wavelet packet.

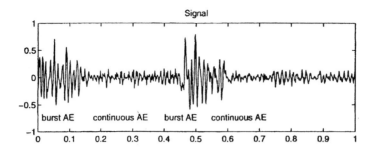

FIGURE 4.9   Continuous- and burst-type AE signals.

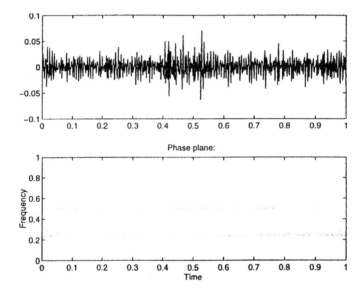

FIGURE 4.10   Normal state of tool.

The acoustic emission signal is usually of two distinct types, continuous and discontinuous or burst (see Fig. 4.9). In the machining process, the continuous-type AE signals are generated in the shear zone, at the tool-chip interface, and at the tool-flank-workpiece interface while the discontinuous burst-type signals are generated due to tool fracture, chipping, and chip breakage [11]. The wavelet packet method has been applied to the AE signals from the turning process to obtain a comprehensive set of time-frequency feature patterns (in the form of phase plane) corresponding to important tool conditions [29, 16, 17]. Figures 4.10, 4.11, 4.12, 4.13, and 4.14 show the time-frequency feature patterns in the phase plane representations after the best-basis wavelet packet transform of the corresponding AE signals obtained for different tool conditions.

Vibration (Acceleration)

Another possible approach to monitoring tool wear in turning is to measure the vibration of the tool shank using an accelerometer. For example, there exists a peak at the natural frequency $f0$ of the tool overhang in spectra of the acceleration signal. Sometimes, another peak can also be found at half of the resonant frequency $f_{1/2}$ as shown in Fig. 4.15(a). However, $f_{1/2}$ does not always exist, as can be seen in Fig. 4.15(b) for the case of the coated tool under the same machining conditions. Studies on the tool shank vibration indicate that although there exists a characteristic trend in the amplitude of the natural

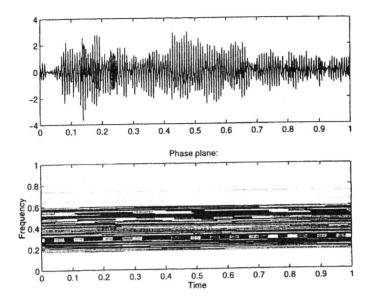

FIGURE 4.11 Intensive tool wear.

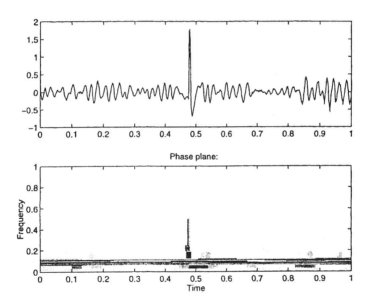

FIGURE 4.12 Tool chipping.

frequency f_0, this trend is not consistent [13, 24]. A more reliable approach is to employ the coherence function of two cross vibration signals.

Coherence Function of Cross Vibration Signals

The coherence function between two cross accelerations from the bending vibration of the tool shank has been found to be suitable for the identification of both tool wear and chatter in turning [23]. The cross accelerations in the horizontal (X) and vertical (Z) directions are measured by two piezoelectric

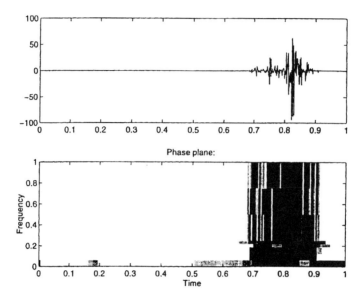

FIGURE 4.13   Tool fracture.

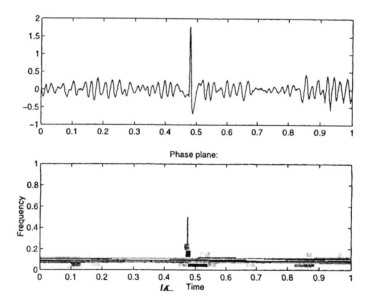

FIGURE 4.14   Chip breakage.

sensors mounted on the tool shank in a CNC lathe, as shown in Fig. 4.16. The coherence function of the two acceleration signals $x(t)$ and $z(t)$ is defined as

$$\gamma^2(f) = \frac{|G_{xz}(f)|^2}{G_x(f)G_z(f)}$$

$$0 \leq \gamma^2(f) \leq 1$$

where $G_x(f)$ and $G_z(f)$ are the respective auto-spectra of $x(t)$ and $z(t)$, and $G_{xz}(f)$ is the cross-spectrum between $x(t)$ and $z(t)$. $\gamma^2 = 0$ when $x(t)$ and $z(t)$ are uncorrelated over the range of frequency f. $\gamma^2 = 1$

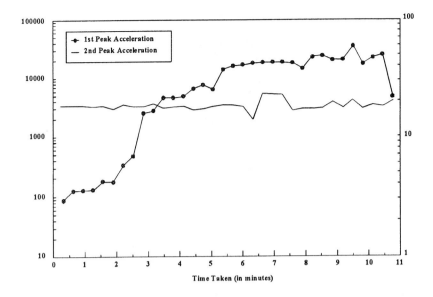

FIGURE 4.15(a) Acceleration of tool overhang (uncoated turning insert).

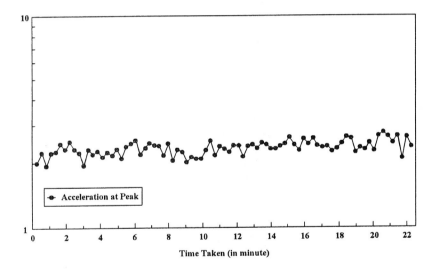

FIGURE 4.15(b) Acceleration of tool overhang (coated turning insert).

when $x(t)$ and $z(t)$ are completely correlated for all f. Otherwise, $0 \leq \gamma^2 \leq 1$ according to the degree of correlation between $x(t)$ and $z(t)$.

Figures 4.17(a) and (b) show the coherence function of the acceleration signals measured during turning of a nickel-based super alloy [4]. As can be seen from the figures below, the value of the coherence function at the chatter frequency reaches unity at the onset of chatter. Its values at the first natural frequency of the tool shank approach unity in the severe tool wear stage. The advantage of using this method is that the thresholds for detecting severe tool wear and chatter can be easily set for the following two reasons: first, the values of coherence function are normalized to a range of between zero to unity and secondly, they are also not so susceptible to changing cutting conditions because the value of coherence function is close to unity at the onset of chatter and severe tool wear.

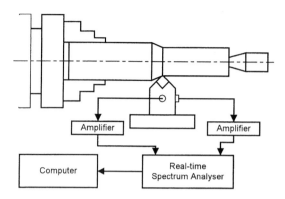

FIGURE 4.16   Setup for measurement of *X*- and *Z*-accelerations of tool shank.

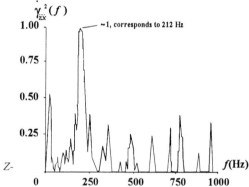

FIGURE 4.17(a)   Coherence function of *X*- and *Z*-accelerations at onset of chatter.

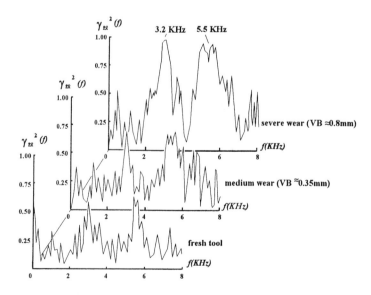

FIGURE 4.17(b)   Coherence function of *X*- and *Z*-accelerations at different tool wear.

4.4 Feature Extraction

Once a preprocessed signal exhibits characteristics of features that can be correlated with the tool conditions, it is necessary to develop suitable techniques to extract those features. The successful identification of the various tool conditions depends on how the features can be reliably extracted from the raw or preprocessed sensor signals for the sensor fusion stage. These features can be expressed in the form of a feature vector from different signal sources. The selected features should comprehensively characterize the different tool conditions. Although redundant features can reinforce the decision process in the determination of the tool condition, a minimum possible set of complementary features may have to be used to reduce computation requirements. For reliable and generalized identification of tool conditions, features used should not be sensitive to machining parameters, such as speed/feed, tool/workpiece materials, tool geometry, etc. In practice, this requirement is very difficult to meet. One method of achieving this is to have some means of normalizing the preprocessed output, such as the use of the aforementioned coherence function. By using a normalized output, threshold setting is confined to a range between zero and unity and is simpler to set. Another is to employ some form of pattern recognition to deduce the state of the tool from characteristic patterns of the processed outputs rather than by some single-value threshold setting.

4.5 Neural Network Architectures

After the feature extraction, an inference process is needed to associate each feature vector to a corresponding tool condition. This classification process usually involves an inference engine to map groups of feature vectors to their associated tool conditions. Such process, especially when crisp logic rule based methodology is employed, can be extremely tedious for feature vectors of large dimension. As discussed in the previous section, more extracted feature components are needed to reinforce the identification decision and complement each other

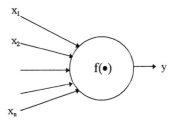

FIGURE 4.18 A simple processing element.

in their limited operational range. A more practical approach is to use an intelligent system which can "learn" itself to correct inference decision via past examples. A solution to such problem is to use an artificial neural network as a classifier [8, 16, 17, 29, 33, 34].

An artificial neural network is an information-processing system inspired by the performance of the human brain. It consists of a collection of very simple processing element called the neuron as shown in Fig. 4.18. The neuron embodies a node (soma) with multiple inputs (dendrite) and an output (axon). The output of the node can be described by

$$y = f(w_1, w_2, \ldots, w_n, x_1, x_2, \ldots, x_n) \tag{4.3}$$

where $w_i, \forall i \in [1, n]$ is the weight of the node associated with each input and $f(\cdot) : \mathbf{R}^n \to \mathbf{R}$ is called a transfer function of the neuron.

The collection of such neurons forms a neural network. A type of neural network is characterized by three fundamental factors as follows:

- Transfer function—the transfer function defines the way the output associated with the inputs and their corresponding weights, and the type of activation function used.
- Architecture—the architecture defines the pattern of connections between neurons and the number of layers of connection.
- Learning algorithm—the method of determining the weights on the connections.

There are numerous types of neural networks developed to suit the needs of their applications [12, 14, 18, 45]. This chapter does not intend to show all types of neural networks but the three common types of neural networks that are used in the case study have examples in the following sections. There are Multi-layer Perceptron (MLP), Kohonen Network, and Adaptive Resonance Theory 2 (ART2).

Multi-Layer Perceptron (MLP)

This is the most common neural network model used by many researchers and is better known as the back-propagation network due to the way it is trained [14]. A simple M-layer MLP is illustrated in Fig. 4.19 with each neuron represented by a circle and each interconnection, with its associated weight, by an arrow. For each neuron, the ith neuron in pth layer, its output, can be described by the equations

$$h_i^{(p)} = \sum_{j=1}^{N_{p-1}} w_{ij}^{(p)} v_j^{(p-1)}, \quad \text{for } p \in [1,M] \tag{4.4}$$

$$v_i^{(p)} = f_p(h_i^{(p)}) \tag{4.5}$$

where $f_p(\cdot)$ is any monotonically increasing activation function. A common choice is to use the sigmoid function for all the hidden layers and a linear function for the output layer. That is:

$$f_p(h) = \begin{cases} \dfrac{1}{1 + e^{-h}}, & \text{for } p \neq M \\ h, & \text{for } p = M \end{cases} \tag{4.6}$$

The number of hidden layers in the network and the number of nodes for each layer are arbitrary. Hecht-Nielsen [15] has proved that one hidden layer is sufficient to approximate a bounded continuous function to an arbitrary accuracy. However, one hidden layer may result in an excessive number of neurons used in the hidden layer. Thus, for practical implementation, one hidden layer is used for simple functions and two hidden layers are used for functions that are more complex.

The learning law employed by this network is the gradient decent based delta rule:

$$w_{ij}^{(new)} = w_{ij}^{(old)} + \Delta w_{ij} \tag{4.7}$$

and

$$\Delta w_{ij} = -\eta \frac{\partial J}{\partial w_{ij}} \tag{4.8}$$

where J it the cost function usually defined as $\sum_{i=1}^{n_y}(\zeta_i - y_i)^2$ and ζ_i is the desired output of the ith node.

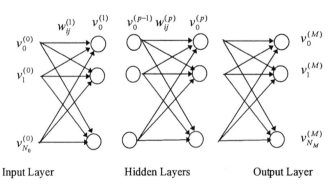

FIGURE 4.19   A M-layers multi-layer perceptron networks.

With the assumption of $f_x(\cdot)$ and $f_y(\cdot)$ as defined in Eq. 4.7 and Eq. 4.8, respectively, the training procedure can be summarized as follows:

- Step 1: Read in the input pattern and the desired output (x, ζ)
- Step 2: Propagate the signal forward by Eq. 4.5 and Eq. 4.6
- Step 3: Calculate the output layer error, δ as

$$\delta_i = \zeta_i - y_i \tag{4.9}$$

- Step 4: Back propagate this error to the hidden layer by

$$\delta_i^{(p-1)} = f'(h_i^{(p-1)}) \sum_j w_{ji}^{(p)} \delta_j^{(p)} \tag{4.10}$$

- Step 5: Calculate Δw_{ij} by

$$\Delta w_{ij}^{(p)} = \eta \delta_i^{(p)} y_j^{(p-1)} \tag{4.11}$$

- Step 6: Update the weights by

$$w_{ij}^{(p)} = w_{ij}^{(p)} + \Delta w_{ij}^{(p)} \tag{4.12}$$

- Step 7: Go to Step 1

Kohonen Networks

Kohonen Network [19] is comprised of a single layer network as shown in Fig. 4.20. This network uses a learning technique called competitive learning. Only one component of the output vector is activated (or ON) at a time. It is a network used for clustering operation.

For each output neuron, the output compete with one another by the equation below:

$$h_i = D(\mathbf{w}_i, \mathbf{u}) \geq 0 \tag{4.13}$$

$$y_i = \begin{cases} 1, & h_i = \min_j\{h_j\} \\ 0, & \text{otherwise} \end{cases} \tag{4.14}$$

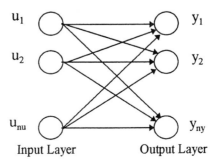

Input Layer   Output Layer

FIGURE 4.20 A Kohonen Network.

where $D(\bullet,\bullet)$ $\mathbf{R}^{n_u} \times \mathbf{R}^{n_u} \to \mathbf{R}$ is an metric function which measures the "distance" vectors and $\mathbf{w}_i \in \mathbf{R}^{n_u}$ is the weight vectors that connect the ith out with the input vector $\mathbf{u}$. For this network, the activation function sets the output node whose weight vectors is closest to the input pattern to one and the rest of the output node to zero. The metric function $D(\bullet,\bullet)$ can be any distance measurement, say the Euclidean norm. However, for computational simplicity, the Hammin distance

$$D(\mathbf{w}_i, \mathbf{u}) = \sum_{j=1}^{n_u} |w_{ij} - u_j| \tag{4.15}$$

is more commonly used.

The learning process for the Kohonen network only involves the weights associated with the winning output nodes. Hence, Kohonen learning is sometime referred to as competitive learning. The Kohonen learning law can be presented as

$$\mathbf{w}_i^{(\text{new})} = \mathbf{w}_i^{(\text{old})} + \eta(\mathbf{u} - \mathbf{w}_i^{(\text{old})})y_i, \quad \forall\, i \in [1, n_y]. \tag{4.16}$$

As shown in Eq. 4.17 that the learning law employed by Kohonen network does not involve the desired output, it does not require past examples to train the network. It is sometime referred as self-organizing network.

ART2 Networks

The adaptive resonance theory (ART) was developed by Carpenter and Grossberg [2] in two forms. One form, ART1 is designed for handling binary pattern whereas ART2 [1] can accept patterns of continuous magnitude. As shown in Fig. 4.21, a typical ART2 network is composed of two successive stages or layers of cells–an input representation layer $F1$ and a category representation layer $F2$. These layers are linked by feedforward and feedback weight connections (w_{ij}, ω_{ji}) that define a pattern specified by a corresponding $F2$ cell. Each $F1$ cell consists of three processing sub-layers with six nodes that enhance the

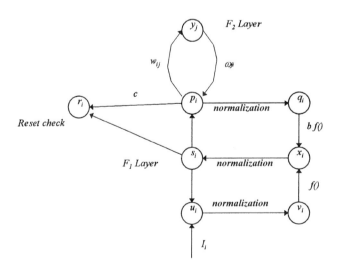

FIGURE 4.21   An ART2 network.

salient feature and suppress noise in the received signals, as shown in Fig. 4.21. Their activities can be characterized by the following equations:

$$u_i = I_i + as_i \tag{4.17}$$

$$v_i = \frac{u_i}{e + \|\mathbf{u}\|} \tag{4.18}$$

$$t_i = f(v_i) + bf(q_i) \tag{4.19}$$

$$s_i = \frac{t_i}{e + \|\mathbf{t}\|} \tag{4.20}$$

$$p_i = s_i + \sum_{j=1}^{M} g(y_j) \cdot \omega_{ji} \tag{4.21}$$

$$q_i = \frac{p_i}{e + \|\mathbf{p}\|} \tag{4.22}$$

where

$$f(x) = \begin{cases} x, & \text{if } x \geq \theta \\ 0, & \text{if } x < \theta \end{cases} \qquad \text{for}(0 \leq \theta \leq 1) \tag{4.23}$$

Every cell in the *F*2 layer competes with the others by the rule denoted in Eq. 4.25 until the only winner remains active:

$$g(y_j) = \begin{cases} d, & \text{if } y_j = \max_{r \in [1, M]}\{y_r\}, \text{ i.e., } j^{th} F2 \text{ cell is active} \\ 0, & \text{otherwise} \end{cases} \tag{4.24}$$

in which

$$y_j = \sum_{i=1}^{N} p_i \cdot w_{ij} \tag{4.25}$$

The degree of match between an input pattern and a responded feedback pattern is measured by the combined normalized feedforward/feedback vector $\mathbf{r} = (r_1, r_2, \ldots, r_N)$ with

$$r_i = \frac{s_i + cp_i}{e + \|\mathbf{s}\| + \|c\mathbf{p}\|} \tag{4.26}$$

where $0 < c < 1$ and $\frac{cd}{1 - d} \leq 1$.

The vigilance parameter ρ, which has a value of between 0 and 1, determines how well an input pattern matches with the feedback pattern of the active $F2$ cell. The closer ρ is to 1, the more sensitive is the system to mismatches. The matching criterion is defined as shown in the table below.

$$
\begin{array}{ll}
\text{Condition} & \text{Match Result} \\
\|\mathbf{r}\| < \rho - e & \text{mismatch} \\
\|\mathbf{r}\| \geq (\rho - e & \text{proper match)}
\end{array}
\tag{4.27}
$$

where

a. For $\|\mathbf{r}\| \geq \rho - e$, $F2$ is reset so that the active cell is deactivated with an output value of zero. Meanwhile, the input pattern is considered to mismatch with the stored pattern.
b. For $\|\mathbf{r}\| \geq \rho - e$, $F2$ is not reset. The active cell supposedly presents the proper class of input pattern during the classification operation, and is activated with an output value of one. Besides, if the active $F2$ cell has not been encoded with any pattern, such as during the initial learning, then a new category is established in the weighted connections with the cell, or the input pattern is merged into the weighted connections with the correctly encoded or stored pattern during the multi-sample learning cycle.

The feedforward and feedback weights are adjusted according to the following equations

$$
w_{ij}(t + 1) = [1 - g(y_j)][w_{ij}(t) - p_i] + p_i
$$
$$
w_{ij}(t + 1) = \alpha g(y_i)s_i + \{1 + \alpha g(y_j)(g(y_j) - 1)\}w_{ij}(t)
\tag{4.28}
$$

$$
\omega_{ij}(t + 1) = \alpha g(y_j)s_i + \{1 + \alpha g(y_j)(g(y_j) - 1)\}\omega_{ij}(t)
\tag{4.29}
$$

Further details of the ART2 algorithm can be found in the description by Carpenter and Gorssberg [1].

4.6 Tool Condition Identification Using Neural Networks

Features extracted from the sensor inputs are synthesized for more reliable and accurate estimation of the state of the process. Basically, tool condition diagnostics approaches can be divided into two categories: [3]

- Statistical Approach. In this approach, estimates of the state variables are obtained by evaluating process models based on the physics of the process. Statistical data concerning the physical properties of the materials are used. Bayesian estimator and Shafer-Dempster reasoning methods are examples of this approach.
- Training Approach. The synthesis occurs through a *mechanism* or a *mode* which first *learns* through a training phase on how the synthesis should occur. Methods based on this approach go through a training phase to capture the behavior or learn to synthesize the data.

This chapter present the works based on the training approach, in which, neural networks are used as the learning mechanism.

The following sections demonstrate three different approaches to the application of the neural network for tool condition identification. The first example uses very simple preprocessing and an MLP network to identify various tool conditions for the turning process. The following three examples use a combination of ART2 networks. Two are based on a single-ART2 network and employ a two-step feature extraction strategy for the identification of tool wear state. The first single-ART2 network uses only

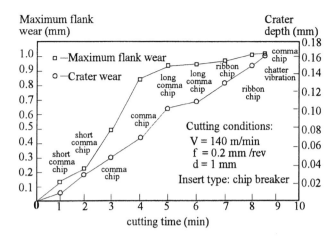

FIGURE 4.22 A typical tool wear experiment.

acoustic emission sensing and a wavelet transform technique in the signal processing. The other uses both force and acoustic sensing. The third neural network consists of multiple ART2 sub-networks and uses three sensors to measure acoustic emission and cross vibrations from the tool shank. The input feature information to the multi-ART2 network is obtained from the distribution of the coherence function between two cross vibration signals and the power spectrum of the acoustic emission signal in the effective frequency bandwidth of these signals.

MLP for Force Sensor with Simple Pre-processing [32]

In modern manufacturing processes, an essential part of a machining system is the ability to monitor and automatically diagnose the faults that occur in the machining process. In practice, such faults are not isolated and they are often co-existing. Figure 4.22 shows the coexistence of various cutting conditions in a typical turning process. It is therefore important to identify these faults from the detected abnormal phenomena during the machining process.

The occurrence of a fault can be due to either system characteristics' changes (true reflection of a fault) or propagation of other undetected faults. It is important that the root of the fault must be detected correctly. Otherwise, this may result in a wrong control action being taken. The process of fault detection in machining processes consists of the detection of abnormality, followed by the identification of the cutting state and then the recognition of the cause of the fault. This section focuses the discussion on the monitoring of tool wear, chatter vibration, and chip breaking in a turning process.

Feature Requirements

From the pattern recognition viewpoint, the necessity for process diagnostics is to extract enough independent governing features for the purpose of identification. In this experiment, the exact value of the dynamic cutting force is not important. It is the sudden variation from its normal zero-mean value that reflects the presence of chip breaking. This phenomenon is reflected as "spikes" in the dynamic cutting force. The magnitudes of the spikes are more distinctive in the worn tool than the fresh tool. These spikes usually have negative values of a few times the magnitude of the normal zero-mean values. Therefore, a subtraction from its maximum value will ensure all data to have positive value and high spikes are the point of chip breaking as suggested by Eq. (4.31). However, these spikes are usually of high magnitude. Therefore, the set of data is divided into ten data subsets defined in Eq. (4.30) to avoid these chip-breaking spikes from dominating the other inherent properties of the signal. Four features are extracted from the cutting force in feed direction with the following operation:

1. The signal (2000 data points) is passed through a low-pass filter to eliminate the high frequency noise.
2. This signal is divided into ten data subsets S_i, for $i \in [0, 9]$ with

$$S_i = \{x(200i + 0), x(200i + 1), ..., x(200i + 199)\} \tag{4.30}$$

where $x(i)$ is the filtered dynamic component of cutting force.

3. For each S_i, the data are transformed into an absolute valued data by

$$y_i = |x(200i + t) - x^i_{max}|, \text{ for } t \in [0, 199], i \in [0, 9] \tag{4.31}$$

where

$$x^i_{max} = \max[x(200i+0), x(200i+1), ..., x(200i + 199)] \tag{4.32}$$

4. For each pre-processes data $y_i(.)$, the following features are extracted.

   • The mean value F_i of y_i is defined by

$$F_i = \frac{\alpha_1}{200} * \sum_{t=0}^{199} y_i(t) \tag{4.33}$$

where α_1 is a scaling factor.

   • Similarly, the variance V_i of y_i is defined as

$$V_i = \frac{\alpha_2}{200} * \sum_{t=0}^{199} [y_i(t) - F_i]^2 \tag{4.34}$$

where α_2 is a scaling factor.

   • To avoid the presence of too many high valued spikes, the ratio R_i of mean and variance is also introduced as

$$R_i = \alpha_3 * V_i / F_i \tag{4.35}$$

   • Another dominating property is the magnitude D_i of the spike which is defined as

$$D_i = \alpha_4 * \max\{y_i(t)\}, \quad \text{for } t \in [0, 199] \tag{4.36}$$

   • One important property is the frequency of chip breaking. This can be measured via its coherent coefficient P_i which is defined by

$$P_i = \left| \frac{\sum_{t=1}^{199} [y_i(t) - F_i] * [y_i(t - 1) - F_i]}{\sum_{t=1}^{199} [y_i(t - 1) - F_i]^2} \right| \tag{4.37}$$

These features form the input to the MLP for the diagnosis process.

An Integrated Fault Diagnosis Scheme

In this case, an intelligent faults diagnosis scheme (IGDS) is proposed. For a fixed interval, the IFDS collects 2000 points of the dynamic cutting force component measured in the feed direction which are sampled over a time span of 0.8 second. This set data is filtered to cut-off the high-frequency noise component. The filtered data are divided into ten subsets of 200 points, in which five features are extracted from each subset to form a feature vector. These ten feature vectors are used in the MLP.

The MLP has an architechture of 5-8-3 and sigmoid function as its threshold function. The five input nodes of the MLP correspond to the five feature components of the extracted feature vector and these will be explained in the next section. The minimum of one hidden layer in a multi-layer perceptron is inevitable due to the linear separability properties of perceptron. This experiment shows that a hidden layer with eight nodes is sufficient for the purpose of identification. The output values of the MLP are between 0 and 1. When a binary decision is needed, output values of less than 0.5 are considered to be 0 or otherwise to be 1. The first, second, and third component of the output layer represent the level of tool wear, chatter vibration, and chip breaking, respectively and can distinguish typical patterns: (0,0,0), (0,0,1), (0,1,0), (0,1,1), (1,0,0), (1,0,1), (1,1,0), (1,1,1). Noting that, (0,1,1) and (1,1,1) are redundant in actual machining process. This is because long chips are not produced when chatter vibration occurs.

Experiment

A lathe was used in this experiment on chip breaking, chatter vibration, and tool wear in turning. Three workpiece of ASSAB 760 steel each with sizes of $\phi 60 \times 1000$, $\phi 130 \times 700$, and $\phi 130 \times 260$ mm. Two types of inserts, AC25 coated and G10E uncoated, were used. The cutting speeds were varied from 50–160 m/min, feed rates from 0.1–0.4 mm/rev. and depth of cut from 0.5–1.5 mm. Three kinds of chip breaking experiments has been carried out under the conditions mentioned above. The first was under cutting state of chatter vibration with fresh tool, and the size of the workpiece was $\phi 60 \times 1000$ mm and AC25 coated inserts were used in the experiments to guarantee that no tool wear occurred in one cutting pass. The second was under cutting state of chatter vibration with a worn tool and the size of the workpiece was $\phi 130 \times 700$ mm G10E uncoated inserts were used to ascertain at least the average flank wear land width V_B, of 0.3 mm in one cutting pass and when AC25 coated inserts were used, no chatter vibration occurred in machining process. The third was under the cutting state of no chatter vibration with a worn tool. The size of the workpiece was $\phi 130 \times 260$ mm and G10E inserts were used.

Discussion of Results

A total of 136 measurement samples (200 data point per sample) corresponding to variance levels of chip breaking and the cutting states (tool wear and chatter vibration) were collected. For each measurement sample, the set of data are divided into ten data subsets in which a feature vector of five features is extracted from data sets. This forms a total of 1360 feature vectors for the experiment. Among these 1360 feature vectors, representative training samples belonging to types of measurement samples were chosen and used to train the MLP. The remaining sets of feature vectors were used for testing and validation.

During the training stage, the target states of the output nodes were fixed at 0 for patterns of fresh tool, no chatter vibration and short comma chips; and 1 for patterns of worn tool, chatter vibration and long chips, respectively. The learning rate of the FDNN is 0.2 and its momentum coefficient is 0.4. The weights were initialized to uniformly distributed random values between -0.1 and 0.1. During the testing stage for each sample, the final output vector of the IFDS is the mean value of the MLP output vectors for the ten subgroup of feature vectors.

Figure 4.23 shows the relationship between each features and the six typical patterns. It is clear that the system will fail when insufficient feature components are used for the identification process. Table 4.1 shows the training results with different number of features used as the input vector. The results in Table 4.1 show that a minimum of four feature components (P, F, V and D) must be used for successful identification. However, experiments show that the addition of the mean ration (R) enhances the convergent rate in MLP training. Hence, all the five feature were used as the input vector to the MLP.

This system has been tested on a completely new set of data and shown an approximately 95% success rate.

TABLE 4.1   Training Results with Different Number of Features

Number of Features	1	2	3	4	5
Features	P; F; V; D; R	PF; PV; PR; PD; FV; FR; FD; VR; VD; RD	PFV; PFR; PFD; PVR; PVD; PRD; FVR; FVD; FRD; VRD	PFVD	PFVDR
Training Results	Failure	Failure	Failure	Success	Success

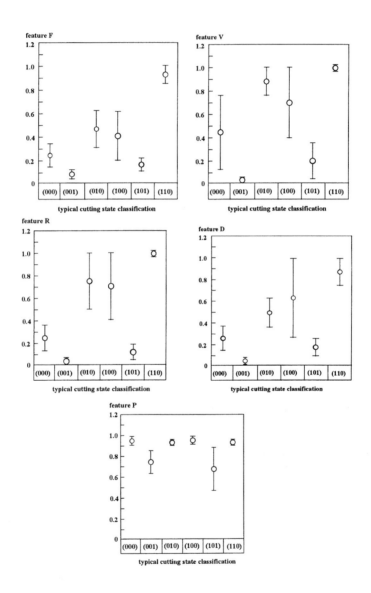

FIGURE 4.23   The relationship between features and typical patterns.

Single-ART2 Neural Network with Acoustic Emission Sensing

As shown in the section entitled "wavelet packet analysis of AE and force signals," using the best basis wavelet packet transform and phase plane representation, a comprehensive set of time-frequency feature patterns corresponding to different tool conditions can be obtained from the AE signals generated during

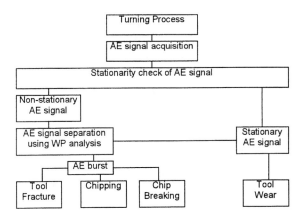

FIGURE 4.24 Multiple tool condition monitoring system.

turning. Based on the processed information, a new strategy for multiple tool condition identification has been established. An efficient separation scheme using the inverse wavelet transform for the separation of burst AE and continuous AE has also been developed. This method plays a key role in the progressive tool wear identification of a tool condition classification approach based on an ART2 neural network discussed in this section. Figure 4.24 shows a graphical presentation of this approach.

Given on AE signal from the turning process, a stationary check is first conducted to judge the state of the signal (stationary or non-stationary). If the signal is stationary, then it will be used for tool wear monitoring. If the signal is non-stationary, it is first separated into burst and continuous signals by using a separation scheme. The burst signal is then used for the identification of transient tool conditions, such as the tool fracture, chipping, etc., and the continuous signal is used for tool wear monitoring.

The characteristics of this approach based on wavelet packet transform and ART2 neural network [29] are:

- Wavelet packet transform is employed for AE signal separation and burst AE signal feature extraction.
- For tool wear monitoring, a two-step feature extraction scheme is employed aiming at obtaining more meaningful feature vector related to tool flank wear process.
- An unsupervised ART2 neural network is employed for multi-category classification of tool conditions. Two separately trained ART2 networks are used for automatic classification of transient tool conditions (fracture, chipping, and chip breakage) and progressive tool flank wear.

Transient Tool Condition Identification

As presented earlier, the time-frequency feature of the tool fracture, chipping, and chip breakage each produces a set of characteristics components in a certain frequency range at the location of the burst. Tool fracture produces the largest magnitude in almost the entire frequency range. Chip breakage produces the smallest magnitude in a very narrow frequency range, and those of chipping have characteristics that are between the other two states. Based on this understanding, a transient tool condition identification approach is proposed as shown in Fig. 4.25.

A feature vector corresponding to frequency band value of the wavelet packet coefficients in the phase plane at the burst location is extracted. In the phase plane, the entire frequency range is divided into 16 equal zones (each zone equivalent to 62.5 kHz), and two time indices t_s and t_e which indicates the start and end point of the burst signal are determined from the separated AE burst signal. At each frequency division, all wavelet packet coefficients within the time interval t_s and t_e are summarized. Thus, a feature vector $\boldsymbol{F}_t$

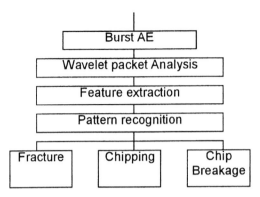

FIGURE 4.25 Transient tool condition identification approach.

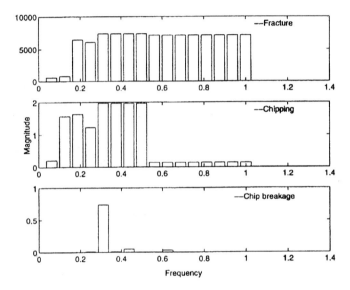

FIGURE 4.26 Feature patterns of different AE burst signals.

containing 16 elements is derived. Figure 4.26 shows the typical feature pattern of fracture, chipping, and chip breakage. The characteristics of the above three typical feature vectors can be summarized from Fig. 4.26:

- For tool fracture and chipping, the last 8 elements of the feature vector (corresponding to 0.5–1 MHz) are generally of the same value, while the first 8 (corresponding to 0–0.5 MHz) have varied values. The difference lies in that the sum of the last 8 elements is no less than the first 8 for tool fracture, while, for chipping, the case is just opposite.
- For chip breakage, there only exists a dominant element somewhere between 0–0.5 MHz while the remaining ones are nearly zero.

From the above feature extraction process, the advantage of wavelet packet analysis over FET is evident. Only with this efficient time frequency analysis can this kind of feature extraction in certain time location be made possible.

An ART2 network with 16-element input is trained to classify these three transient tool conditions. A total of 20 train patterns is employed to train the ART2 network (including 3 patterns from the fracture, 2 from the chipping, and 15 from chip breakage). Then the trained ART2 network is used for the transient tool condition identification from sampled data of AE signal. Fracture, chipping, and chip breakage incidents that have occurred during the capture of the sampled AE signal during the machining test have been successfully identified.

Tool Wear Monitoring

As mentioned previously, the acoustic emission signal from the flank wear process originates mainly from rubbing between the workpiece and tool flank face. If this rubbing process is even and continuous in machining, we expect a consistent increase in the continuous AE signal. However, in practical machining, the tool flank wear is not a uniform process so that fluctuation in the AE signal is normally encountered. Nevertheless, the energy of the AE signal pertinent to the tool flank wear is generally related to severity of rubbing between the workpiece and tool flank. The experimental signal shows that although there is much variation in the AE signal energy during flank wear, there exists a progressive increase in the AE energy in a certain frequency range (300 kHz–600kHz) with tool flank wear, as shown in Fig. 4.27.

In view of the above reason, a special feature analysis approach has been developed. As shown in Fig. 4.28, this approach essentially applies a two-step feature extraction (the primary and secondary feature extraction) strategy, coupled with an ART2 neural network for condition identification. Figure 4.29 shows a sample of primary and secondary AE features. A feature vector $\mathbf{F}_w$ is derived from primary and secondary AE features and used as input to ART2 neural network. Preliminary results show that for un-grooved tool inserts, good classification results have been directly achieved. For grooved tool inserts, there is significant interference from chip breakage. As a result, the burst signal separation scheme has to be applied before a correct identification result can be obtained. Table 4.2 shows the results of the ART2 classification results of experimental test data obtained for different machining conditions involving coated and uncoated tools and different workpiece materials.

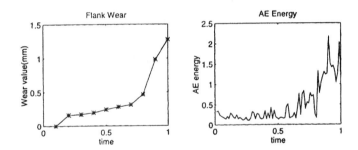

FIGURE 4.27 AE energy characteristics with tool flank wear.

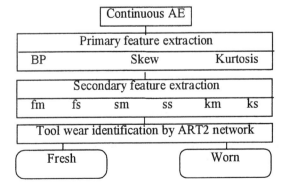

FIGURE 4.28 Tool wear identification from the continous AE signal.

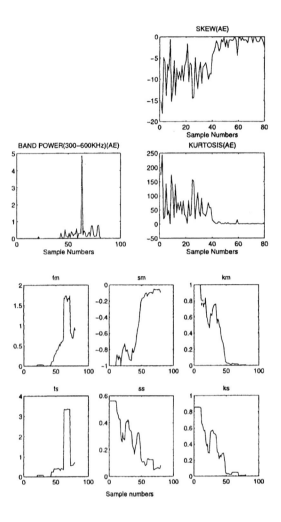

FIGURE 4.29   Primary and secondary AE features for tool wear identification.

Single-ART2 Neural Network with Acoustic Emission and Force Sensing

Based on acoustic emission and force sensing, an intelligent sensor system Fig. 4.30 has been developed to integrate multiple sensing, advanced feature extraction, and information fusion methodology [28]. Such a system employs

- More than one sensor to extend the effective range of the sensing system.
- Signal processing techniques to extract the compact feature vector sensitive to the monitored tool condition.
- Information fusion methodology to make correct decisions about the condition of cutting tool.

A two-step feature extraction strategy is proposed. Spectral, statistical, and dynamic analysis have been used to determine primary features from the sensor signals. In the primary feature determination, three features from the acoustic emission, the frequency band power (300 kHz–600 kHz) by FFT and the skew and the kurtosis by statistics, are obtained. One prominent feature from the tangential force signal, the natural frequency component resulting from the tool overhang [20], is also derived. Figure 4.31 shows an example of the behaviour of these features with the tool flank wear. It can be seen that the frequency band power exhibits increasing activities while the skew and the kurtosis have decreasing activities as the

TABLE 4.2   Identification Results by ART2 Network for Different Cutting Conditions

Case	Speed (m/min)	Feed (mm/r)	DOC (mm)	Work	Tool	Identification Result
TEST 1	170	0.2	2.0	A	X	
TEST 2	170	0.25	2.0	A	X	
TEST 3	180	0.3	2.0	B	X	
TEST 4	180	0.4	2.0	B	X	
TEST 5	170	0.2	2.0	A	Y	
TEST 6	170	0.3	2.0	A	Y	
TEST 7	230	0.3	2.0	B	Y	
TEST 8	200	0.3	2.0	A	Z	

DOC-Depth of cut; A-ASSAB760; B-ASSAB705.

X-uncoated ungrooved; Y-coated ungrooved; Z-coated grooved

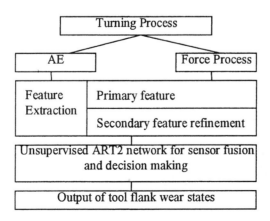

FIGURE 4.30   Intelligent sensor system for tool wear monitoring.

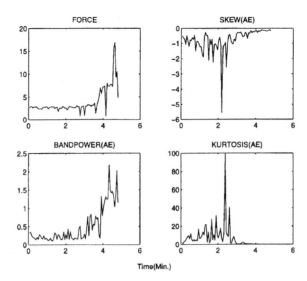

FIGURE 4.31   Primary features.

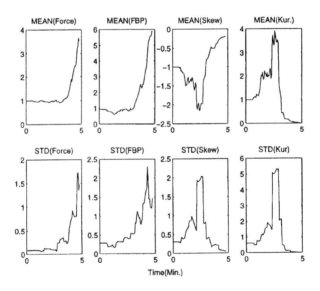

FIGURE 4.32   Secondary features.

flank wear progresses. As the values of the skew and the kurtosis are negative and positive, respectively, they provide complementary functions that have the effect of enhancing feature contrast. The tangential force components at the natural frequency of the tool overhang display an accelerated increase trend with respect to the tool wear before falling rapidly preceding the onset of tool failure. Another important characteristics of these four features, as experimental results show, is that they are nearly independent of the cutting conditions. Although changes in the cutting conditions may expedite or impede the tool wear process, the nature of the aforementioned features generally does not change.

Although the trends of the four primary features correlate well with the tool flank wear, they cannot be reliably used, either as an individual monitoring index or collectively as a feature vector for the decision making process. This is because of the severe variation of the features during the tool wear process (as shown in Fig. 4.31). Experimental evidence shows that even for the least varied force feature, there are still occasional fluctuations along the process of the tool flank wear, let alone the other three greatly flucturated AE features. For the purpose of getting more meaningful features, further refinement of the primary features is necessary for reliable tool flank wear identification.

Therefore, a secondary feature refinement is further applied to the primary features in order to obtain more correlated feature vector to the tool flank wear process. For each of the primary features, two refined features are extracted, namely the mean and the standard deviation within a moving window (Fig. 4.32). For a feature series (it can be frequency band power, skew, kurtosis, or tangential force component at the natural frequency of tool overhang) $P(n)$ $(n = 1,2,...,N)$, where N is the number of data sample, the secondary features in the moving window $(n, n + l - 1)$ can be represented as follows (l is the window size):

1. Mean value

$$\text{mean} = \frac{1}{l} \sum_{i=n}^{n+l-1} P(i) \qquad i = n,...,n+l-1 \qquad (4.38)$$

2. Standard deviation

$$\text{std} = \sqrt{\frac{1}{l-1} \sum_{i=n}^{n+l-1} (P(i) - \text{mean})^2} \qquad i = n,...,n+l-1 \qquad (4.39)$$

From the above process, a feature vector **F** is obtained:

$$\mathbf{F} = \{fm, fs, pm, ps, sm, ss, km, ks\}$$

- fm, pm, sm, km correspond to the mean values of the four primary features (the tangential force component, the frequency band power, the skew, and the kurtosis);
- fs, ps, ss, ks are the standard deviation values of the four primary features.

Figure 4.33 shows a typical representation of the above feature vector. It can be seen from Fig. 4.33 that the refined features are more meaningful to the tool flank wear than the primary features. The general feature vector patterns which reflect the fresh and worn states are given in Fig. 4.34. It can be

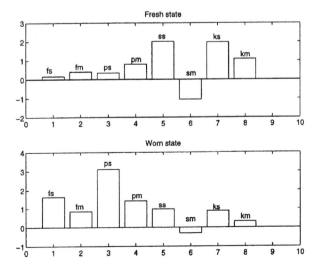

FIGURE 4.33 Pattern of feature vectors for fresh and worn tools.

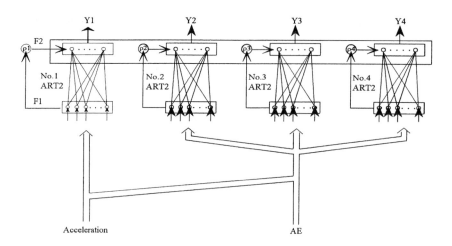

FIGURE 4.34   Parallel multi-ART2 neural network for tool condition monitoring.

seen from Fig. 4.34 that the fresh state feature pattern is quite different from that of the worn state. In the fresh state, the standard deviation and mean of the force feature and frequency band power are smaller than those of the skew and kurtosis. While in the worn state, the case is just the opposite. One thing worth noting is that the standard deviations from the four primary features are more prominent than their corresponding mean values. Therefore, the feature vector thus obtained can distinctively reflect the tool wear state. For a neural network, the more meaningful features we have, the more reliable and faster the classification can be achieved.

Note that the window size l is determined experimentally and the first l samples are used to calculate the first feature vector. Therefore, it is assumed that tool is fresh within the first l sample. This assumption is practically acceptable because the time used to obtain the first l sample is very short. The flank wear of the tool is very unlikely to occur in such a short time.

After obtaining a meaningful feature vector **F**, it is then used as input to an ART2 neural network for the fusion of AE and force information and decision-making of the tool flank wear state. As shown in Table 4.3, the experimental results confirm that the developed intelligent sensor system can be reliably used to recognize the tool flank wear state over a range of cutting conditions. The cutting conditions of the turning tests used to verify the proposed system include various types of inserts and workpiece materials under different machining conditions. The ART2 recognition results are listed in the last column of Table 4.3 (the X-axis represent cutting time in minutes). Table 4.4 shows the flank wear values at identified intermediate tool wear states (uncoated tool only) and worn state of tool (both the uncoated tool and coated tool). From the results obtained, it can be seen that:

1. For both the uncoated and coated tools, the fresh and worn states of the turning tool are all successfully recognized. The worn tools are identified around 0.3 mm of the measured flank wear value which is very close to those used in industry.

2. For uncoated tools inserts, the tool flank wear states have been classified into three categories: the fresh state (0), intermediate wear state (1) and worn state (2). The worn state occurs around the tool flank wear threshold value of 0.3 mm. The existence of the intermediate tool wear state may be caused by the fuzziness of the feature information between the fresh state and the worn state of the tool, since there is no clear cut information between the fresh and worn state in the wearing process of the uncoated insert tool. The intermediate state (1) can be used as a warning index in tool wear monitoring of the uncoated tools.

3. For coated tool inserts, the tool flank wear states have been correctly classified into two categories: the fresh (0) and the worn (1) states. Unlike uncoated tools, no intermediate flank wear state has been produced. This is because the wear process of the coated tool inserts is quite different. The

TABLE 4.3 Cutting Conditions and Identification Results of Test Data

Process	Speed	Feed	Depth of Cut	Workpiece	Tool	Identification Result
TEST 1	180	0.3	2.0	ASSAB705	A	
TEST 2	180	0.4	2.0	ASSAB705	A	
TEST 3	200	0.4	2.0	ASSAB705	A	
TEST 4	200	0.3	2.5	ASSAB760	A	
TEST 5	180	0.4	2.0	ASSAB760	B	
TEST 6	230	0.3	2.5	ASSAB760	B	

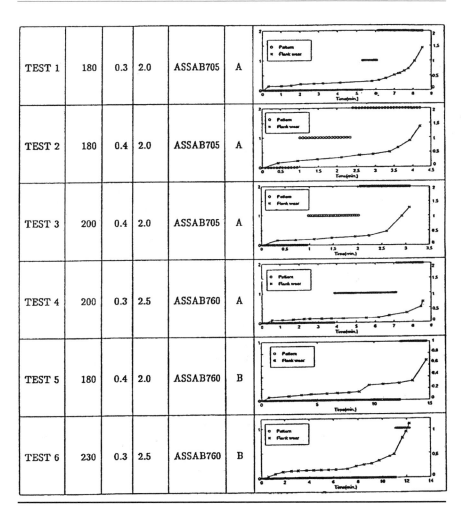

Note: A-SNMN-120408; B-SNMG120408

TABLE 4.4 Tool Flank Wear Value Identified by ART2

	Intermediate Wear States (mm)	Worn States (mm)
TEST1	0.30	0.45
TEST2	0.25	0.40
TEST3	0.20	0.35
TEST4	0.15	0.30
TEST5	—	0.30
TEST6	—	0.48

tool flank wear increases very slowly until it reaches a point where the coated materials near the tool edge are worn off. Then the tool flank wear increases significantly because in this case the tool insert is similar to an uncoated one. So, distinct information between the fresh and worn states has been produced.

4. The proposed approach is less dependent on cutting conditions, since the features selected are only those relevant to the nature of the tool wear. The change of cutting conditions may expedite or slow down the tool flank wear. However, when tool wear occurs the characteristic of the primary features are fixed. Therefore, cutting conditions have little influence on this approach, as the above results indicate (Table 4.5).

Multi-ART2 Neural Network with Force and Vibration Sensing

The value of ρ determines the granularity in which the input patterns are classified by the network. For a given set of patterns with different classes, a larger value of ρ may result in finer discrimination between patterns of the same class and a smaller value of ρ may lead to the merging of some of the classes into a single category. In a complex situation, the input patterns can be presented in different classes (such as tool failure and chatter), with each class having several cases (e.g., the various types of tool condition, the different frequencies or different magnitudes of chatter as well as the different cutting conditions). Due to the varied boundaries and dispersions of the various classes of input patterns, it is difficult to choose an accurate value of ρ for a single-ART2 network to classify all the patterns correctly, which is even less possible in a complex situation. In view of the varying cases, the network may not effectively represent each class of input patterns with only one cell in the $F2$ layer of the ART2 network. It then becomes necessary to encode or enroll more cells in the $F2$ and $F1$ layers in order to handle the different classes as well as their cases in complex patterns. The corresponding increase in the computation time for the expanded single-ART2 network then makes it less feasible for on-line use.

Due to the aforementioned limitations of the single-ART2 network, a novel parallel multi-ART2 neural network (as shown in Fig. 4.34), has also been developed to recognize tool conditions and machining chatter in turning operations [22]. With the multi-ART2, each subnetwork is individually designed with its own network parameters and is fed with the appropriate feature vector(s) relevant to a targeted machining condition (that is, only one of the input categories) so that the setting of the threshold is easier, compared with that for the single-ART2 network. In Figure 4.34, the input vector I consists of multiple features extracted from both vibration and AE signals. The input vector γ, also with multiple elements, contains features based on the vibration signals only. The neural network consists of four parallel ART2 subnetworks with the binary outputs denoted by $Y1$, $Y2$, $Y3$, and $Y4$, respectively. These sub networks are employed to identify the following four categories of tool states: machining chatter, tool failure, simultaneous severe tool wear, and chatter. According to the simultaneous outputs of the multi-ART2 network (i.e., $Y1$, $Y2$, $Y3$ and $Y4$), the patterns of the four machining states can be indicated by the following:

1. *chatter*: if $Y1 = 1$, $Y2 = 0$, $Y3 = 0$, $Y4 = 0$.
2. *tool failure (severe wear and breakage)*: if $Y1 = 0$, $Y2 = 1$, $Y3 = 0$, $Y4 = 0$.
3. *simultaneous chatter and severe tool wear*: if $Y1 = 0$, $Y2 = 0$, $Y3 = 1$, $Y4 = 0$.
4. *normal*: if $Y1 = 0$, $Y2 = 0$, $Y3 = 0$, $Y4 = 1$.

During the learning process, each of the four ART2 sub-networks is trained to represent one category of input patterns through arranged sample learning by grouping the test samples according to the respective categories. The $F2$ cells within an ART2 subnetwork characterize the typical cases of each category. Hence, if needed, more categories can be achieved by adding more ART2 subnetworks while more cases of each category can be handled by enrolling more $F2$ cells in the respective category representation sub-layer. To minimize the possibility that an input pattern does not match the existing patterns, the multi-ART2 network can be retrained to represent a new pattern according to the following two cases:

1. The input pattern does not belong to any of the existing categories and needs to be oriented. In this case, an additional category can be incorporated by adding a new ART2 subnetwork with the new input data, without any changes in the other ART2 subnetworks.

2. The input pattern is considered to be a specific case of one of the existing categories. In this case, the new pattern can be merged into the weighted connections with a newly added cell by the new input data, with no changes in the other cells.

Furthermore, four finer vigilance thresholds $\rho_1, \rho_2, \rho_3,$ and ρ_4 for the four ART2 subnetworks are employed in place of a single vigilance threshold ρ which is set to classify all categories for single-ART2 network. By appropriate setting of their threshold values, $\rho_1, \rho_2, \rho_3,$ and ρ_4 are directed at the respective four categories of input patterns (a)–(d) so that four specialized classifiers are formed on their feedforward/ feedback weight connections. The values of the other parameters, such as $a, b, c, d,$ and θ, are also chosen for every $F1$ cell of each ART2 sub-network so as to enhance the contrast of the information that is correlated with the corresponding category.

When a new input pattern is presented for classification, the multi-ART2 network carries out parallel processing and fast searching among the four ART2 subnetworks. The operation of the parallel multi-ART2 neural network

1. *Competition within each subnetwork*: each of the four ART2 subnetworks operates for the single-ART2 network. Hence, only one cell in an $F2$ sub-layer of a subnetwork is activated, which represents one of the four patterns (a)–(d);
2. *Competition among the subnetworks*: when two or more $F2$ cells from different subnetworks happen to be activated after Step 1, the $F2$ cell with the maximum degree of match is the one selected to represent the corresponding one of the four patterns (a)–(d) as the final result classified by the multi-ART2 network.

Feature Information

With reference the coherence function for the acceleration signals:

$$\gamma(f) = |G_{xz}(f)| / \sqrt{G_x(f).G_z(f)} \tag{4.40}$$

If the concerned frequency range is divided into $N1$ appropriate bands, $\Delta f_i = f_i - f_{i-1}(i = 1, 2,..., N1)$, the maximum value of the coherence function in each band is

$$\gamma_i = max.[\gamma(f), f \in (f_{i-1}, f_i)](i = 1,2,...,N1) \tag{4.44}$$

An $N1$-feature vector is thereby extracted from the acceleration signals as

$$\gamma = (\gamma_1, \gamma_2,..., \gamma_{N1}) \tag{4.45}$$

Ten frequency bands are used, (i.e., $N1 = 10$). Five are distributed over the chatter range from 0 to 500 Hz with $\Delta f_i = 100$ Hz. One is from 500 Hz to 3 kHz. The remaining four ranges are around the first natural frequency of the cantilever shank, from 3 kHz to 5 kHz with $\Delta f_i = 500$ Hz, for monitoring the tool condition. Figure 4.35 presents an example of the increase in the maximum frequency-band value pertaining to the coherence function of the two acceleration signals corresponding to the onset of chatter. Different distribution characteristics have been observed for the severe tool wear as well as the simultaneous onset of chatter and severe tool wear [23].

It has also been found from a careful study of the power spectrum that the AE signal emitted during the turning operation is more sensitive to chipping, breakage and severe wear of the cutting tool in the characteristic frequency range of 200 kHz–500 kHz. Hence, the spectrum in this band embodies the input patterns corresponding to the conditions of the tool. Meanwhile, the AE spectrum in the same frequency band does not exhibit a consistent relation to the occurrence of chatter. If the power spectrum

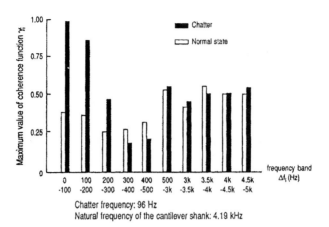

FIGURE 4.35   Frequency band distribution of maximum coherence function (Chatter) Depth of Cut - 2 mm; Feed-45 mm/min; Speed-72 m/min; Workpiece-S45C; Tool-P10.

of AE signal is $G(f)$ and the chosen frequency range is separated into $N2$ equivalent frequency bands, $\Delta f_j = f_j - f_{j-1}$ $(j = 1, 2,..., N2)$, then the relative weighted power ratio in each band is

$$\delta_j = \int_{f_{j-1}}^{f_j} \lambda G(f)\, df \Big/ \int_{f_{j-1}}^{f_j} G(f)\, df \qquad (j = 1, 2,..., N2) \qquad (4.41)$$

with

$$\lambda = \begin{cases} 1.0, & G(f) > T_2 \\ 0.5, & T_1 < G(f) \le T_2 \\ 0.25, & G(f) \le T_1 \end{cases} \qquad (4.42)$$

λ is the weight, T_2 and T_1 $(T_2 > T_1)$ are two threshold constants which can be estimated during sample runs. T_2 and T_1 indicate the lower levels of the AE spectrum in normal and abnormal tool conditions, respectively. Thus, the $N2$-feature vector extracted from the AE signal is

$$\delta = (\delta_1, \delta_2,..., \delta_{N2}) \qquad (4.43)$$

$N2$ is set to 10 for the frequency range of 200 kHz–500 kHz, so that $\Delta f_j = 30$ kHz. Based on the aforementioned vectors of γ and δ, the N-input feature vector I from both acceleration and AE signals is assigned as follows:

$$I = (I_1, I_2,..., I_N)$$
$$= (\gamma, \delta)$$
$$= (\gamma_1, \gamma_2,..., \gamma_{N1}; \delta_1, \delta_2,..., \delta_{N2})$$

where $N = N1 + N2$ is chosen to be 20 in this study.

   With the above input feature information from the maximum coherence function of two acceleration signals and the relative weighted frequency-band power ratio of an acoustic emission signal, the multi-ART2 network has been found to identify various tool failure and chatter states in turning operations with more than 93% success rate over a wide range of cutting conditions, compared to that of 71.4%–89.3% obtainable with the single-ART2 neural network (Tables 4.6 and 4.7).

TABLE 4.5 Machining Conditions

	Cutting Speed (mm/min)	Feed Rate (mm/min)	Depth of Cut (mm)	Workpiece Material	Tool Type (Insert)
C1	134	50	1.7		
C2	134	70	2.5	S45C	P10
C3	95	70	1.7		
C4	95	50	2.5		
C5	95	100	1.7		
C6	134	70	2.0		
C7	95	70	2.5		
C8	64	50	3.1		
D1	85	27	1.0		
D2	85	35	2.0	Aluminum	K6
D3	70	35	1.0	alloy	
D4	70	27	2.0		
D5	70	50	1.0		
D6	85	27	1.5		
D7	70	35	1.5		
D8	62	35	2.0		

TABLE 4.6 Experimental Results based on Multi-ART2 Network

Pattern Categories	(Sample Size/Machining Conditions*)		Recognized Rate (%)
	Training	Classification	
Chatter	2/C2, C4	16/C2, C4, C7, C8	100
Tool Breakage	4/D1-D4	8/D1-D8	100
Severe Tool Wear	4/C1-C4	8/C1-C8	87.5
Chatter & Tool Wear	2/C2, C4	8/C2, C4, C7, C8	87.5
Normal State	8/C1-C4, D1-D4	16/C1-C8, D1-D8	100
Total (Σ)	20/C1-C4, D1-D4	56/C1-C8, D1-D8	96.4

TABLE 4.7 Experimental Results Based on Single-ART2 Network

Pattern Categories	(Sample Size/Machining Conditions*)		Recognized Rate (%)
	Training	Classification	
Chatter	2/C2, C4	16/C2, C4, C7, C8	93.8
Tool Breakage	4/D1-D4	8/D1-D8	87.5
Severe Tool Wear	4/C1-C4	8/C1-C8	62.5
Chatter & Tool Wear	2/C2, C4	8/C2, C4, C7, C8	62.5
Normal State	8/C1-C4, D1-D4	16/C1-C8, D1-D8	81.3
Total (Σ)	20/C1-C4, D1-D4	56/C1-C8, D1-D8	80.4

(*Refer to Table 4.1 for specific machining conditions.)

4.7 Conclusions

Tool conditions monitoring is an important component in modern manufacturing environment, in particular, the machining processes. The chapter presented some simple neural networks architectures for the application of tool condition monitoring. The approaches presented in the examples illustrates the use of neural network to identify the tool condition from the input feature vectors. The chapter also illustrates the important of sufficiently rich feature data to be extracted for successful identification.

References

1. Carpenter, G.A. and Grossberg, S., "ART2: Self-organization of stable category recognition codes for analog input patterns," Applied Optics, Vol. 26, No. 23, (1987) pp. 4919–4930.
2. Carpenter, G.A. and Grossberg, S., "A massively parallel architecture for a self-organizing neural recognition machine," Computer Vision, Graphics, and Image Processing, 37, 54–115, 1987.
3. Chryssoloursis, G. and Domroese, M., "Sensor synthesis for control of manufacturing process," J. Engineering for Industry, Trans. ASME, May 1992, pp. 158–174.
4. Cohen, L., "Time-frequency distributions—A review," Proceedings of the IEEE, Vol. 77, 1989, pp. 941–981.
5. Coifman, R., Meyers, Y., Quake, S., and Wickerhauser, M.V., "Signal processing and compression with wavelet packets," Progress in Wavelet Analysis and Applications, Meyer and Roques Ed., Editions Frontieres, Toulouse, France, 1992, pp. 77–93.
6. Colgan, J., Chin, H., Danai, K., and Hayashi, S. R., "On-line tool breakage detection in turning: A multi-sensor method," Trans. ASME Journal of Engrg. for Industry, Vol. 116, February 1994, pp. 117–123.
7. Dornfeld, D. A., "In-process recognition of cutting states," Int. Journal JSME, Series C, Vol. 37, No. 4, 1994, pp. 638–650.
8. Dornfeld, D. A., "Neural network sensor fusion for tool condition monitoring," Annals of CIRP, 1990, pp. 101–105.
9. Du, R., Elbestawi, M. A., and Yan, D., "Time-frequency distribution of acoustic emission for tool wear detection in turning," Proc. of 4th World Conference on Acoustic Emission, Boston, M.A., 1991, pp. 269–285.
10. Editors, "Future view: Tomorrow's manufacturing technologies," Manufacturing Engineering, January 1992, pp. 76–88.
11. Emel, E., and Kannatey-Asibu, E., "Tool failure monitoring in turning by pattern recognition analysis of AE signals," Trans. ASME J. Eng. Industry, Vol. 110, 1988, pp. 137–145.
12. Fausett, A., "Fundamentals of neural networks, architectures, algorithms, and applications," Prentice-Hall International Ltd., 1994.
13. Gui, J.S., "On-line tool failure monitoring of carbide inserts," B. Eng. Thesis, Department of Mechanical and Production Engineering, National University of Singapore, 1994/95.
14. Hecht-Nielsen, R., "Neurocomputing," Addison-Wesley Publishing Company, 1989.
15. Hecht-Nielsen, R., "Theory of backpropagation neural network," Proc of the Int. Conf on Neural Networks, I, 593–611, IEEE Press, New York, June 1989.
16. Hong, G.S., Rahman, M., and Zhou, Q., "Using neural network for tool condition monitoring based on wavelet decomposition." International Journal of Machine Tools & Manufacture: Design, Research and Application, 36, no. 5 (1996): 551–556.
17. Hong, G.S., Rahman, M., and Zhou, Q., "Tool condition monitoring using neural network." In Proceedings of the 26th International Symposium on Industrial Robots, pp. 455–460. Singapore: International Federation of Robotics, 4 October 1995.
18. Khanna, T., "Foundations of neural networks," Addison-Wesley Publishing Company 1990.
19. Kohonen, T., "Self-organization and associative memory" (3rd Edition), Berlin: Springer-Verlag.
20. Lee, K.S., and Gan, C.S., "On the correlation between dynamic cutting force and tool wear," International Journal of Machine Tools & Manufacture: Design, Research and Application, Vol. 29, pp. 295, 1989.
21. Lee, K.S., Lee, L.C., and Teo, C.S., "On-line tool wear monitoring using a PC," Journal of Materials Processing Technology, Vol. 29, 1992, pp. 3.
22. Li, X.Q., Wong, Y.S., and Nee, A.Y.C., "A comprehensive identification of tool failure and chatter using a multi-ART2 neural network," To be published in ASME Journal of Manufacturing Science and Engineering.

23. Li, X.Q., Wong, Y.S., and Nee, A.Y.C., "Tool wear and chatter detection using the coherence function of two crossed acceleration signals," International Journal of Machine Tools & Manufacture: Design, Research and Application, Vol 4, 1997, pp. 425–435.

24. Lim, C., "Tool wear in milling," B., Eng. Thesis, Department of Mechanical and Production Engineering, National University of Singapore, 1995/96.

25. Lim, Y.H.C., "Surface coating for cutting tools," Ph.D. Thesis, Department of Mechanical and Production Engineering, National University of Singapore, 1996/97.

26. Matsushima, K., Bertok, P., and Sata, T., "In-process detection of tool breakage by monitoring the spindle motor current of a machine tool," ASME Winter Annual Meeting, Phoenix, Arizona, Nov. 1982, pp. 145–153.

27. Nee, A.Y.C., Wong, Y.S., and Chan, K.Y., "Force pulsations in milling," Society of Manufacturing Engineers, Technical Paper, No. MRR78–09, U.S.A. (1978).

28. Niu, Y.M., Wong, Y.S. and Hong, G.S., "An intelligent sensor system approach to reliable tool flank wear recognition." To be published in The International Journal of Advanced Manufacturing Technology.

29. Niu, Y.M., Wong, Y.S., Hong, G.S., and Liu, T.I., "Neural-based multi-category classification of tool conditions using wavelet packets and arts network." To be published in ASME Journal of Manufacturing Science and Engineering.

30. Niu, Y.M., Wong, Y.S., and Hong, G.S., "A comprehensive review on tool condition monitoring techniques," TRME-002-CON96, Dept of Mech & Prod Eng, National University of Singapore, 1996.

31. Owen, J.V., "Feedback from the cutting edge," Manufacturing Engineering, January 1993, pp. 39–45.

32. Rahman, M., Hong, G.S., and Zhou, Q., "On-line cutting state recognition using neural network." The International Journal of Advanced Manufacturing Technology (October 1995): 87–92.

33. Rangwala, S., Dornfeld, D. A., "Sensor integration using neural network for intelligent tool condition monitoring," J. Engineering for Industry, Trans. ASME, Vol 112, Aug., 1990, pp. 219–228.

34. Tansel I. N., Wagiman, A., and Tziranis, A., "Recognition of chatter with neural networks," International Journal of Machine Tools & Manufacture, Vol. 31(4), pp. 539–552, (1991).

35. Tonshoff, H.K., "Development and trends in monitoring and control of machining processes," Annals of CIRP, 1988, pp. 611–622.

36. Trent, E.M., "Metal Cutting," 3rd Edition, Butterworth-Heinemann, Oxford, 1991.

37. Trent, E.M., "Tool wear and machinability," Journal of the Institute of Production Engineers, Vol. 38, pp. 105–130.

38. Venkatesh, V.C., and Sathithanandam, M., "A discussion on tool life criteria and total failure causes," Annals of the CIRP Vol. 29/1/1980, pp. 19–22.

39. Waschkies, E., Sklarczyk, C., and Hepp, K., "Tool wear monitoring in turning," Trans. ASME Journal of Engrg. for Industry, Vol. 116, Nov. 1994, pp. 521–524.

40. Wright, P.K. and Bagchi, A., "Wear mechanisms that dominate tool-life in machining," Journal of Applied Metal Working, Vol. 1, No. 4, pp. 15–23, 1981.

41. Wright, P.K. and Trent, E.M., "Metallurgical appraisal of wear mechanisms and processes on high-speed-steel cutting tools," Metals Technology, Vol. 1, pp. 12–23, 1974.

42. Chryssoloursis, Domroese, M., and Beaulieu, P., "Sensor integration for tool wear estimation in machining," Sensors and Controls for Manufacturing, Vol. 33, ASME Winter Annual Meeting, pp. 115–123.

43. Zhou, Q., Hong, G.S., and Rahman, M., "New tool life criterion for tool condition monitoring using neural network". Engineering Applications of Artificial Intelligence, 8, no. 5 (1995): 579–588.

44. Zhou, Q., Hong, G. S., and Rahman, M., "A neural network approach for the on-line diagnosing of tool wear." In Proceedings of the IEEE International Conference on Neural Network Applications to Signal Processing, 17–20 August 1993, pp. 132–137. Singapore, 1993.

45. Zurada, J.M., "Introduction too artificial neural systems," Info access Distribution Pte Ltd., 1992.

5

Intelligent Real-time Expert System Environment in Process Control

Grantham K. H. Pang
The University of Hong Kong

Raymond Tang
Esso Petroleum Canada

Stephen S. Woo
Esso Petroleum Canada

This chapter discusses the fundamental issues in a real-time expert system environment for process control. A new methodology which integrates Petri Net, fuzzy logic, and real-time expert system called Continuous Fuzzy Petri Nets (CFPN) is presented. This methodology has already been applied to the monitoring of an oil refinery processing unit. The major advantage of CFPN is that it provides a novel approach for engineers to carry out system modeling, operational analysis, process monitoring, and control. The developed system can relieve the operator from monitoring sensor data information and allow him to concentrate on the higher level interpretation of process event occurrences.

5.1 Introduction

The process control industries embody a major sector of activities in our modern society. They include petrochemical, pharmaceutical, pulp and paper, rubber and plastics, food and beverage, glass, metal industries etc. Efficient and safe operation of these process control systems is mandatory in the advancement of any society or country. Automation of process control plants very often involves automating a process or a combination of a number of subprocesses that need to be controlled. At the top level of operation, a corporate computing system could have all the information of the process, including all the business and manufacturing aspects. At the middle level, a supervisory computer carries out the network communications and maintains the database and application. At the lowest level of operation, it consists of the control and instrumentation layer with instruments to monitor, sense, and manipulate the process variables. There are many opportunities to introduce artificial intelligence (AI) or expert systems (ES) into process control, especially at the supervisory host computer level. This level is tightly integrated into the control layer.

In recent years, many interesting applications for expert systems have been developed and introduced in process control systems. The applications of intelligent systems could appear in many areas. First, alarm management is one of them. In a plant environment, either a large quantities of alarms go off at the same time or no alarms go off. An expert system which would intelligently figure out what is really causing the problem and what to do about it would be of tremendous benefit in the safety and operational functioning of a plant. The second important use of an expert system is in scheduling. It is a very tough problem because it affects all of the other aspects of the plant environment. The problem is particularly complicated in the batch processing industries where a finite run of one item is followed by a finite run of another item. Proper scheduling affects the productivity and inventory control of the plant. Thus, it could have important financial consequences to the operation. The third area is in supervisory control and optimization. Supervisory control and optimization adjusts the parametric coefficients in the various traditional control structures throughout the plant. It would require an intelligent system to identify that there is a need to tune, to reason about which tuning method to employ, and then to go ahead and apply the new tuning coefficients.

5.2   An Expert Systems Approach

The AI/ES approach to process control is both viable and necessary in many situations. The first is the existence of manufacturing processes whose behavior does not make them suitable for conventional techniques. For example, there are system properties which are intractable using a model-based approach but can be overcome using a rule-based approach. That is, a collection of formulas cannot totally model the process and allow you to arrive at a solution functionally and analytically. These properties can include nonlinearities in the dynamics of the process and unreliable or scarce measurements of process variables. Under such circumstances, a control strategy which relies on mathematical models is often ineffective and valid in only a narrow range of operating conditions.

Another situation which points to the use of expert systems is when the control of the process mimics certain qualities that are normally associated with the way human beings function. It may exhibit some reasoning, based on broad understanding, to focus experience, skill, and knowledge on an aspect of a problem at just the right time. Also, there is a need to search, which is somewhat related to not being able to find a closed form analytical solution. Very often, the problem at hand could deal with objects and relationships between objects. There may be many non-numerical data and associations involved in the process.

The environmental and safety concerns are crucial in the control of many industrial processes. Three outstanding incidents that have raised the public awareness are the Three-Mile Island incident, the methyl isocyanate release at Bhopal, and the nuclear failure at Chenobyl. In some cases, it has been shown that traditional control schemes have had success in optimizing raw material and energy efficiencies but cannot fulfill the role of monitoring safety and environmental aspects of a process simultaneously. It has become evident that the human operator plays an extremely important role in the mitigation of process plant accidents. However, a typical process plant may include hundreds of interacting units of various types. A human operator of such a plant could be provided with computer displays of the plant status. The operator may be responsible for measurement readings and alarm indicators of the order of thousands, or even ten thousands. Also, the information is presented at a lower level than individual measurements or trend plots of measurements. The interpretation of these readings and displays is the operator's responsibility. Hence, it is clear that the operator suffers from an overload of information.

In process control plants, the human operator is responsible for choosing the operating setpoints which can affect the quality, safety, and economy of the plant. In addition, he is in charge of the plant startup, shutdown, load change, and production changes. Of particular significance is the handling of accidents and emergency conditions. A skillful operator may have years of experience in observing process behavior and the results of control action. He should be aware of the process equipment failure modes and the pattern of behavior they induce. The failure of sensors and the consequences on control and alarm behavior of the control plant is also very important. Very often, there is a trend in the measurement readings or a pattern of events which leads to an emergency situation. An experienced operator would be able to detect such a trend or sequence of events and apply the correct actions before the crisis occurs. In the face of a flood of low-level measurements and alarms, it is very difficult to filter the information and make the appropriate judgment. An expert systems approach can bring along great benefits in such a situation.

5.3   Real-time Control and Petri Nets

The field of Artificial Intelligence (AI) now embodies a broad range of tools and techniques that permit the representation and manipulation of knowledge. With the advancement of both computer hardware and software, AI will continue to have a profound effect in areas such as process control. In this chapter, a novel approach called Continuous Fuzzy Petri Net (CFPN) for real-time process control and modeling is described. Continuous Fuzzy Petri Net combines several paradigms and technologies–Fuszzy control, Petri nets and real-time expert systems. These three areas are integrated to produce a powerful tool in the area of real-time process control supervision.

Petri Nets were selected because of its inherent quality in representing logic in an intuitive and visual way. Brand and Kopainsky [4] discussed the principles and engineering method of process control with Petri nets. Techniques of fuzzy control was incorporated because it is well known that operators and engineers often provide inexact knowledge in the form of rules, heuristics, or even conflicting knowledge. This problem is especially prevalent in the process control industries where plant operation is sometimes more of an art than a science. Fuzzy logic provides a framework for amalgamating these uncertainties and ad-hoc techniques into a mathematically-sound method for logical inferencing. Hence, fuzzy control is appropriate for the process control area of application.

Real-time expert systems form the backbone of this CFPN approach. The issues of real-time expert system play an important role in setting constraints and providing goals for the proposed Continuous Fuzzy Petri Net. In particular, knowledge base validity maintenance over time is an important issue. A method for the aging of assertions based on temporal distance from the present state is adopted as a means of automatic truth maintenance.

The Fuzzy Petri Net method presented in this chapter was developed for ESSO (Imperial Oil Ltd.) in Canada. The objective was to provide advanced process monitoring at an oil refinery process in Sarnia, Ontario. The tool was implemented using the G2 [11] real time expert system, from Gensym Corporation. G2 is a very powerful expert system development environment for building intelligent real-time applications. It is an object-oriented and graphical tool, together with a structured natural language. The physical and abstract aspects of the application can be represented using the G2 objects. The object-oriented environment allows the user to create new object instances by cloning existing objects or by defining objects and their properties and behaviors. All objects are organized in a hierarchical class structure and multiple inheritance is also allowed. The objects have built-in connectability which allows the developer to connect objects graphically to represent the data and logic flows. This feature allows G2 to represent the Fuzzy Petri Net concept. The set of objects created for CFPN is shown in Fig. 5.1.

Another fundamental feature of G2 is the ability to capture knowledge by creating generic rules, procedures, formulas, and relationships, that apply across entire classes of objects. The basis of developing expert systems is that knowledge can be captured and represented in the form of rules. The rules of G2 work in real time and mimic the human ability to reason on specific situations. The rules can be data-driven (or event-driven) and get triggered when new measurements are obtained from the process.

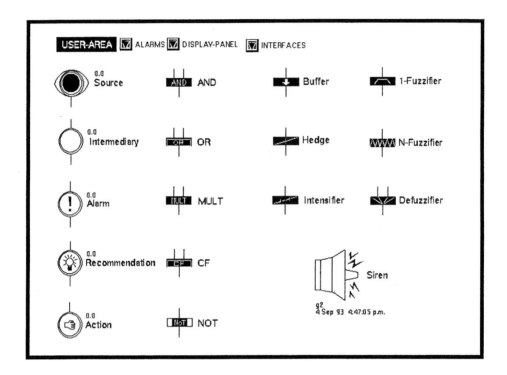

FIGURE 5.1   CFPN Symbol.

Alternatively, rules can be data seeking through backward chaining to invoke other rules, procedures, or formulas. This goal-driven feature is also implemented in CFPN and it is useful as a trouble-shooting and diagnostic tool. When the process has deviated from its desirable performance, CFPN helps to analyze the possible causes of the problem. The graphical objects and generic rules embody the knowledge of the process and its structure. The appropriate rule would be invoked automatically to infer the causes of the variance in the system operation and the possible causes of the problem and their remedy procedures.

As a tool for intelligent control, Fuzzy Petri Net can be used in many ways. It can be used to improve operator decision support. This, in turn, can improve the performance of the plant. For example, the aim could be an increase of octane content of the petroleum products, which means an increase of yield and profit. Another aim could be to optimize operations like continuous catalytic reformer (CCR). The Fuzzy Petri Net can also be used as a diagnostic tool. It can act an assistant to the operator and help to monitor the operation and trend of hundreds or thousands of sensors and actuators. If it senses an impending incident (the possibility of an undesirable event is high), it can alert the engineer with a color coded alarm. Remedy procedures can also be suggested to the operator. Thus, it is seen as an "extra pair of eyes" against problems.

5.4   Overview of Fuzzy Logic

In 1965, Lotfi A. Zadeh developed the idea of fuzzy logic and fuzzy sets [27]. Developing his ideas further in 1973, the concept of linguistic variables was introduced [28, 29, 30, 31]. Linguistic variables have values which are words rather than numbers. For instance, a linguistic variable "speed" may have values "fast," "slow," and "not very fast" and so on. Essentially, fuzzy logic provides a mathematical approach to dealing with the world which is full of imprecision and vagueness. In combination with fuzzy **IF-THEN** rules, the concept has been found useful in dealing with uncertainty in real world tasks.

Fuzzy Expert Systems

Many areas of research, such as medical diagnosis and process control, have taken advantage of the use of expert systems. However, imprecision and uncertainty were found to be an important aspect of these fields. Thus, traditional expert systems have been evolved to fuzzy expert systems to tackle these problems.[12]

The method for handling imprecision must be excellent in order for an expert system to succeed in becoming a useful tool. It has to be simple and natural so that a domain expert, such as a plant operator, can transform his knowledge into an expert system knowledge-base without difficulty.

Visualization is a powerful method in creating an environment which facilitates the translation from expert knowledge to expert system. One fuzzy expert system tool which is particularly relevant to this chapter is CASNET [26]. CASNET is a semantic net-based expert system where each node represents a state and has an associated certainty factor. There are forward and reverse weights associated with each node that give the strengths of causation between nodes. These weights correspond to the following interpretations: sometimes, often, usually, always, etc. Rules attached with a confidence value between −1 and 1 are used to link observations with states. The CFPN also takes advantage of the visual qualities in a net-based approach to expert system knowledge representation.

Fuzzy Control

One application area which has caught the attention of fuzzy logic researchers was control systems. Thus, fuzzy logic control was born and has made fuzzy logic an important and popular research topic. One of the first industrial applications of fuzzy logic control was F. L. Smidth Corp.'s cement kiln which became operational in Denmark during 1975 [13]. In a conventional control scheme, what is modeled is the system or process being controlled, whereas, in a fuzzy logic controller, the focus is on the human operator's behavior. Fuzzy controllers adjust the system parameters through a fuzzy rule-base expert system. The rule-base forms a logical model of the thought processes a human operator might go through in manipulating the system. In conjunction with the two key processes of *fuzzification* and *defuzzification*, the link between vagueness and exactness is established to make fuzzy logic a practical tool in process control.

The process of *fuzzification* is used to evaluate an exact or crisp value to its linguistic equivalent. The *membership function* plays a key role in this process. The membership function does not necessarily represent a probability density function. In fact, it is more of a similarity measure. Although probabilities may be a good starting point, the shape of the membership function can be quite arbitrary. In practice, it may often be obtained by gathering information from an experienced operator's opinion. The membership functions in practical fuzzy control may need to be fine tuned further to achieve better control. The reciprocal process of *defuzzification* brings the vague and imprecise logic of a fuzzy rule-base expert system back to the exact world of the system it is controlling. The fuzzy rule-base reaches some conclusions and makes recommendations on how to change the system parameters. Membership functions from various recommendations are combined and weighted, and through defuzzification, a crisp control value is obtained. There are various methods for defuzzification. The centroid method is a popular approach, where the centroid of the various weighted membership functions of all recommendations is taken as the crisp control value.

5.5 Overview of Petri Nets

Petri Net, as an approach to discrete event system modeling and analysis, has found great diversity and versatility. Since its creation in 1962 by Carl A. Petri [20], a rich body of knowledge concerning Petri Net theory and applications has been developed. Petri Nets have been used with varying success in modeling logic systems in many diverse fields, such as flexible manufacturing systems [16, 32, 6, 33], software engineering [14, 23, 24, 3], as well as process control and monitoring [22, 4, 1].

The most basic type of Petri Net is referred to as Ordinary Petri Net. It consists of components called places, transitions, tokens, and arcs. Hence, it is also known as Place/Transition Nets [5,2] Petri Nets have been applied in many diverse areas, which have prompted many variations and extensions of the Ordinary Petri Net. Some of the major variations are Color Petri Nets [15], Timed Petri Nets [21], Stochastic Petri Nets [19], Fuzzy Petri Nets [18, 25, 6, 7, 10] and Hybrid Petri Nets [17].

Fuzzy Petri Nets

Fuzzy Petri Nets [18, 25, 6, 7, 8, 9, 10] have been developed to model uncertainty in a knowledge base system. Tokens in these approaches represent uncertain assertions with a truth value between zero and one.

Looney [18] first defined the concept of fuzzy Petri net and applied it for rule-based decision making. Since that time, various models have been developed. Cardoso et al. [6] have attempted to introduce uncertainty in the marking of a Petri net with application in the monitoring of Flexible Manufacturing Systems. They aimed to reduce the combinatorial explosion of the complexity of the Petri net by considering a larger set of transitions as enabled by a marking with uncertainty.

In the paper by Chen et al. [7], they have developed a fuzzy Petri Net (FPN) which can work as a tool for real-time expert system modeling. However, CFPN offers more, as it is also useful as a tool for real-time expert control of both continuous and discrete systems. The fuzzification and defuzzification elements in the CFPN allows it to do this. The FPN does not include these elements. The FPN of Chen et al. is capable of dealing with uncertainty, but does not address the final step which is to bring that uncertainty to a crisp action.

Garg et al. [8] have also made modifications to the original Petri net model which is used for the representation of a set of fuzzy formulas. The resulting model can be used for automated reasoning and decision making in a fuzzy environment. However, the model is simple as it only represents knowledge in the form of propositional or first-order logic. Bugarin and Barro [9] have also proposed a fuzzy Petri net model for the knowledge representation of fuzzy production rules. Algorithms for the execution of the fuzzy Petri nets are developed and the process is carried out for the sup-min compositional rule of inference. A continuous fuzzy Petri net tool was described by Tang et al. [10]. An object-oriented description of the approach was presented. In this chapter, a formal definition of the Continuous Fuzzy Petri Net is given.

Hybrid Petri Nets

In the paper by Le Bail, Alla and David [17], a type of Petri Net which combined continuous Petri Nets with traditional Petri Nets, call Hybrid Petri Nets was introduced. Their approach was based on dividing a token into smaller units. Therefore, a continuous firing of transitions will build up sufficient of those smaller units to form one token unit.

In CFPN, the token at a place does not get accumulated. The certainty factor associated with the token represents the certainty of the assertion associated with the place. The CFPN approach is meant to provide continuous real-time inferencing. It has the ability to continuously make new inferences as sensor readings are updated.

5.6   The Continuous Fuzzy Petri Net Concept

In this section, a brief discussion of Continuous Fuzzy Petri Net is given. Continuous Fuzzy Petri Net as a modeling tool combines the paradigms of Fuzzy Logic and Petri Nets, each having different characteristics and advantages in one integrated tool. Therefore, a new paradigm that takes advantage of both approaches, through visual programming, is formed.

The Continuous Fuzzy Petri Net extends from an ordinary Petri net which consists of places, transitions, arcs, and tokens. In CFPN, places can be used to denote fuzzy propositions or other declarative knowledge.

The presence of a token represents the actual assertion and the degree of which an assertion holds true. The certainty of an assertion is color coded to provide added visual stimulus to the user. Transitions are used as functional nodes which can be used to represent linguistic hedges, fuzzification and defuzzification procedures, as well as logical operations such as AND, OR, and NOT. Arcs are used to interconnect the elements to form the logical structure of the Continuous Fuzzy Petri Net. Together, they form an intuitive visual representation of a fuzzy expert control system.

This Continuous Fuzzy Petri Net approach also introduces two important extensions to the Petri Net concept. These are the addition of a time based pattern matching algorithm for fuzzification and negative certainty values in the fuzzy logic paradigm.

The implication of the use of *continuous* to describe this approach is to contrast the difference between some of the previous work in the area of Fuzzy Petri Nets, where the Fuzzy Petri Net was used as a one-shot inferencing mechanism for knowledge representation or deals with strictly discrete event systems. This approach is meant to provide continuous real-time inferencing for the purpose of process control and modeling. Hence, it has the ability to continuously make new inferences as sensors are updated in real-time.

5.7 Definition of a Continuous Fuzzy Petri Net

A generalized Continuous Fuzzy Petri Net structure can be defined as a 12-tuple:

CFPN = $(P, T, \mathcal{P}, I, O, \Phi, \Theta, \Psi, \tau, \delta, \nu, \sigma)$, where

$P = \{p_1, p_2, ..., p_n\}$ is a finite set of *fuzzy places.*

$T = \{t_1, t_2, ..., t_m\}$ is a finite set of *fuzzy transitions.*

$\mathcal{P} = \{\rho_1, \rho_2, ..., \rho_n\}$ is a finite set of *propositions.*

$I : P \times T \to \{0, 1\}$ is an input function that defines the set of directed arcs from P to T.

$O : T \times P \to \{0, 1\}$ is an output function that defines the set of directed arcs from T to P.

$\Phi : P \to \{Source, Intermediary, Action, Alarm, Recommendation\}$
    is a mapping of fuzzy place to fuzzy place subclass.

$\Theta : T \to \{AND, OR, MULT, CF, NOT, Buffer, Hedge,$
    Intensifier, $1 - Fuzzifier$, $N - Fuzzifier$, *Defuzzifier*$\}$
    is a mapping of fuzzy transition to fuzzy transition subclass.

$\Psi : P \to \mathcal{P}$ is a bijective mapping from fuzzy places to propositions.

$\tau : T \to [0, 1]$ is a mapping of fuzzy transition to a real value between zero and one, denoting a threshold value.

$\delta : T \to \{d\}$ is a mapping of fuzzy transition to time delay d, expressed as a time interval.

$\nu : T \cup I \cup O \to [-1, 1]$ is a mapping of fuzzy transition and arc to a real value between zero and one, denoting a weighting.

$\sigma : P \to [-1, 1]$ is a mapping of fuzzy place to a real value between zero and one, and denotes the certainty value of the place.

P and T define a set of fuzzy-places and fuzzy-transitions, respectively. Ψ maps each fuzzy-place to a fuzzy proposition. (e.g., "*temperature is high*"). τ and δ define, respectively, the threshold and time delay associated with each fuzzy-transition. These establish a set of basic functions that a fuzzy-transition is to perform.

The functions I and O establish the presence of fuzzy-arc connections between fuzzy-places and fuzzy-transitions. ν defines the weighting of these fuzzy-arcs.

The associated mapping functions Φ and Θ determine the particular subclass of fuzzy place and fuzzy transition which each node ($P \cup T$) belongs. These subclasses define the nature of each fuzzy-place and fuzzy-transition. The symbol $\mathcal{F}(\cdot)$ will be used to represent a formula associated to a particular fuzzy-transition such as min, max, or fuzzify, which will be explained in a later section. These characteristics of the fuzzy entities provide us with the reasoning power in the CFPN approach.

σ maps each fuzzy-place to a certainty value (or certainty factor, which we will use interchangeably). When this σ function assigns a certainty value to a fuzzy-place, the corresponding fuzzy proposition becomes a fuzzy assertion. The certainty value tells us the degree of belief or disbelief we have of the fuzzy propositions defined in the knowledge base embedded in the Continuous Fuzzy Petri Net. So, in other words, a fuzzy proposition is simply a logical statement (e.g., "*the humidity is low*"), about which we have no idea of its truth or falseness. However, a fuzzy assertion occurs when a fuzzy proposition is assigned certainty value (e.g., "*the humidity is low with a degree of truth of 0.7*"). The σ function also assigns a marking to the Continuous Fuzzy Petri Net and puts a *token* in a fuzzy-place to denote when a fuzzy proposition becomes a fuzzy assertion.

Figure 5.1 illustrates the symbols used to represent each type of fuzzy-place (mapped by Φ) and fuzzy-transition (mapped by Θ) within the CFPN approach.

Execution of a Continuous Fuzzy Petri Net

In the CFPN approach, which is similar to Looney's approach [18], tokens are not consumed when they are *fired*. This is in contrast to how they are commonly used in other Petri Net approaches. The presence of a token in a fuzzy-place denotes a fuzzy assertion. A fuzzy assertion does not become invalid after a fuzzy-transition fires; therefore, tokens remain within fuzzy-places regardless of the firing state of a fuzzy-transition. This is necessary because we want to represent the truth states of the system we are modeling and/or controlling rather then simply providing a reasoning mechanism.

The certainty value of each fuzzy-place is determined by whether or not it is being *substantiated*. A fuzzy-place is substantiated by any tokens that are fed into that particular fuzzy-place. Tokens are modeled as a continuous entity since the CFPN is trying to deal with continuous real-time inferencing. Tokens are created inevitably at a source place as sensors are updated. Tokens are fed into a fuzzy-place by any of its input fuzzy-transitions. A fuzzy-transition fires if the absolute value of the certainty factor, evaluated by the formula governing the behavior of the fuzzy-transition, equals or exceeds the threshold value of that particular fuzzy-transition. Hence, a fuzzy-place is substantiated by a token if:

$$\nu(t_i) \times \mathcal{F}(x(t)) \geq \tau(t_i)$$

where $\nu(t_i)$ is the certainty value modifier of the fuzzy-transition, (t_i) feeding the fuzzy-place, and, $\mathcal{F}(x(t))$ is the formula governing the behaviour of the fuzzy-transition. This function represents the mathematical or logical function performed by a fuzzy-transition such as min, max, fuzzify, etc. $x(t)$ is the set of inputs to the fuzzy-transition at time, t, representing the certainty value of tokens feeding a fuzzy-transition from all of its inputs. However, note that transitions such as buffer, not, intensifier, hedge and the fuzzifier have only one input. $\tau(t_i)$ is the threshold of the fuzzy-transition t_i.

While a fuzzy-place is being substantiated, the certainty value of a fuzzy-place p_j is given by:

$$\sigma_j(t + d) \;=\; \nu(t_i \times p_j) \nu(t_i) \times \mathcal{F}(x(t))$$

for one or more fuzzy-transition inputs. $\nu(t_i \times p_j)$ represents the modifier of the fuzzy-arc connecting a fuzzy-transition t_i to the fuzzy-place p_j, and d is the delay of the transition t_i, with the rest defined as above.

When a fuzzy-place p_j is not substantiated by any fuzzy-transition output, the certainty of p_j is given by:

$$\sigma_j(t) \;=\; \text{Age}(\sigma_j(t_o), t - t_o)$$

where Age($\cdot$) is a formula governing the aging process of the fuzzy-place p_j and t_o was the time the fuzzy-place p_j was last substantiated. The current time t will always be greater then or equal to t_o. The formula Age($\cdot$) is defined such that $\sigma_j(t) = \sigma_j(t_o)$, for $t = t_o$ and $\sigma_j(t) \leq \sigma_j(t_1)$, for $t \geq t_1$. That is Age($\cdot$) is a non-increasing function of time. In effect, the Aging function determines the certainty value of a token that has been put into a fuzzy-place. While a fuzzy-place is being substantiated, we can view the situation as a token continuously replacing an old one, thereby resetting t_o to the current time continuously. (e.g., $t = t_o$)

CFPN Places

The mapping function Φ determines the nature of a fuzzy-place within the Continuous Fuzzy Petri Net framework. Each place has a certainty factor associated with it and this represents the degree of truth or disbelief we have about a certain fuzzy assertion or state. The CFPN approach assigns a color to the associated token within each of these places to reflect the certainty to take advantage of the visual quality of Petri Nets. A different color coding scheme is used for each type of fuzzy-place. For example, intermediary places may have a color scheme that range from white through grey to black, while an alarm place may have a color scheme that range from green through yellow to red.

Source Place

This type of fuzzy-place is used to collect real time sensor data. Its certainty indicates the freshness of the data. As sensor readings are obtained, the certainty of a fuzzy-source is set to fuzzy-source places are the input interfaces of the CFPN to the external world.

Intermediary Place

The intermediary place is used to denote general fuzzy assertions and intermediary fuzzy assertions which may be used in combination to make conclusions. The certainty indicates the degree of truth of the assertion.

Action Place

The action place is used to store the results from a defuzzifier transition and specifies a particular control output value that should be sent to an actuator. This type of place is the output interface of the CFPN to the external world. This completes the final step in the aim of using CFPN to encompass fuzzy control from sensor gathering, through inferencing, to control decisions.

Alarm Place

The alarm place is used to assert fuzzy propositions that can be construed as an alarm condition (e.g., "*the boiler is overheating*"). The certainty factor indicates the degree of truth of the alarm condition. Also, a graded belief of an alarm condition can be used as a sort of early warning system, such that an alarm condition can be seen to develop (and action can be taken immediately) rather than wait until a full fledged alarm condition has occurred.

Recommendation Place

The recommendation place is used to make recommendations (e.g., "*set valve high*") to a defuzzifier transition. The certainty indicates the confidence of the particular action we are recommending. These objects are usually connected to a defuzzifier transition. As a tool for a practical application, it was found that it was necessary to define a reasonable action for all possible logical states. For example, we have a recommendation with a positive certainty to "set value high." If the statement is "do not set value high," it does not necessarily mean we actually want it to be set to anything in particular. That is, the certainty factor of that case is zero, while a negative certainty factor would have meant some particular action to take place. Each recommendation place also has a membership function associated with it. The membership function determines the range of valid control values for that recommendation and is required for the defuzzification process.

CFPN Transitions

The function $\mathcal{F}(\cdot)$ governing the behavior of a fuzzy transition is determined by the mapping function Θ. The actual functions for each of the possible mappings are described in this section.

AND Transition

This type of fuzzy-transition is used to form logical AND constructs for a fuzzy rule base. It uses the minimize function to model this behavior. That is, the lowest *certainty* value of all input **fuzzy-places** is obtained as the resultant certainty.

The AND transition function is defined as:

$$\mathcal{F}(x(t)) = \min(x(t))$$

where $x(t)$ is the set of inputs to the fuzzy-transition at time, t, representing the certainty value of tokens feeding a fuzzy-transition from all of its inputs. Each component of $x(t)$, x_i, is defined as $\sigma(p_i) \times v(I(p_i \times t_j))$ with p_i being an input fuzzy-place to the fuzzy-transition t_j.

OR Transition

The OR transition is used to form logical OR constructs for a fuzzy rule base. The maximize function is used to model this behavior. The highest certainty value of all input **fuzzy-places** is obtained as the resultant certainty.

The OR transition function is defined as:

$$\mathcal{F}(x(t)) = \max(x(t))$$

with $x(t)$ defined as previously.

MULT Transition

The MULT transition is used to multiply all input certainty values which can be used to represent conditional probabilities. This is an alternative definition for the logical AND condition. Depending on the purpose of their application (i.e., to base the conclusion on either the conditional probabilities or the MYCIN approach), one would choose between these two types of AND transitions.

The MULT transition function is defined as:

$$\mathcal{F}(x(t)) = \Pi_{i=1}^{n} x_i(t)$$

where $x_i(t)$ is the certainty value of each input fuzzy-place weighted by their connecting fuzzy-arcs. Each $x_i(t)$ is defined as a component in $x(t)$ and n is the number of components.

CF Transition

This transition is used as a certainty factor (CF) combiner (from which its name is derived). It is used to combine positive and negative evidence for a fuzzy assertion. The algorithm adopted allows any number of fuzzy conditions whose certainty can range from -1 to 1, to be combined to form a resulting fuzzy assertion with certainty within a range of -1 to 1. All positive evidence strengthens the measure of belief (as do negative evidence for the measure of disbelief).

Just as the MULT transition is an alternative model for a logical AND operator, the CF transition can be used as an alternative model for a logical OR. This type of fuzzy-transition is appropriate in modeling fault detection rules. Often, a fault condition depends not only on the strength of a single condition but upon several highly coupled conditions. One condition may add confidence to another while others may reduce the degree of confidence to form a combined level of confidence.

The CF transition function is defined by the following algorithm:

1) $MB = 0$
2) $MD = 0$
3) for all $x_i(t) \geq 0$ $\{MB = MB + x_i(t) - MB \times x_i(t)\}$
4) for all $x_i(t) < 0$ $\{MD = MD + |x_i(t)| - MD \times |x_i(t)|\}$
5) $CF = (MB - MD)$

where $x_i(t)$ is the certainty value of tokens of each input fuzzy-place weighted by their connecting fuzzy-arcs. Each $x_i(t)$ is defined as $x(t)$ in the AND-transition. *MB* and *MD* represent the intermediate level of belief and disbelief of a fuzzy assertion. The difference between these two values is the actual certainty value (or degree of truth) of the fuzzy assertion we are deducing.

Buffer Transition

The buffer transition allows only one input place. This transition simply passes the certainty value it receives from an input fuzzy-place to its outputs after applying the usual modifiers and fuzzy-transition actions.

The buffer transition function is defined as:

$$\mathcal{F}(x(t)) = x(t)$$

where $x(t)$ is defined as in the AND-transition. It can be used to model simple time delay or certainty factor modifier like an arc.

NOT Transition

This type of transition is used to model logical NOT constructs for a fuzzy rule base. It also allows only one input place.

The NOT transition function is defined as:

$$\mathcal{F}(x(t)) = (1 - |x(t)|) \times \text{sign}(x(t))$$

where $x(t)$ is defined as in the AND-transition.

Intensifier Transition

An intensifier transition is used to apply a contrast intensifier upon input certainty values. This serves to increase the strength of the certainty value when beyond a threshold value (i.e., 0.5), but decrease it when below this threshold. Essentially, it serves to create a steeper slope on membership functions, bringing it closer to traditional bi-level logic. That is, it has the effect of reducing the fuzziness of an assertion. Like the Buffer transition, an intensifier allows only one input fuzzy-place.

The intensifier transition function is defined as:

$$\mathcal{F}(x(t)) = 2 \times x(t)^2 \times \text{sign}(x(t)), \qquad \text{for } 0.0 \geq |x(t)| \leq 0.5$$
$$\mathcal{F}(x(t)) = (1 - (2 \times (1 - x(t))^2)) \times \text{sign}(x(t)), \quad \text{for } 0.5 \geq |x(t)| \leq 1.0$$

where $x(t)$ is defined as in the AND-transition.

Hedge Transition

The hedge transition is used to apply linguistic hedges (e.g., "*very*," "*slightly*") to fuzzy assertions. This is accomplished by multiplying the magnitude of the incoming certainty value by a power factor. This transition allows only one input fuzzy place.

The hedge transition function is defined as:

$$\mathcal{F}(x(t)) = |x(t)|^\alpha \times \text{sign}(x(t))$$

where $x(t)$ is the certainty value of each token in each input fuzzy-place weighted by their connecting fuzzy-arcs. α is the power factor attribute of the fuzzy hedge transition, where $\alpha > 0$.

1-Fuzzifier Transition

A 1-fuzzifier transition contains a fuzzy membership function which can be used to map a single sensor data value to a certainty factor.

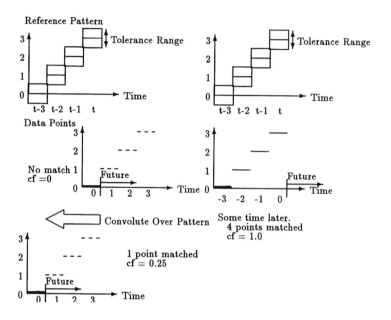

FIGURE 5.2 N-Fuzzifier Pattern Matching Procedure.

The 1-Fuzzifier transition function is defined as:

$$\mathcal{F}(x(t)) = \mu_F(\text{Val}(p_i)) \times x(t)$$

where $\mu_F(\cdot)$ is a membership function of a fuzzy set F, $\text{Val}(p_i)$ is the data value of the input fuzzy-source place p_i. $x(t)$ is defined as in the AND-transition and it can be used to model sensor uncertainty.

N-Fuzzifier Transition

These transitions contain a *pattern* and a *tolerance* value for each data point in a reference pattern. Each point of the reference pattern is compared with our historical data, with each point being evaluated for its similarity to this reference pattern and combined with equal weighting to form a single membership rating from 0 to 1. This evaluation also checks for partial matching of the reference pattern to detect conditions that maybe building up to the situation which we are trying to detect with our reference pattern. This is a very important function, as it allows the CFPN to provide a sort of early warning method, by producing a certainty factor evaluation which gradually increases as the data pattern becomes a closer match to the reference pattern. Figure 5.2 illustrates this procedure.

The **N-Fuzzifier** transition function is defined as:

$$\mathcal{F}(x(t)) = \max_{j=0}^{t}\left(\frac{\sum_{i=t-n+1}^{t} \mu_{q(i)}^{(w(i-j))}}{n}\right) \times x(t)$$

where $\mu_{q(i)}(\cdot)$ is a membership function for each reference point $q(i)$. The pattern q has n data points. $w(i)$ is the $i'th$ data point of the input fuzzy-source place. The max operator is used to represent the operation of shifting the reference pattern over the data points. Both the pattern and data points are stored in an array. The array is matched as is, and then shifted by one point to check for gradual matching up to the present time t.

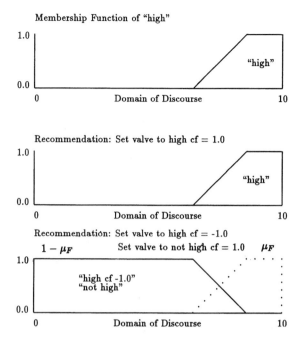

FIGURE 5.3 Defuzzifying with Negative Certainty Values.

Defuzzifier Transition

The defuzzifier transition is used to defuzzify recommendations for actions deduced from a fuzzy rule base. Fuzzy assertions are converted into crisp control output values through a defuzzifer transition. Defuzzifiers must have at least one fuzzy-recommendation and one fuzzy-action place connected to it in order to operate properly.

A special case occurs when $x_i(t)$ is a negative number. In order for the CFPN to provide a concise method for representing control actions such as *"set valve to not low,"* as well as reconciling the incorporation of a measure of disbelief, negative certainty factors were used to model this behavior. We propose that when a defuzzifier transition receives a recommendation with a negative certainty, the formula $1 - \mu_F$ should be used in place of the normal membership function μ_F defined for an input fuzzy-recommendation. This provides us with a logical and consistent method for extending the fuzzy logic paradigm to include -1 to 1 logic. Figure 5.3 illustrates this procedure.

The control output value is calculated as:

$$\frac{\int_D y \times g(y)\,dy}{\int_D g(y)\,dy}$$

where D is the domain of discourse, which is the continuous range of possible real values from which crisp values are to be mapped onto fuzzy sets. $g(y)$ is defined as $\Sigma(\mu_{Fi}(y) \times x_i(t))$, over all input fuzzy-recommendation places. $\mu_{Fi}(y)$ is the membership function of a recommendation such as "set power high" and $x_i(t)$ is the certainty of the $i'th$ recommendation, thus it is used as a weighting factor on the membership function associated with that recommendation. $x_i(t)$ is defined as in the AND-transition.

5.8 Examples

A Simple Control Example

This example illustrates how one might develop an application from input sensor values to control output, by combining various logic elements of the CFPN approach to form a fuzzy rule-base. Objects such as the **Buffer** transition and the **OR** transition are introduced. **Fuzzy-recommendation** places, a **Defuzzifier** transition and an **Fuzzy-action** place are also used in the example to illustrate the control aspect of the Continuous Fuzzy Petri Net.

The CFPN constructed for this example represents the following two rules:

> IF the temperature is low OR
> the pressure is increasing as (0,1, 2, 3, 4, 5) then
> set power to high

and another rule

> IF the temperature is high then
> set power to low

Referring to Fig. 5.4, the readings of the temperature and pressure sensors are simulated by a sawtooth waveform which varies from 0 to 10 units. The temperature sensor (**source place**) feeds into two different **1-fuzzifiers**: *temperature high* and *temperature low*. A simple linear mapping of the temperature values to a certainty factor is done by each of these two fuzzifier transitions. The **N-fuzzifier** transition attempts to match the pressure sensor reading to a reference pattern which varies as (0, 1, 2, 3, 4, 5). Each of the **recommendation** places, *power-low* and *power-high,* contains a membership function which determines what is considered as a low and high power setting, respectively. For the power-low setting, a triangular membership function where the power setting of 0 has a certainty of 1.0 and a power setting of 10 has

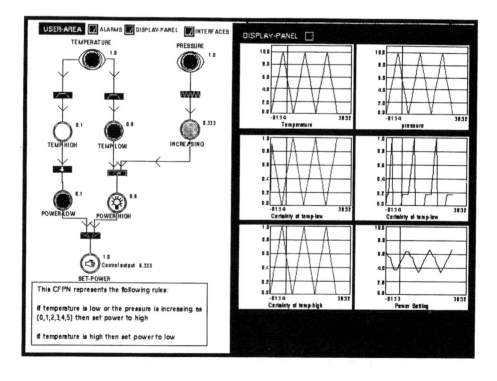

FIGURE 5.4   A Simple Control Example.

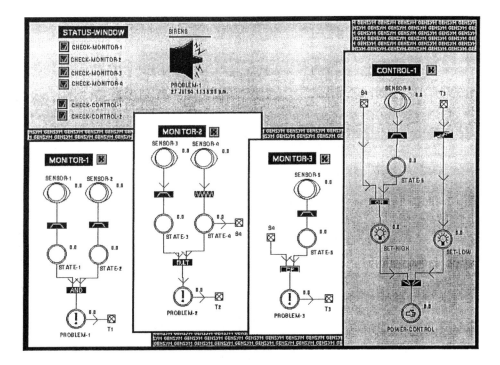

FIGURE 5.5   Sub-Windows for large CFPN networks.

a certainty of 0.0 is defined. The membership function of the power-high setting is simply the reverse with a power setting of 0 having a certainty of 0.0, and a setting of 10 having a certainty of 1.0.

The **DISPLAY-PANEL** shows the temperature and pressure sensor readings, along with their corresponding certainty values for each fuzzy assertion. The last graph in this example shows the control output (*power setting*) produced by this fuzzy rule-base.

Dealing with Large CFPN Networks

For a large industrial process, one may be faced with the problem of constructing a large CFPN to model the complex process control and diagnostic knowledge-base. To combat this problem, a separate window can be used to hold each fuzzy rule.

Within the G2 framework, a large Continuous Fuzzy Petri Network can be subdivided into several workspaces. Various "windows" can be used to access these workspaces. Some may be used to hold alarm conditions and *sirens* or to access other sub-windows. Figure 5.5 illustrates how this windowing structure operates. In this example, the CFPN is divided into six sub-workspaces or "windows" (four of which are shown). In addition, a **STATUS-WINDOW** is used to access these sub-workspaces and to hold any *alarms* created by the fuzzy rule-base represented by the CFPN. The CFPN has also been used for the monitoring of a Waste Effluent Treatment System (WETS) was developed at an ESSO oil refinery site in Sarnia, Ontario, Canada. The objective was to provide an enhanced detection and response system so as to avoid any reportable environmental incidents. A component of WETS, an Oily Water System, is shown in Fig. 5.6.

5.9   Conclusions

In this chapter, an intelligent real-time expert system environment based on the Continuous Fuzzy Petri Net (CFPN) is described. The CFPN concept has been implemented using the G2 real-time application development tool. G2 is a powerful tool ideal for the application due to its object-oriented and graphical tools, as well as the features of concurrent real-time execution, real-time rules, and structured natural language that support rapid prototyping. A formal definition of the Continuous Fuzzy Petri Net is

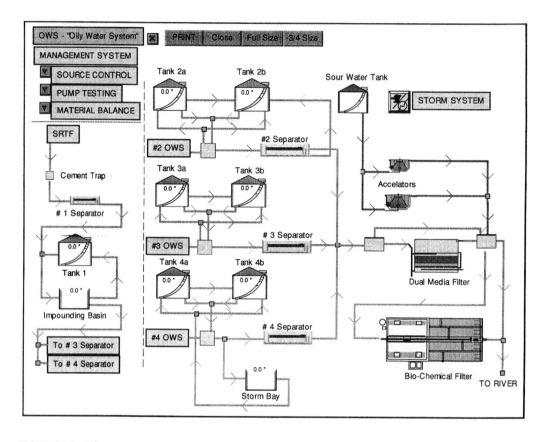

FIGURE 5.6   Oily water system.

presented. The execution and the function of the various components of the Continuous Fuzzy Petri Net are described. A time based pattern matching algorithm for fuzzification is also given. In addition, negative certainty values in fuzzy logic has been introduced as well as a method for handling negative certainty values in the defuzzification process.

A simple control example, using a Continuous Fuzzy Petri Net, is presented to illustrate the concepts developed in this chapter. ESSO Canada Ltd. has adopted the CFPN approach and has integrated it successfully with their refinery process monitoring system in Sarnia, Ontario, Canada. The Continuous Fuzzy Petri Net has addressed areas not covered by previous work on Fuzzy Petri Nets and it operates on a different level of logic compared to David and Alla's Continuous Petri Nets. Hence, the development of CFPN has made a contribution to the field, especially in that it has been used in a real industrial setting. Indeed, CFPN has provided significant insight into the understanding of the dynamics of the system at ESSO and has helped them to develop a process diagnostic system for the monitoring of an oil refinery.

The Continuous Fuzzy Petri Nets approach is a new direction in Petri Net development. It combines the flexibility of fuzzy logic and the graphical nature of Petri Nets to form a tool which can be useful for monitoring, diagnosis, decision support, and intelligent control.

References

1. K-E. Årzén, *Grafcet for intelligent real-time systems*, In *Proceedings of 12th IFAC World Congress*, Vol.4, pp. 497–500, Sydney, Australia, July 1993.
2. R. Y. Al-Jaar and A. A. Desrochers, *Petri Nets in Automation and Manufacturing*, In G.N. Saradis (ed), *Advances in Automation and Robotics: Knowledge-Based Systems for Intelligent Automation*, Vol. 2, pp. 153–225, Greenwich, Connecticut, JAI Press, Inc., 1990.

3. M. Baldassari and G. Bruno, *PROTOB: An Object Oriented Methodology for Developing Discrete Event Dynamic Systems*, In K. Jensen and G. Rozenberg (eds), *High-level Petri Nets: Theory and Application*, Springer-Verlag, pp. 624–648, 1991.

4. K. P. Brand and J. Kopainsky, *Principles and engineering of process control with petri Nets*, In *IEEE Transactions on Automatic Control*, Vol. 33 No. 2, pp. 138–149, Feb., 1988.

5. W. Brauer, W. Reisig, and G. Rozenberg (eds.), *Advances in Petri Nets 1986*, (Parts I and II), *Petri Nets: Central Models and Their Properties* and *Petri Nets: Applications and Relationships to Other Models of Concurrency. Proceedings of an Advanced Course*, Bad Honnef, Germany, Sept. 8–19, 186. Lecture Notes in Computer Science, Vol. 254 and 255, Springer-Verlag, New York, 1987.

6. J. Cardoso, R. Valette, and D. Dubois, *Petri nets with uncertain markings*, In G. Rozenberg (ed.), *Advances in Petri Nets*, Lecture Notes in Computer Science, Vol. 483, Springer-Verlag, pp. 65–78, 1990.

7. S. M. Chen, J. S. Ke, and J. F. Chang, *Knowledge Representation Using Fuzzy Petri Nets*, In *IEEE Transactions on Knowledge and Data Engineering*, Vol. 2, No. 3, Sept. 1990.

8. M. L. Garg, S. I. Ahson, and P. V. Gupta, *A fuzzy Petri net for knowledge representation and reasoning*, Information Processing Letters, 39, pp. 165–171, 1991.

9. A. J. Bugarin and S. Barro, *Fuzzy reasoning supported by Petri nets*, IEEE Trans. on Fuzzy Systems, Vol. 2, No. 2, pp. 135–150, May 1994.

10. Tang, R., G. K. H. Pang, and S. S. Woo, *A Continuous Fuzzy Petri Net Tool for Intelligent Process Monitoring and Control*, IEEE Trans. on Control Systems Technology, Vol. 3, No. 3, pp. 318–329, Sept. 1995.

11. *G2 Reference Manual*, G2 Version 3.0, Gensym Corp., 125 CambridgePark Drive, Cambridge MA 02140, U.S.A, 1992.

12. L. O. Hall and A. Kandel, The Evolution from Expert Systems to Fuzzy Expert Systems, In A. Kandel (ed.), *Fuzzy Expert Systems*, CRC Press, Boca Raton, 1992.

13. P. Holmblad and J-J. Stergaard, *Control of a Cement Kiln by Fuzzy Logic*, In M. M. Gupta and E. Sanchez (eds.), *Fuzzy Information and Decision Processes*, Amsterdam, North-Holland, pp. 398–399, 1982.

14. G. S. Hura, H. Singh, and N. K. Nanda, *Some design aspects of databases through Petri Net modeling*, In *IEEE Transactions on Software Engineering*, Vol. 12 No. 4, pp. 505–510, April 1986.

15. K. Jensen, *Coloured Petri Nets and the invariant method*, In *Theoretical Computer Science*, Vol. 14, pp. 317–336, Springer-Verlag, 1981.

16. E. Kasturia, F. DiCesare, and A. Desrochers, *Real time control of multilevel manufacturing systems using colored petri nets*, In *IEEE Int'l Conference on Robotics and Automation*, pp. 1114–1119, 1988.

17. J. Le Bail, H. Alla, and R. David, *Hybrid Petri Nets*, In *Proceedings of 1'st European Control Conference*, Grenoble, France, July 1991.

18. C. G. Looney, *Fuzzy Petri Nets for Rule-Based Decision making*, In *IEEE Trans. on Systems, Man, and Cybernetics*, Vol. 18, No. 1, pp. 178–183, Jan/Feb 1988.

19. S. O. Natkin, *Les reseaux de Petri stochastiques et leur application a l'evaualtion des systems informatiques*, These de Docteur-Ingenieur, Conservatoire National des Arts et Metiers (CNAM), Paris, June 1980.

20. C. A. Petri, *Kommunikation mit Automaten*, Ph.D. dissertation, University of Bonn, Bonn, West Germany, 1962.

21. C. Ramachandani, *Analysis of asynchronous concurrent systems by timed Petri Nets*, Ph.D. dissertation, Dept. of Electrical Engineering, MIT, Cambridge, MA, Sept. 1973.

22. W. H. Ray, *Advanced Process Control*, McGraw-Hill, New York, 1981.

23. W. Reisig, *Petri Nets for software engineering*, In *Petri Nets: Applications and Relations to Other Models of Concurrency*, Springer-Verlag, Berlin, pp. 63–96, 1986.

24. W. Reisig, *Petri Nets in Software Engineering*, In W. Brauer, W. Reisig and G. Rozenberg (eds), *Petri Nets: Applications and Relationships to Other Models of Concurrency*, Springer-Verlag, New York, pp. 63-96, 1987.

25. R. Valette, J. Cardoso, and D. Dubois, *Monitoring manufacturing systems by means of Petri nets with imprecise markings*, In *Proceedings of the IEEE Int'l Symposium on Intelligent Control 1989*, pp. 233–238, Albany, New York, Sept. 1989.

26. S. M. Weiss and C. A. Kulikowski, *Representation of expert knowledge for consultation: the CASNET and EXPERT projects*, In P. Szolovits (ed), *Artificial Intelligence in Medicine, AAAS Symp. Series*, Westview Press, Boulder, CO, 1982.

27. L. A. Zadeh, *Fuzzy Sets*, In *Information and Control*, V. 8, pp. 338–353, 1965.

28. L. A. Zadeh, *Outline of a new approach to the analysis of complex systems and decision processes*, In IEEE Transactions on Systems, Man, and Cybernetics, Vol. 3, pp. 28–44, 1973.

29. L. A. Zadeh, *The concept of a linguistic variable and its applications in approximate reasoning—Part 1*, In information Sciences, Vol. 8, pp. 199-249, 1975.

30. L. A. Zadeh, *The concept of a linguistic variable and its applications in approximate reasoning—Part 2*, In information Sciences, Vol. 8, pp. 301-357, 1975.

31. L. A. Zadeh, *The concept of a linguistic variable and its applications in approximate reasoning—Part 3*, In information Sciences, Vol. 9, pp. 43-80, 1975.

32. W. X. Zhang, *Representation of assembly and automatic robot planning by petri net*, In *IEEE Trans. on Systems, Man, and Cybernetics*, Vol. 19 No. 2, pp. 416–422, March/April 1989.

33. M. C. Zhou, F. DiCesare, and D. Rudolph, *Control of a flexible manufacturing system using petri nets*, In *1990 IFAC Congress*, Vol. 9, pp. 43–48, Tallinn, USSR, 1990.

6

Adaptive Neuro-Fuzzy Control Methods for Milling Operations in Manufacturing Systems

Y. S. Tarng
National Taiwan University of Science and Technology

N. T. Hua
National Taiwan University of Science and Technology

G. J. Huang
National Taiwan University of Science and Technology

In this chapter, an adaptive neuro-fuzzy control system is developed and then applied to control milling processes with non-linear and time-varying cutting characteristics. First, a neuro-fuzzy logic controller is employed to obtain a constant milling force under varying cutting conditions. To obtain optimal control performance, a learning algorithm is used to tune the weights of the fuzzy rules. It is shown that the developed neuro-fuzzy logic controller can achieve an automatic adjustment of feed rate to optimize the production rate with a constant cutting force in milling operations.

6.1 Introduction

Increasing the productivity of machine tools is a principal concern for the manufacturing industry. In recent years, computer numerical control (CNC) has made great progress to increasing the productivity of machine tools [1]. However, a common drawback for CNC technology is that the selection of cutting parameters in the numerical control program is not straightforward, depending greatly upon the well-experienced programmers. To avoid the occurrence of a broken tool, poor surface accuracy, or varying cutting conditions, conservative cutting parameters are usually chosen in the whole machining cycle so as to reduce the productivity of CNC machine tools. Therefore, the use of an adaptive control system for adapting the cutting parameters to the cutting conditions is required [2]. However, machining processes contain complicated, non-linear, and time-varying characteristics due to the interaction of the dynamics of the chip-removal process, the structural dynamics of the machine tool, and the dynamics of the machine tool driver. Hence, the design of the adaptive controller in machining operations with high control performance is a difficult task even using various forms of modern adaptive control algorithms [3-6].

During the past decades, fuzzy control has been proven to be a powerful tool for dealing with complicated, non-linear, and time-varying systems [7]. This is because the fuzzy control action is based directly on the linguistic rules acquired from the knowledge of experts and expressed mathematically through the theory of fuzzy sets [8]. As a result, the fuzzy control can simulate the control action that a human expert would take when controlling the given process. The fuzzy control has also been applied to the control of milling operations [9]. However, there are some drawbacks to the approach using the fuzzy control. First, the fuzzy control design has relied on a priori knowledge of human experts and, thus, the controller performance is dependent on the quality of this expertise. Second, a reliable linguistic rule for the controlled process may not always be obtainable. Third, some significant process changes may be outside the operator's experience and the design procedure appears to be limited by the elucidation of the heuristic rules. Furthermore, evaluation and tuning of the fuzzy logic controller are typically done by a time-consuming trial and error manner. To solve these problems, an adaptive neuro-fuzzy logic controller has been proposed in this study. The adaptive neuro-fuzzy controller has a learning algorithm and is capable of modifying linguistic rules based on an evaluation of the system performance [10]. As a result, the proposed controller can start from an empty linguistic rule base. The modification of linguistic rules is achieved by assigning a credit to the control action based on the present control performance. Milling processes with varying depths of cut [11] are the controlled plant used in this study. It is shown that an optimal neuro-fuzzy logic controller can be obtained by a learning process to achieve an on-line adjustment of feed rate in milling operations with a constant cutting force.

In the following sections, an overview of the adaptive control of milling operations using the adaptive neuro-fuzzy control is described first. A milling process model used as the controlled plant is described next. Then, the development of the adaptive neuro-fuzzy control in milling operation with a constant cutting force is described. Finally, computer simulation and experimental verification of the adaptive neuro-fuzzy control system in milling operations are shown.

6.2   Adaptive Control System for Milling Operations

Basically, the use of an adaptive control system with a constant cutting force in milling operations is to achieve an automatic on-line adjustment of feed rate for optimizing the production rate. Usually, during the milling process, the radial or axial depth of cut increases; correspondingly, the cutting force increases also or even exceeds the preset constant cutting force. It is expected that the control system senses this increase and immediately generates a smaller feed rate to avoid tool breakage. On the other hand, as the radial or axial depth of cut decreases, the cutting force will decrease below the preset constant cutting force. Under this condition, the control system senses this decrease and automatically generates a larger feed rate to maintain the preset constant cutting force with a higher cutting efficiency.

In reality the definition, such as that for the larger or smaller feed rate, contains a certain degree of uncertainty and vagueness. Furthermore, machining processes contain highly non-linear, time-varying and complex characteristics. Therefore, designing the adaptive control system for machining operations is a challenging task. It has been shown that the adaptive neuro-fuzzy control not only has the better potential for controlling non-linear, time-varying, and complex system, but also is a very effective tool for dealing with an uncertain and vague system [12]. Therefore, an adaptive neuro-fuzzy control system has been proposed and developed for the control of milling operations in the present study.

6.3   Adaptive Neuro-Fuzzy Control of Milling Operations

The overall block diagram of the adaptive neuro-fuzzy control system in milling operations is shown in Fig. 6.1. It is shown that the measured cutting force F_c is compared with the reference cutting force F_r from which the error of the cutting force E and the change of the cutting force CF are obtained. These two signals, E and CF, are then sent into the neuro-fuzzy logic controller for generating the change of the feed rate ΔU. In addition, these two signals are also used to tune the structure of the neuro-fuzzy logic controller for improving the control performance in milling operations. The feed rate U is calculated based on the change of the feed rate ΔU. Therefore, an integral control action is generated to eliminate the steady state force error and then the measured cutting force F_c is equal to the reference cutting force F_r. However, the feed rate U must be constrained to a finite value to prevent an excessive cutting force breaking the cutting tool. Therefore, the maximum feed rate U_{max} is used if the feed rate U is greater than U_{max}. The determined feed rate, which is called the command feed rate U_{com}, is sent directly to the machine tool driver system for producing the real feed rate V_f. The cutting tool then proceeds with the real feed rate V_f to cut a workpiece and to generate the measured cutting force F_c again. Hence, an on-line adjustment of feed rate to maintain the constant measured cutting force F_r under varying cutting conditions, can be achieved based on this control system.

Milling Controlled Process

For the machine tool driver shown in Fig. 6.1, the transfer function between the command feed rate U_{com}, and the real feed rate V_f can be expressed as:

$$\frac{V_f(s)}{U_{com}(s)} = \frac{K_n}{(s/\omega_n)^2 + 2\zeta(s/\omega_n) + 1} \tag{6.1}$$

where K_n is the gain of the servo; ω_n is the natural frequency of the servo; ζ is the damping ratio of the servo; and s is the variable of the Laplace transform.

The feed per tooth f_t in the milling process can be expressed as:

$$f_t = \frac{V_f}{Nm} \tag{6.2}$$

where m is the number of teeth and N is the spindle speed (rpm).

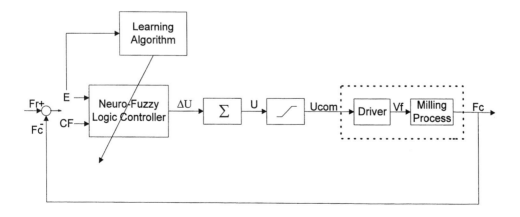

FIGURE 6.1   Block diagram of the adaptive neuro-fuzzy control system in milling operations.

The chip thickness h on each cutting edge, varying not only by the rotation position of the cutting edge but also by the cutter run-out, can be expressed as:

$$h = \begin{cases} f_t \sin \theta_i + (r_i - r_{r-1}) + h_i & \text{if } f_t \sin \theta_i + r_i > r_{i-1} \\ 0 & \text{if } f_t \sin \theta_i + r_i \leq r_{i-1} \end{cases} \tag{6.3}$$

where θ_i is the rotation angle of the i-th cutting edge, r_i is the radius of the i-th cutting edge, and h_i is the over-cut chip thickness for the i-th cutting edge, which can be expressed as:

$$h_i = \sum_{j=1}^{k} f_t \sin \theta_i + r_{i-j} - r_{i-j-1} \quad \text{until } f_t \sin \theta_i + r_{i-k} - r_{i-k-1} > 0 \tag{6.4}$$

The tangential force and the corresponding radial force acting on the cutting edge can be expressed as:

$$F_t = k_s bh \tag{6.5}$$

$$F_r = k_r F_t \tag{6.6}$$

where b is the axial depth of cut and k_r is the ratio of the tangential force and radial force.

The cutting forces acting on the cutting edge in the X and Y directions can be obtained by decomposing the tangential force F_t and radial force F_r into the X and Y directions:

$$F_x(\theta_i) = F_t \cos \theta_i + F_r \sin \theta_i \tag{6.7}$$

$$F_x(\theta_i) = -F_t \sin \theta_i + F_r \cos \theta_i \tag{6.8}$$

Then, the cutting forces in the X and Y directions with multiple cutting edges can be expressed as:

$$F_X = \sum_{i=1}^{m} \delta(i) F_x(\theta_i) \tag{6.9}$$

$$F_Y = \sum_{i=1}^{m} \delta(i) F_y(\theta_i) \tag{6.10}$$

and

$$\delta(i) = \begin{cases} 1 & \text{if } \theta_s \leq \theta_i \leq \theta_e \\ 0 & \text{otherwise} \end{cases} \tag{6.11}$$

where m is the number of teeth on the cutter, θ_s is the start angle of cut, and θ_e is the exit angle of cut. Basically, the start angle of cut θ_s and the exit angle of cut θ_e are a function of the radial depth of cut and the geometry of workpiece. Finally, the resultant cutting force F_c can be expressed as:

$$F_c = (F_X^2 + F_Y^2)^{1/2} \tag{6.12}$$

From Eqs. (6.1) through (6.12), it can be seen that the transfer function of the controlled process between the resultant cutting force F and the command feed rate U_{com} contains complicated, non-linear, and time-varying characteristics. The change of cutting parameters such as axial depth of cut b, spindle speed N, and radial depth of cut varies the open-loop gain of the controlled process. Since the performance of the control system (Fig. 6.1) is characterized by the open-loop gain of the controlled process, any variation in these parameters will directly affect the control system response and might even cause instability. Therefore, the use of the neuro-fuzzy logic controller for improving control performance under variations of these cutting parameters will be discussed next.

Neuro-Fuzzy Logic Controller

The neuro-fuzzy logic controller developed in this study is composed of a number of fully interconnected nodes and well organized into four layers (i.e., input layer, membership layer, rule layer, and output layer). Nodes on the first layer are used to receive and normalize the physical inputs of the controller by the scaling factors. The normalized physical inputs are then fuzzified by the membership functions in the second layer. Next, the third layer performs a fuzzy reasoning on control rules to generate an inference output on each rule. Each inference output has its corresponding weight in the output layer. Finally, the weighted inference outputs are summed to produce the output of the controller. The structure of the two-input-one-output neuro-fuzzy control system used in milling operations is shown in Fig. 6.2. In the following, the function of the neuro-fuzzy logic controller is described as follows.

The measured resultant cutting force F_c is compared with a reference cutting force F_r and converted into two controller inputs, the cutting force error E and the cutting force change CF, that is:

$$E(i) = F_r - F_c(i) \tag{6.13}$$

$$CF(i) = F_c(i) - F_c(i-1) \tag{6.14}$$

where i is the index of time increment for sampling the cutting force.

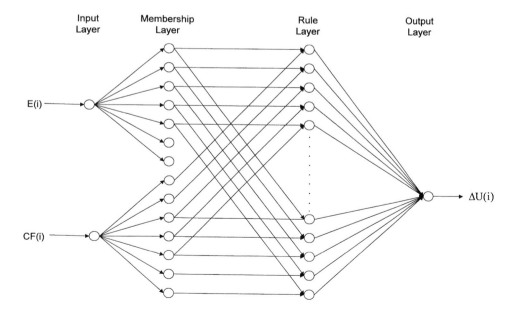

FIGURE 6.2 Structure of the two-input-one-output neuro-fuzzy logic controller.

Therefore, there are two nodes in the input layer of the controller. The two inputs are normalized in the closed interval $[-1, 1]$ by multiplying the corresponding scaling factors, GE and GCF, that is:

$$e(i) = GE \, E(i) \tag{6.15}$$

$$cf(i) = GCF \, CF(i) \tag{6.16}$$

where $e(i)$ and $cf(i)$ are the two normalized inputs.

The normalized inputs are then mapped into suitable linguistic values by using the membership function of the fuzzy sets. A simple triangular shape of the membership function for grading the membership of the class from members to non-members is used in this study. The triangular membership function $\mu_A(x)$ for a linguistic (fuzzy) set A where $-b \leq x \leq b$ is defined as:

$$\mu_A(x) = \begin{cases} \dfrac{x + b}{a + b} & -b \leq x \leq a \\ \dfrac{x - b}{a - b} & a \leq x \leq b \end{cases} \tag{6.17}$$

In the membership layer, seven linguistic sets are defined as follows: NB - negative big; NM - negative medium; NS - negative small; ZE - zero; PB - positive big; PM - positive medium; PS - positive small. Figure 6.3 shows the shapes of seven linguistic sets for the two normalized inputs. As discussed before, each normalized input is mapped into seven linguistic sets; therefore, the membership layer has fourteen nodes due to the two normalized inputs (Fig. 6.2). As to the rule layer, 49 (7×7) linguistic rules are constructed because seven linguistic sets of the normalized input e are fully interconnected with seven linguistic sets of the normalized input cf. It also means that 49 nodes are available in the rule layer. Let the j-th linguistic (fuzzy) control rule be described as follows:

if e is A_j and cf is B_j then z is C_j.

By taking the product compositional operation [7], the fuzzy reasoning of the control rule yields an inference output. Suppose $e = e_o$ and $cf = cf_o$ are the two normalized inputs at time i. The membership function for the j-th rule can be expressed as:

$$\mu_{c_j}(i) = (\mu_{A_j}(x_o) \cdot \mu_{B_j}(y_o)) \tag{6.18}$$

The inference output $\mu_{c_j}(i)$ has its corresponding weight $w_j(i)$ in the output layer. The weighted inference output is equal to the product of the inference output $\mu_{c_j}(i)$ and its corresponding weight $w_j(i)$.

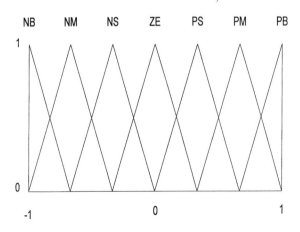

FIGURE 6.3 Shapes of the seven linguistic sets.

The weighted inference outputs are summed to produce the output value of defuzzification $\Delta u(i)$, that can be expressed as:

$$\Delta u(i) = \sum_{j=1}^{n} \mu_{c_j}(i) w_j(i) \tag{6.19}$$

where n is the total number of fuzzy rules in the rule layer.

The output value of defuzzification $\Delta u(i)$, multiplied by the output scaling factor of the controller GU results in the change of feed rate $\Delta U(i)$, that is:

$$\Delta U(i) = GU\Delta u(i) \tag{6.20}$$

Finally, the feed rate U can then be expressed as:

$$U(i) = U(i-1) + \Delta U(i) \tag{6.21}$$

Learning Algorithm for the Neuro-Fuzzy Logic Controller

It is known that the use of the developed neuro-fuzzy logic controller in milling operations is to achieve an on-line adjustment of feed rate with a constant cutting force. To reach the constant cutting force F_r as soon as possible, the cutting force error E and the cutting force change CF must quickly approach zero. Therefore, a performance index that is a function of the normalized force error $e(i)$ and the normalized force error change $cf(i)$ during cutting must be minimized. The performance index PI can be defined as:

$$PI = \frac{1}{2} \sum_{i=1}^{k} (e(i)^2 + \rho \cdot cf(i)^2) \tag{6.22}$$

where k is the total number of sampling data and ρ is a weighting factor.

The negative gradient of the performance index for the optimal control performance [13] can be expressed as:

$$-|\nabla PI| = (-|e(i)| - |cf(i)|) \tag{6.23}$$

To simplify the learning process, only the weight connected with the maximum inference output is adjusted. A change of weight for the j-th rule with the maximum inference output $\mu; (i)$ can be expressed as:

$$\Delta w_j(i) = \eta(-|\nabla PI|) \begin{bmatrix} e(i) \\ cf(i) \end{bmatrix} \tag{6.24}$$

where η is the learning rate.

The new weight $w_j(i)^{\text{new}}$ corresponding to the maximum inference output $\mu_j(i)$ can then be expressed as:

$$w_j(i)^{\text{new}} = w_j(i)^{\text{old}} + \Delta w_j(i) \tag{6.25}$$

6.4   Computer Simulation and Experimental Verification

Computer Simulation

In the simulation, a 12 mm diameter HSS end mill with four flutes, rotating 400 rpm, machining 6061 aluminum blocks (k_s = 1500 N/mm^2, k_r = 0.6) was used. The reference force of 300 N and the maximum feed rate U_{max} of 75 mm/min were selected in the control loop (Fig. 6.1). The peak resultant cutting force in a revolution is the criterion to be controlled in the adaptive neuro-fuzzy control system. The scaling factors of the controller, GE = 0.0033, GCF = 0.0017, GU = 0.1, were chosen, respectively. Since this control system is a sampled-data system, the machine tool driver [Eq. (6.1)] needs to be identified by using Z-transform.

The cut geometry with the changes of the axial depth of cut in milling operations is shown in Fig. 6.4. Figure 6.5 shows the variation of the *PI* value with the number of learning (cutting) cycles. The weighting factor ρ = 6 and the learning rate η = 0.25 were used in the learning process. It can be seen that the *PI* value decreases as the number of learning cycles increases. The initial learning convergence of the *PI* value is very fast, followed by a period of slower convergence to the minimum value. Hence, it can be clearly shown that control performance of the milling process becomes better and better through the learning process.

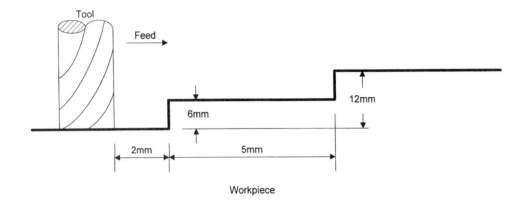

FIGURE 6.4   Cut geometry with the changes of the axial depth of cut.

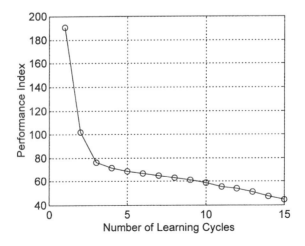

FIGURE 6.5   Effect of the *PI* value on the learning cycle.

The learning process was terminated after 15 cutting cycles because the improvement rate of the *PI* value became insignificant. After the learning process is completed, the weights in the output layer are no longer changed and the neuro-fuzzy logic controller is ready to be applied for machining tests.

Experimental Results and Discussion

To verify the adaptive neuro-fuzzy logic controller, cutting tests with the same cutting conditions as mentioned before were performed on the CNC machining center. The cutting force signal was obtained from a dynamometer (Kistler 9255B) mounted under the workpiece and the feed rate signal was directly measured and extracted from the CNC controller. Both signals were recorded on a PC-486 through a data acquisition board (DT2828).

Figure 6.6 shows the measured feed rate and measured cutting force during machining of the cut geometry with the changes of the axial depth of cut. As shown in Fig. 6.4, the end mill starts 2.0 mm from the workpiece. Since the end mill has not entered the workpiece, the maximum feed rate should be used in order to save the machining time (Fig. 6.6(a)). Once the cutting tool starts to engage the workpiece with the axial depth of cut of 6 mm, a sudden overshoot of the cutting force is generated (Fig. 6.6(b)). The cutting tool is quickly slowed down to prevent the excessive cutting force breaking the cutting tool. After the cutting force returns to the level of the reference force F_r, the cutting tool proceeds at a constant feed rate. A similar cutting phenomenon is also shown in the change of the depth of cut from 6 mm to 12 mm.

A more complicated pocket workpiece with a spiral-out tool path (Fig. 6.7) was used in the experiments. In this pocket machining, the radial depth of cut varies not only by way of different traveling paths but also by different positions of a path [14]. Figure 6.8(a) shows the cutting force and machining time using the constant feed rate generated by a commercialized CAD/CAM system. The use of the constant feed rate generated by the CAD/CAM system is the most common approach in the industry. Figure 6.8(b) shows the cutting force and machining time using the adaptive feed rate generated by the adaptive neuro-fuzzy controller. It is shown that machining time saved is about 23% through this approach.

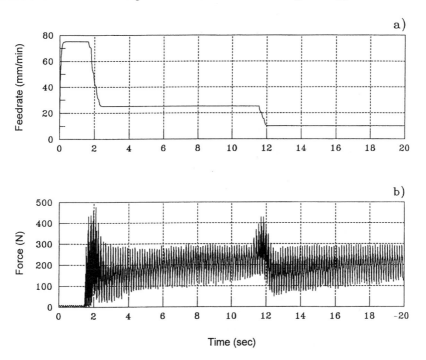

FIGURE 6.6　Experimental results with the changes of the axial depth of cut: (a) measured feed rate; (b) measured cutting force (reference force = 300 N; spindle speed = 400 rpm).

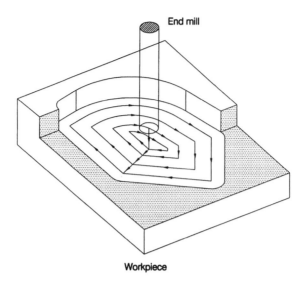

FIGURE 6.7   Pocket machining with a numerical control tool path.

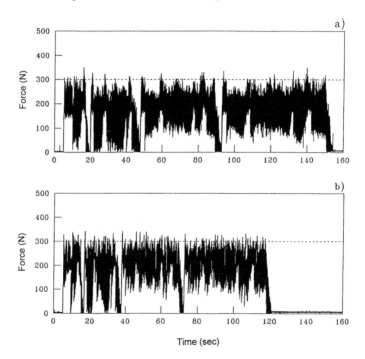

FIGURE 6.8   Measured cutting forces for the pocket machining: (a) without the adaptive neuro-fuzzy control; (b) with the adaptive neuro-fuzzy control (reference force = 300 N; spindle speed = 400 rpm).

6.5   Conclusions

An adaptive neuro-fuzzy control system in milling operations has been described in this paper. The main advantage of this approach is that the control performance of the neuro-fuzzy logic controller can be improved through the learning process. As a result, the proposed adaptive controller starts from an empty rule base and the design cycle time for the neuro-fuzzy control system can be greatly reduced.

Computational simulations and experimental cutting tests have been performed in milling operations to confirm the proposed method. Good control performance of feed rate and cutting force in milling processes have also been shown. Hence, a powerful technique for constructing the adaptive control system in milling operations has been demonstrated in this study.

Acknowledgements

The authors wish to thank Associate Professor Z. M., Yeh, Institute of Industrial Education and Technology, National Taiwan Normal University, Taipei, Taiwan, for his help in the course of this work. This research was supported by the National Science Council of the Republic of China, Taiwan under grant number NSC82-0422-E-011-017.

References

[1] Groover, M., *Automation, Production Systems, and Computer Integrated Manufacturing*, Prentice-Hall, Englewood Cliffs, NJ, 1987.

[2] Koren, Y., *Computer Control of Manufacturing Systems*, McGraw-Hill, New York, 1983.

[3] Elbestawi, M. and Sagherian, R., Parameter adaptive control in peripheral milling, *International Journal of Machine Tools and Manufacture*, 27(3), 399–414, 1987.

[4] Launderbaugh, L. and Ulsoy, A., Model reference adaptive force control in milling, *ASME Journal of Engineering for Industry*, 111, 13–21, 1989.

[5] Elbestawi, M., Mohamed, Y., and Liu, L., Application of some parameter adaptive control algorithms in machining, *ASME Journal of Dynamic Systems, Measurement, and Control*, 112, 611–617, 1990.

[6] Tarng, Y. S. and Hwang, S. T., Adaptive learning control of milling operations, *Mechatronics*, 5(8), 937–948, 1995.

[7] Pedrycz, W., *Fuzzy Control and Fuzzy Systems*, John Wiley, New York, 1989.

[8] Zadeh, L., Fuzzy sets, *Information Control*, 8, 338–353, 1965.

[9] Tarng, Y. S. and Cheng, S. T., Fuzzy control of feed rate in end milling operations, *International Journal of Machine Tools and Manufacture*, 33(4), 643–650, 1993.

[10] Yeh, Z. M., Tarng, Y. S., and Nian, C. Y., A self-organizing neural fuzzy logic controller for turning operations, *International Journal of Machine Tools and Manufacture*, 35(10), 1363–1374, 1995.

[11] Tarng, Y. S., Cheng, C. I., and Kao, J. Y., Modeling of three-dimensional numerically controlled end milling operations, *International Journal of Machine Tools and Manufacture*, 35(7), 939–950, 1995.

[12] Brown, M. and Harris, C., *Neurofuzzy Adaptive Modeling and Control*, Prentice-Hall, Englewood Cliffs, 1994.

[13] Anstrom, K. J. and Wittenmark, *Adaptive Control*, Addison-Wesley, Reading, MA 1989.

[14] Tarng, Y. S. and Shyur, Y. Y., Identification of the radial depth of cut in numerical control pocketing routines, *International Journal of Machine Tools and Manufacture*, 33(1), 1–11, 1993.

7

Instrumental Robots Design with Applications to Manufacturing

R.C. Michelini
University of Genova

G.M. Acaccia
University of Genova

M. Callegari
University degli Studi di Ancona

R.M. Molfino
University of Genova

R.P. Razzoli
University of Genova

7.1 Introduction

Instrumental robotics are developed to provide functionally oriented equipment, having duty adapted activity performance figures in order to accomplish sets of required tasks, with proper autonomy ranges. Basically, robots characterize the domain of intelligent automation, supplying the "active" adaptation of the actuating, handling, grasping, or machining dynamics. Active behavior innovates conventional feedback automation, making it possible to modulate the dynamics as a case arises, while tasks are performed. It is usually understood that adaptivity has to be related to outfit and reset actions, enabled with "knowledge" of the ongoing duty sequences and surrounding influences. Furthermore, task-driven equipment is usually concerned with "uncertainty," since it's end-effector is interfaced to the structured "external" world. The uncertainty is overridden by knowledge-intensive techniques, that mainly exploit system hypotheses to drive the manipulation dynamics with due account for instance, of the modeled nonlinear inertial couplings and observation data to modify the current behavior while counteracting the external off-setting influences.

The domain of instrumental robotics is characterized by its task-dependence by the detailed specification of operation duties, and by the current recognition of execution charges. The robot behavior is mainly assumed to evolve according to structured patterns, i.e., distinguishing the related developments from those that are typically investigated in the field of the artificial intelligence applications and aiming at autonomous agents interfaced to unstructured environments. In this field, the operation autonomy presumes on-line task-planning, performed by exploiting goal-oriented bent and self-learning abilities to generate (turn by turn), proper activity patterns. Autonomy under-emphasizes robots as rigs that do jobs. Matching a "generic" robot to a task requires costly interfaces, a complex program, and cannot reach "optimal" schedules. Functional bent due to instrumental turn, on the other hand, reaches effectiveness which resorts to off-process task-programming by acknowledging "optimal" activity modes, prearranged to secure the accuracy, dexterity, efficiency, and versatility figures required to exactly fill out the wanted set of tasks (and not aim at some generic goals).

The idea behind functional bent is equivalent to a paradigm in robotics stating: *"as soon as a task is acknowledged into a series of instructions, then equipment can be devised to do it."* Actually, the ability to recognize operation models might become restrictive: a person could be able to perform "undefined" tasks, filling gaps with skill; better plans could be stated moving the "intelligence" from manufacturing to artifacts' redesign. The two issues are different and the second has entrepreneurial value. When instructions lack a quantitative basis, it cannot attempt to perform an audit for quality data or artifacts would be delivered with unpredictable specification; and, when product-and-process are simultaneously poised by reengineering, economic and technical criteria could be stated quantitatively to recognise good or bad artifacts. Economic considerations make understanding that the functional bent might become misleading, if built as an unnecessaryly complex technology-driven solution. Artifacts really need to be offered with "fit-for-use" properties and should be designed with bound operation range and application scope. Activity outlook, rather than sophistication, is the winning alternative of instrumental robotics, shifting the concern on adapting products to improving the manufacturing effectiveness and the client's satisfaction.

The Design Cycle for Instrumental Robots

The integrated approach to instrumental robotics attempts to simultaneously define activity modes and functional devices. The design of task-driven robots is a very exacting request to achieve the required operational performance with a return-on-investments benefit. It is incumbent on the designer to set operational rules that assure the fulfilment of the charges as specified by the activities model. It is understood, that outside such range a task-driven robot does not operate properly. It might possibly acknowledge a set of feasible tasks and progress as far as possible or, more correctly, generate "help" or "warning" messages. The integrated design cycle in instrumental robotics, Fig. 7.1, will iterate the following five steps:

1. Specification of the behavioral requirements of the tasks need be executed, with choice sets of competing activity modes so that technology-driven solutions will demonstrate their consistency (through prototypal implementations or the like) and could be classified in terms of cost-driven assessments.

2. Model the operational performance of the hypothesized robotic equipment, in connection with the acknowledged task-oriented specifications. The setting of consistent manipulation

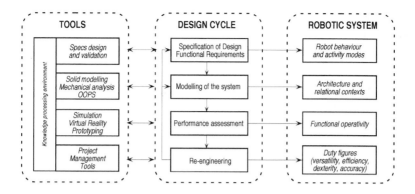

FIGURE 7.1 The design cycle of instrumental robots.

architectures and the (parametric) generation of the related structural and functional contexts are linked to each others.

3. Assess the characteristic features of each solution, with checks on the functional operativity (activity mode and process lay-out) and on the discriminating patterns (technological details and performance achievements) to reduce the development costs. This step exploits simulation and virtual reality checks while it uses prototyping only for critical components.

4. Reconsider the duty figures (versatility, efficiency, dexterity, and accuracy) and of the preset constraints to evaluate hidden merits and dearths. The step requires careful testing on the specifications and on the models, particularly when the prospective instrumental robot suggests 'new' operation modes as compared to conventional (anthropocentric) robotics.

5. Iteration of the cycle, whenever appropriate.

The steps cannot be undertaken unless the behavioural specifications, the relational contexts, the performance assessments, and the reengineering options are quantitatively recognized. Design steps profit from computer aids, supported at different levels of details and the functional and structural models of the proposed robotic fixtures; the addition of "helps" manage the decision loops at the users' interface. Functional bent is a powerful means for the choice of efficient instrumental robots when the design starts conceptualizing with all available data. Due to task dependency, processing of the supporting knowledge is needed along the life-cycle of the fixture: to program duty updating, control fitting, activity modes planning, and to transfer the pertinent data each time the robot's charges are modified (as a different class of tasks has to be accomplished). The computer aids are essential to the robot, since the final set-up's effectiveness depends on the duties actually enabled.

To design instrumental robots is a challenging process using technical and economical targets. Technical concern leads to overemphasizing robotic capabilities, increasing sophistication, and pushing the functional bent to make complex tasks feasible. The return-on-investments criteria need to redefine artifact's construction and manufacturing actions so that robot and duty can be balanced within the reach of "lean" engineering. A clear division between robot and task does not exist. That is why one should be in the position to assess advantages of "advanced" options in order to start reengineering and establish the solution that matches any particular request. Basically the design of instrumental robots could split according to two tracks:

- One design investigates technology-driven options and, mainly by computer simulation, looks for "advanced" setups fit to perform any prospected task no matter how complex, joining "interfacing" equipment and software to achieve the goals, and the issues transferring "exaggerated" ability to robots. This can be recognized on the condition that the adequacy for manufacturability is previously fully analyzed.

- Another explores how fit-for-use artifacts could be designed to favor manufacturing and, simultaneously, how processes could be modified to produce them with the available equipment. The issues possibly slow down innovation, setting aside the features capable of upgrading product quality or improving plant productivity.

The second track faced by manufacturing engineers is at the shop-floor level, aimed at flexible automation with the economy of scope rules. An example of typical issues in "intelligent" settings is given by the authors in the chapter "Techniques in computer-integrated assembly for cost effective developments" [MAC98]. The first track is mainly dealt with in this chapter. The organization of the matter bears the following scheme:

- A section recalls the main features of the issues achieved by computer simulation. The presentation avails itself of an extended programming aid (the 'SIRIxx' set of packages) properly generated to offer opportunities for experimenting sophisticated solutions.
- A section considers alternatives for upgrading functional bent to emphasize process and economical dependence and exploring simultaneous engineering and hardware-software modularity (i.e., standard mechanical parts, normalized command, and/or mobility).
- Further (two) sections give details about example developments aiming at robots with advanced control opportunities (dynamical nonlinearity compensation, redundant position/force control, etc.) and robots with sophisticated manipulation architectures (multiple robot configuration, redundant mobilities robots, etc.).

7.2   The Design of Function-Oriented Robots

Robots got their name from a Slavish root meaning "heavy labor," thus, they are developed to replace manual workers. They should be endowed with handy skills and training ingenuity "sufficient" to accomplish the considered tasks. A robot and it's duty are inseparable, but quite soon the difficulty of quantitative and deterministic job descriptions that humans were able to perform appeared. Designers started to equip the robot with further capability to fill the gaps in the job description. This innovation moved to end fixed automation and special-purpose equipment and to use computer intelligence for flexible automation and multi-task equipment. The achievements are technology-driven issues and business-driven economic patterns that appear as constraints to slow down replacements until return-on-investment (ROI) is verified.

The problem, however, is not just economical or technical. To assess its "fitness-for-purpose," the fixture ought to exist and be tested in the proper surroundings, to verify potential and effectiveness. The checks should cover, besides instrument functional appropriateness, the soundness of robot duties in terms of scheduled goals and task usefulness and increase process reliability and product quality. The design of function-oriented devices, on these grounds, splits into a series of accomplishments (see Fig. 7.2); of course, the separation by conceptual stages or detail levels is mainly done for academic purposes. The practical development of "new" robots, before "new" applications, is undertaken with the know-how of previously tested equipment. A design cycle can be started at any point, neglecting noncritical details or postponing underspecified phases. Phases in re-engineering, nevertheless, can only be started with "sufficient" knowledge on both robot technology and simultaneous engineering. In this chapter, attention is focused on robot technology; however, for instrumental robotics, the design will move from an "effectiveness" model, with "price-time" figures, to a return-on-investment model by monitoring process-added value and productivity performance. If a robot costs too much or takes too long to do the tasks, it will fail in the marketplace. The "effectiveness" model leads to recognition of the close binding of robotics and design and the critical support provided by CAD opportunities.

Conceptual Design of Task-Driven Robot Arms

The design of task-driven equipment aims at robots with operation capabilities and planning options to allow feasibility of the desired charges (regardless of the complexity or product-process consistency). The subject has been tackled from different standpoints related to the handling architectures [BeP97],

To acknowledge the robotic fixtures consistent with the task to be performed by means of a design cycle, undertaken at different levels of detail	
task-level assessment	to specify the corresponding operation details and to explore the consistency of the detailed sequences
task-level programming	to select series of instructions for their accomplishment and to single out manipulation architectures making the programmed charges feasible
operation-level assessment	to specify the commands making coordinated motion feasible and to investigate actual governing strategies
operation-level verification	to evaluate actual performance and robot's fit-for-use according to the tasks to be performed
To acknowledge task sequences consistent with the goals of the processes where robots are included as instrumental fixtures, by iterated design cycles	
goal-level re-engineering	to evaluate products' alternatives preserving or improving customer's satisfaction with 'simplified' technical specifications
process-level re-engineering	to explore the design-for-manufacturing changes, which increase the effectiveness of the production cycles
equipment-level re-engineering	to find out competing instrumental solutions, with better return on investment figures
programming-level re-engineering	to establish software tools, which would expand robot's operation or efficiency ranges

FIGURE 7.2 Conceptual stages and detail levels of the robot design.

[LII96b], [LYK97], [MAC93], [RoB97]; to the manipulation dynamics [Asd88], [AsH79]; or to the path planning [ACH95], [Alg97], [CCS91], [Whi69b], [ZOY96a], [ZOY96c] requirements, and a lot of practical know-how is available. Example results provide hints on the evolution of the design in robotics, from advanced technology-driven fixtures, to (just technical) fit-for-purpose solutions. Topics are reviewed following the authors' experiences and are presented, without presuming completeness, considering example issues, starting from abstract users' interface requests and entering into technical details that make possible the planned tasks with increasing performance.

- *Path planning and architectural analysis.* Path planning deals with the mapping of the end-effector set-points defined in the work-space, into actuation commands fed to each joint. In general, the effector should possess six degrees of freedom (three angular and three displacement co-ordinates) and the arm needs to combine the related number of powered mobilities. The robot is thus, a multi-variable system requiring adapted inputs for steering the tip location and attitude. Six powered mobilitiy manipulators are basic references for establishing kinematic path-planning problems.

The forward (to the work-space) and the backward (to the joint-space) kinematics can be stated after the manipulation architecture is selected. These (algebraic and nonlinear) transforms are:

$$x = f(q) \quad q = g(x) \quad q \in Q^n \quad x \in X^m \tag{7.1}$$

where x-work-space coordinates and q-joint-space coordinates. The kinematic inversion is generally unsolved, unless a few geometric restrictions are introduced to specialize an algebraic solution (as exemplified by the later mentioned SIRI-CA package).

The availability of the forward and backward kinematics (7.1) can immediately be used to explore the considered architecture's worthiness to accomplish the given set of tasks. The analysis has to be done as the initial job of the design cycle; sometimes less than six mobilities could be sufficient and "reduced" degrees-of-freedom robots are "good" choices. Sometimes, obstacles in the work-space or in the joint-space are properly avoided with additional powered joints and "redundant" mobilities robots are a "better"

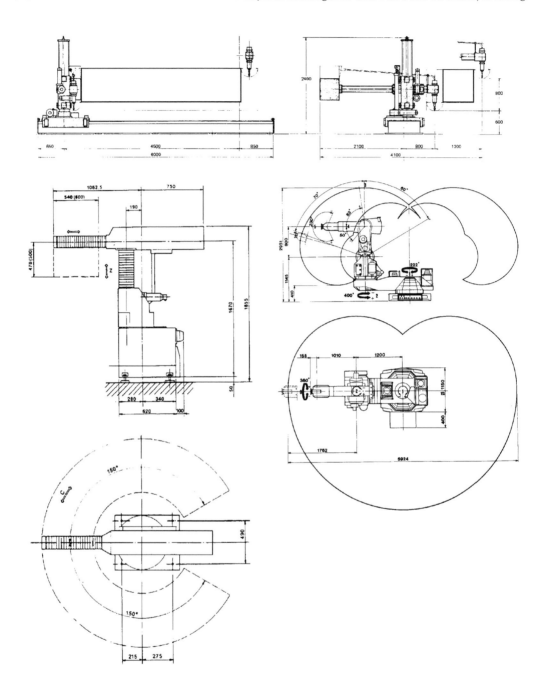

FIGURE 7.3 Example of robot architectures and related work-domains: (a) cartesian structure; (b) cylindrical structure; and (c) articulated structure.

option. The six-mobilities case is, however, basic reference to begin analyses and to obtain practical descriptions useful for CAD opportunities. The forward kinematics is usually defined with bounded spans of the joint coordinates; the mapping leads to closed domains of the work-space with shapes, (Fig. 7.3), which are related to the robot architectures according to patterns that can be used for classifying purposes.

To simplify the generation of computer solutions, linearized models are considered, assuming that velocity vectors, in both spaces, should steer the tip to smoothly approach the target. For the reference

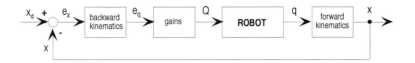

FIGURE 7.4   The work-space incremental control scheme.

architecture of an arm with six sequential mobilities, the so-called 'Jacobian approximation' holds:

$$\dot{x} \approx \frac{dx}{dt} \approx \left[\frac{\partial f}{\partial q}\right]\frac{dq}{dt} \approx \left[\frac{\partial f}{dq}\right]\dot{q} \approx J\dot{q} \qquad \dot{q} \approx \left[\frac{\partial f}{dq}\right]^{-1}\dot{x} \approx J^{-1}\dot{x} \tag{7.2}$$

$$\delta x \approx \dot{x}\delta t \approx J\dot{q}\delta t \approx J\delta q \qquad \delta q \approx J^{-1}\delta x \tag{7.3}$$

Jacobian matrices mapping end-effector's angular velocity (instead of Euler angles' rates of change) can be easily worked out by exploiting the previous relations (7.2). At each point, the Jacobian matrix is computed (as incremental ratio) and (numerically) inverted to enable a rate control, Fig. 7.4 (after the mapping is calibrated).

The approach is not without drawbacks for mapping accuracy and useful tricks are, therefore, explored. The traditional task of programming has been based on "teach-by-doing" and by that way:

- The static calibration is performed with (external) work-space references.
- The path planning may introduce rate compensation (depending on localized trajectory anomalies, on speed biasing influences, etc.).
- The setting of the instructions is trimmed to the specialized charges of each application, without explicitly acknowledging the actual robot behavior.
- The operation scheduling does not require models of the physical world (provided that teaching is fulfilled exactly replicating every work condition).

Time delays always occur with digital controllers related to sampling and processing rate. Unstable or swaying paths could be established, unless commands are fulfilled with no closed-loop meddling. Computer compensation of the time-delays is possible, if robot behavior and coupled surroundings models are stated (with possible account of uncertainty through fuzzy logic), or if sensorized interfaces operate on joints and at the work-space end. The compensation, however, is fixture-dependent and, assuming the approximated mapping (7.2), traceability of calibration tests are run and deviations assessed (by sensitivity analysis) for any of the allowed tasks.

- *Control planning and performance analysis.* Control planning is concerned with the choice of feedback loops to be applied at each actuator to make "dynamics shaping" so that the effector executes the assigned tasks with the "best" effectiveness. Again considering six-mobilities arms as reference architecture, the joint-space dynamics for unconstrained motion maneuvres is given by:

$$Q = A(q)\ddot{q} + B(q) + C(q,\dot{q}) + D(q,\dot{q}) \tag{7.4}$$

where Q is joint actuation force or torque; $A(q)$ is mass matrix of the robot; $B(q)$ is centroid off-set unbalances; $C(q,\dot{q})$ is transport and Coriolis terms; and $D(q,\dot{q})$ is friction and damping terms.

The reference dynamics (7.4) are nonlinear with combined effects of the carried links modifying the driving actuation needed at each joint and to impart a co-ordinated motion for the direct control of the tip. High performance robots are pushed to operate at high speed and acceleration. The dynamic coupling nonlinearities could generate undesired troubles, when feedbacks were established for linearized approximations only.

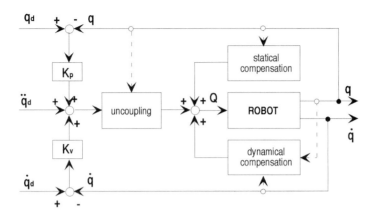

FIGURE 7.5   Joint-based actuation with linearizing and uncoupling feedback.

The free motion joint-based models dynamics has, thus, to deal with the actuation errors:

$$Q = A(q)u \quad u = E_p(q_d - q) + E_v(\dot{q}_d - \dot{q}) + \ddot{q}_d \tag{7.5}$$

where q, $\dot{q}_d$, and $\ddot{q}_d$ are the desired position, speed, and acceleration; E_p and E_v are the gains of proportional and derivative feedbacks.

The suppression of unacceptable back-nuisances is easily performed if the dynamics is properly modeled (7.4) and used as modulation gain, while closing the control loop in the joint-space. This means to exploit the canonical transform which is known to exist for (series of) rigid bodies joined by pivots. The compensation, Fig. 7.5, is easily done on condition, of course, that the architecture of the arm is assessed with known masses, centers of mass, mass quadratic moments, principal directions, etc. After compensation, the actuated joints are uncoupled and can be designed to behave as properly controlled linear second-order blocks. The scheme is a valid option when the coordinated motion control remains a hidden attribute. For instance, tasks are programmed on-line via "teach-by-doing" with due account of the desired performance.

Task programming is presently preferably done off-line supported by intelligent interfaces. The setting refers to the work space and selection of the properly controlled behaviour of each joint needs explicitly deal with the forces and torques transmitted by the distal link. Then, the transforms of the generalized forces can be stated as the dual approximation of the kinematics mapping, leading to:

$$\delta Q = J^T \delta F \quad \delta F = [J^T]^{-1} \delta Q \quad \text{where: } J^T = \left[\frac{\partial f}{\partial q} \right]^T \tag{7.6}$$

The compensation scheme, Fig. 7.6, can be modified accordingly, providing a clear guess on the operation conditions with, however, additional computations to be performed in real time. The command setting which follows is known as 'transposed local adaptation' since it is operated in the work space, then mapped in the joint space. The strategy gives useful results, when only joint displacements are observed, while the tip location in the workspace, not directly measured, cannot be used for off-setting the task errors.

A different approach tries to avoid real-time reckoning with the local relations (7.2) and (7.5); instead, it exploits current measurements for closing the control loops. As a matter of fact, all robots have encoders to obtain joint (absolute) displacements and to compute joint velocities; with addition of accelerators, (Fig. 7.7), alternative evaluations of the inertial couplings (7.4) are possible, to enable compensation

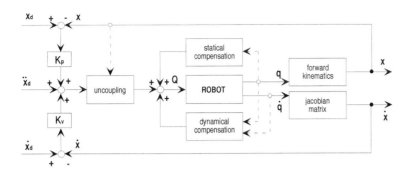

FIGURE 7.6 Work-space commands with compensation of nonlinearities.

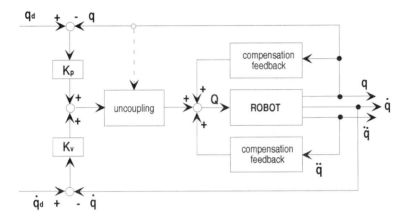

FIGURE 7.7 Sensor-driven compensation of nonlinearities.

feedbacks, or to trim the parametric choices of $A(q)$, $B(q)$, $C(q,\dot{q})$, or $D(q,\dot{q})$. Further opportunities are provided by measurements in the workspace. The direct observation of the tip linear and angular motion has been based on optical devices (cameras, laser sensors, etc.) with, nevertheless, data fusion drawbacks. Relevant advantages are expected by force/torque sensors located at the end-effector for measuring the interactions with the external world. The use of the information, however, needs more sophisticated models, with inclusion of the compliance attributes of the arm (joints, links, etc.) and of the interfaced bodies.

Conceptual Design of Work-Constrained Robot-Arms

The previous paragraph has mainly considered how to drive an arm so that its tip moves along a given path with prescribed attitude. Coordinate measurement machines are good examples of such an approach. Instrumental robots more usually characterized by tasks, with sharply changing charges, and splits the path planning (and control design) to deal with three different models, (Fig. 7.8), for usual duty conditions. Generally, these are reduced to *unconstrained* or *constrained* manueuvers. The third one, dealing with transient constraint manueuvers, needs be accounted for in front of special applications by using computer-aids.

Robot behavior under transient and work-constrained conditions is over-specified when the analysis is limited to rigid body degrees-of-freedom. For practical purposes, investigations are undertaken trying to separate the different influences. In order to be concerned only by a relevant phenomenon (any other input being reduced to be a disturbance); example developments; for instance, are

un-constrained motion maneuvers (free) *navigation path*	The actuation is accomplished on the arm, as isolated system, to bring the end-effector from an initial to a final position and attitude; alternatively, the manipulator is driven in the joint-space without contact with the environment, to make the tip track a trajectory with prescribed orientation and kinematic rules
transient constrained motion manoeuvres *engagement path*	The actuation faces the discontinuity arising between the navigation and the operation motion; the (collision-like) effects appear as (transient) disturbances and are most of the time neglected by robot designers, provided that suitable over-all stability conditions are granted along the operation work-cycle
work constrained motion manoeuvres (coupled) *operation path*	The actuation deals with the required tasks, mastering the end-effector constrained motion while subject to external influence or disturbance; the arm, driven in the joint-space, is manoeuvring in a constrained environment and needs transmit appropriate interfacing forces and torques, in the work-space

FIGURE 7.8   Characteristic duty conditions during robot maneuvers.

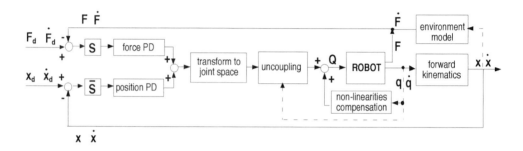

FIGURE 7.9   The combined position/force controller.

- Manipulators, having links and/or joints with (lumped or distributed) compliance, carrying masses (sensors, etc.) and sweeping a region with prescribed paths;
- Manipulators, having rigid (compliant) parts, colliding against compliant (rigid) walls, with different paths and bouncing coefficients;
- Manipulators, with rigid (compliant) parts, performing actions (machining, assembly, etc.) with controlled position and force of the tip.

The conceptual design of robots is said to be concerned by "duty programming" (with "duty" monitored about "task" departure). The addition of "advanced" abilities is still not essential for most of current industrial robot applications. It follows that topics in sophisticated activity modes are well acknowledged, while practical achievements lag far behind. A reason might be the lack of field experience, with clear assessment of actual opportunities. Few concepts are, therefore, recalled to provide hints on how the "duty charges" should be tackled (with respect to simple "tasks") through actuation 'flexibility' (besides "rigid" work-cycles). It can be started by noting that trained and skilled operators, when assembling artifacts, exploit force data to steer the joining motion; thus, performing position control and force modulation. Then, six-axis force-torque sensors are basic rig for anthropomorphic assembly and redundant control strategy applies, Fig. 7.9, by steering the separate feedback loops according to needs.

When one tries the "peg in the hole" task with such a sensorized robot, it may happen that instabilities arise unless the feedback time delay is chosen to match the combined "wrist, peg, and contact zone" stiffness. The force modulation should not be thinner than the measured contact force error. These kinds of instabilities do not depend on the path followed during the co-ordinated motion. By further analyzing the same task, the normal (indenting) and the tangential (friction) components of the contact forces are assessed to understand how insertion progresses along the prescribed directions (e.g., the sequence of Fig. 7.37). Current data is then, available to assess friction and damping effects and to feed engineered compliance so that assembly is fulfilled with delicacy (out of measurement uncertainty) unachievable by people. This means that position/force control is a simple affair of performance: when the combined accuracy, dexterity, efficiency, and versatility figures will be useful for transferring given quality data to the industrial artefact, the option could be checked in terms of return on investment.

Deburring tasks also highly profit by redundant control to grind out weld beads, to finish precision casts, or to honey machined surfaces aiming at constant quality artifacts (not depending on the ability of skilled operators or on the attention to accomplish the job), with cost reduction if burrs removal is granted with accuracy and delicacy. The application is later recalled, with introductory details on the process from a conceptual design viewpoint. The description of tip-to-burr interactions is given by a locally linearized impedance, binding the (generalized) force and displacement components, according to the (approximated) model:

$$\delta \tilde{F}_E \; = \; G(s)\,\delta \tilde{x} \quad \text{with:} \; G(s) \; = \; Ms^2 + Hs + K \tag{7.7}$$

where the reduced inertia M, damping H, and stiffness K depend on the solid zone to be ground and on the actual interfacing characteristics of the powered arm.

Here again, stability problems need, be addressed, due to the contrasting requirements on the normal and on the tangential stiffness figures. The removal of the dithering behavior cannot be extended to avoid the "worst case burr" everywhere in the work space when the deburring robot operates on a fixed artifact because of the contact rigidity between grinding wheel and piece. In spite of that, industrial applications exist (from Japan) with the force-torque sensor technology embedded into six mobilities robots. A better option is possible, carrying the piece by means of a six mobilities rig, so that the contact rigidity is properly modified, process-adapting normal and tangential stiffness. As mentioned, details of the option are later recalled.

The introductory comments of the paragraph give hints about possible "advanced" solutions. The analysis of existing applications to manufacturing (still covering the largest share for instrumental robotics) shows, most of the time quite monotonical replication of manipulation architectures and activity modes. The replication may depend on the functional orientation (meaning that once the task bent leads to an effective rig, it is no use looking for a different one), still the capability of investigating nonconventional architectures and/or behavioural options might suggest how to get out from assessed habits, aiming at 'unexpected' upgrading, by means of alternatives (in particular, if these have already been checked by virtual reality experiments). This is the reason to look for sophisticated models, capable of duplicating the dynamics and control strategies of the instrumental robots, with consistency of details up to the sought technical charges. The models are explored by means of computer aids, leading to comparative assessments and providing, as decision support, a choice of the function-oriented structural settings (component, facility-configuration and command, CFC, frame) and the operation-befit activity modes (monitoring, decision-manifold and management, MDM, frame).

Computer Aids Based on Functional Modeling and Simulation

Computer aids, providing virtual reality experimentation, are a powerful means for fostering the innovation in the field, carefully checking feasibility and effectiveness of new CFC frames and/or new MDM frames, while helping to evaluate the expected return on investment. Computer simulation is a critical means, as the instrumental robotics does not move from anthropocentric functional models, rather it acknowledges task-oriented solutions for the setup of more effective activity modes. Certainly in many cases, robots have been conceived for replacing man, giving rise to replication of 'anthropomorphic devices'; in other cases, they have been developed to perform tasks out of mens' potentialities, giving rise to new fields of 'instrumental robotics,' depending on the conditioning applications, on the transferred level of autonomy and intelligence, and on the actualy achieved performance (accuracy, dexterity, efficiency, and versatility).

The design of operation-oriented robots is a fascinating technical challenge by aiming at the "best" fixtures no matter how complex, providing the conformance to specification is reached and the desired tasks are properly accomplished. The subsequent sections of this chapter try to mitigate the challenge by looking for balanced solutions by having the complexity relieved by the fitness for purpose of the 'economy of scope' approach. By now, the design cycle, Fig. 1, according to the mentioned four steps, is iterated to bring forth pace-wise (performance-pulled and knowledge-pushed) betterments, until 'best' setups are obtained. Iteration progresses with CAD modules supplying virtual-reality display of robot actual dynamics, interfaced with expert modules, as decision support for addressing improved solutions. CAD codes, moreover, help to enter into the details of the predicted functional behavior, to assess the standards to be preserved according to specification, and the entities to be monitored for pro-active maintenance diagnostics.

Several CAD opportunities are available to help the designer, principally,

- General purpose software (such as Pro/ENGINEER series), granting background tools for the buildup of any personalized CAD instance.
- Computer packages suitably arranged into a virtual reality environment to provide systematic support for comparative assessments between competing equipments.
- Computer programs, purposely developed in the framework of a particular project, to give efficient account of the peculiarities of the application.

All three opportunities are useful. The first deserves growing interest since the offered software covers larger and larger CAD details and is endowed with friendly interfaces (costs are the main drawback). The third is largely exploited as soon as the robot equipment is chosen at the suitable level of specification and is the intermediate opportunity best suited to explore for innovation. Typical aspects are considered, hereafter referring to the work carried on by the Industrial Robot Design Research Group at the University of Genova, Italy, which has prepared and used the CAD environment SIRIxx, Fig. 7.10, to develop instrumental robots properly tailored to individual applications.

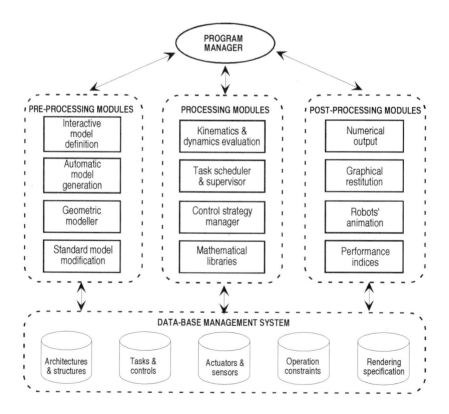

FIGURE 7.10   The SIRIxx robots' design environment.

These packages could be further distinguished by application range; on one side, operation bent is faced as general priority:

- Path planning and architectural analysis are linked with the aid of a set of modules, programmed to generate forward and backward kinematics, for several robots' families.
- Control planning and performance analysis are assessed with the help of a library of modules, programmed to generate example steering strategies, at different levels of complexity.

At higher levels of details, product-and-process matching is explored addressing to less conventional settings and/or more demanding situations:

- Duty planning and function fitting are investigated with the aid of animated displays or robot co-operation to weigh task requirements departures.
- Work constraints and process conditions are tackled by means of control modulation or special effects assessment in front of exacting effects.

This CAD support has been developed over the past ten years with a modular base in order to expand the covered subjects, while preserving computational efficiency. The results have been collected systematically and used as pre-set data for the later discussion of knowledge-based architecture SIRI-XE [ACM88]. The main characteristics of the packages are presented in several papers, with example developments. We defer to the references for details since only a short overview is given hereafter.

The 'SIRIxx' CAD Environment: The Basic Modules

The usefulness of recurrent design-cycle is better explained with examples of the SIRIxx series of packages [AMM87], [AMM90], [ACM91b] which provides useful hints on opportunities and issues underlying systematic investigations. The first group of packages is organised as general purpose CAD support.

- SIRI-CA: providing the usual path-planning objectives. It is built on the availability of forward and backward kinematics of 32 families of open-chain manipulators [ACM86].
- SIRI-AD: assuring the automatic generation of the nonlinear dynamics of open-chain manipulators [AMM84b]. It exploits a step-wise recurrent formulation propagating the dynamical behavior [MMA83].
- SIRI-CL: performing the path-planning and generating the dynamics of three families of closed cinematic chain manipulators. The package presents an oriented structure, exploiting the specialization of the internal constraints [ACM96c].
- SIRI-SC: assessing the robot dynamical performance with competing control strategies [AMM84a], [BMM85]. A modular and extended library of options is provided to help robot control planning.

With the recalled packages, the design cycle evolves, according to the four logic steps of Fig. 7.1, to give rise to sequences of phases, Fig. 7.11, such as:

1. The robot topology is provided by a first specification, employing the application area functional-data and the workspace general-constraints.
2. The payload and mobility requirements provide the main structural properties of the manipulator members and joints.
3. The productivity, with related speed and accuracy figures, makes possible an initial selection of the actuating devices.
4. The task complexity and the functional performance are detailed jointly with the observation schemes to define the control strategies.
5. The architectural consistency is equalized by assessing the robot dynamical behaviour for the set of allocated tasks.

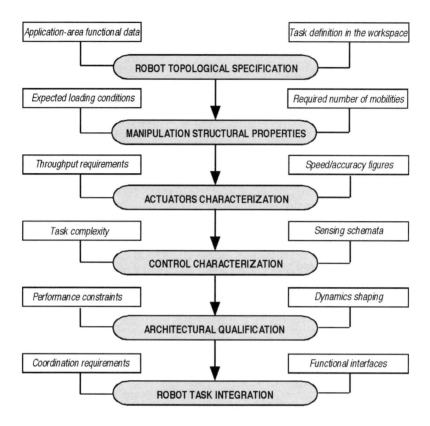

FIGURE 7.11   Basic development phases in the robot design.

6. The job integration of the (software/hardware) resources is performed with concern to the robot communication constraints.

To develop a suitable robotic equipment, an extended background is required which joins the results of the experimentation on existing devices with the investigation on the (structural and behavioral) properties of feasible solutions obtained by a functional description based on the current dynamics. The CAD-environment helps close a set of information loops, Fig 7.12. The designer needs explicit access to all the parameters that may significantly affect the robot operativity. The architectural analysis is required to set the *configuration data*, giving topology and geometrical bounds. In terms of the *structural attributes*, the basic choices concern the

* Actuation data to comply with pay-loads and throughput;
* Observation data to define the appropriate sensing and monitoring scheme;
* Regulation data to specify the control strategies granting the wanted performance.

In parallel, the designer should fix the *conditioning bounds*, specifying the

* Execution data for the management of the scheduled job agendas;
* Co-ordination data for specifying communication and synchronization requests;
* Organization data for prescribing action modes depending on the assigned tasks.

Along with the development cycle, the designer faces interlaced problems, namely,

* The parameters of the functional models (inferred from presumed system hypotheses) should be adapted to improve robot performance with respect to the selected tasks.

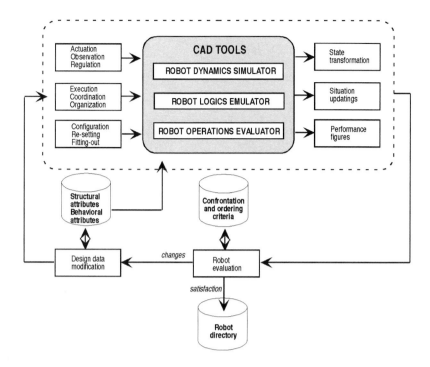

FIGURE 7.12   CAD environment for robot design.

- The models, identifying the task implementation, shall be modified (updating system hypotheses) improving the reference knowledge with account of the conditioning functional criteria.

Typical situations faced by a designer for developing robotic equipment are covered briefly.

- *Architectural analysis*—The setup of the robot layout is, logically, one of the final issues (it is at the fifth phase in Fig. 7.11); production engineers, however, shall directly define a preliminary architecture for their own application by simply considering existing robots (with, possibly, special-purpose rigging). The SIRI-XE framework is available for that purpose. It is based on the X-ARS package [ACM88] that essentially comprises: a *general data-base*, where investigated robots are orderly catalogued into frames with a hierarchical presentation of the available information: a *data-base management block* employed for creating, deleting, or modifying the frames and/or their contents; a *user interface* for interactive operations (through a nested menus sequence) or for assisted operation (through an "expert" block); and an *expert manager* with a rule-based procedural knowledge providing the inference mechanism based on heuristics for the selection of application-consistent robots.

The possibility to broaden, update, and modify robot records is a noteworthy option of the framework. Robot technology is a rapidly evolving area: new devices have to be added; classification criteria, updated; and functional abilities, modified, etc. It is important, moreover, to have an instrument that may be personalized according to the application needs. A database management block is required to help code new knowledge on robot performance under actual running conditions and to expand the structural and behavioural data. The SIRI-XE framework exploits this database management block to implement the ordered recording of existing robots in terms of combined architectures-tasks data. Then the X-ARS procedural knowledge can be employed as expert consultant: in this mode a set of consistent functional features (configurational conditions and/or programming resources) is recommended and the user can accept, reject, or modify the suggestions, updating the robot directory with these 'new' options.

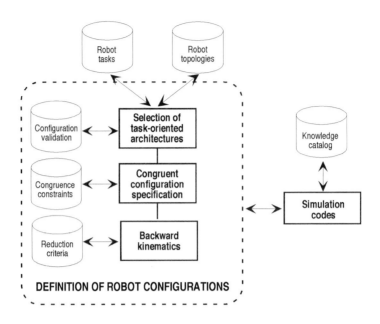

FIGURE 7.13   Selection of path-consistent robot topologies.

- *Activity-modes analysis*—The robot functional characterization depends on the set of tasks to be performed in the workspace once the robot topologies are selected, Fig. 7.13. Tasks are basically described in terms of activity modes, plotting the path and the attitude of the end-effector. The requirement is investigated solving the backward-kinematics problem (referred to trajectory planners), to plot the mobilities (or joints co-ordinates) trends. The solutions, referred to six-degrees-of-freedom robots, can present singularities and the selection of the correct branching needs complex tests when numerical procedures are implemented.

The hindrance is removed by the SIRI-CA framework: with resort to the operation constraints, that characterize actual manipulators, a set of modular elements are chosen; and then, restricting the study to six mobilities configurations, expressed by sequences of sliding or revolving joints, analytical solutions are worked out [ACM86]. The SIRI-CA framework contains a library for generating task-consistent sequences of activity modes, (i.e., three mobilities are basically employed for covering the work volume and three for the local position trimming and the attitude setting). The availability of analytical formats for workspace to joint space mapping provides a direct check on the robot congruence with regard to both the fixed and moving obstacles in the task domain.

- *Dynamic nonlinearities analysis*—Factory automation, with the increased versatility of the resources, moves toward high performance (in speed and accuracy) fixtures. Their dynamics should be generated, fully describing the inertial crosscoupling effects, to quantify the actual properties [ACM96d]. With heavy pay-loads, member compliance effects should be considered [AMP89], [Kov97]. Several programming facilities already exist. The package SIRI-AD was originally developed by the authors and employed as a service kernel of the simulation environment [AMM84b], modularly built for the development of robotic equipment. It can be expanded to cover different actuation possibilities.

This package is based on the recurrent modeling of the dynamics of the supported rigid bodies going back from the distal member to the fixed base. A preprocessing block (incorporating a 3-D geometric modeler) for shaping the robot arms and computing the related structural parameters (center of gravity positions, bulk quadratic moments, etc.) is included. The user can display the robot topology and check the configurations, all along any given task, each time calling on the graphic routines for visual presentations. The dynamics depends on the actuation laws and the solutions are available in the joint space and in the work-space. Referring to the development stages of Fig 7.2, the SIRI-AD framework provides the pertinent data for setting the actuators, once a reference configuration is obtained with the help of the SIRI-CA package.

- *Steering strategies analysis*—Robot design, (Fig. 7.11), needs the phases of selecting the control strategies; of verifying the task-congruence effectiveness; and of integrating the equipment in the manufacturing process. The availability of high efficiency processing devices enables families of feasible strategies with comparatively high sophistication-level control-schemes. Dynamic nonlinearities can be accounted for if the appropriate simulation facilities are employed starting with the ideation phase.

The SIRI-SC framework is, accordingly, built as standard CAD reference with modular layout; it can be extended to include all the different control options in use or proposed [ACM93], Fig. 7.14, such as,

- Point-to-point and path-continuous control with kinetically-balanced feedback;
- Position-follower control with force feedback and partial kinematics compensation;
- Piece-wise control with local (numerical) inversion of the (full) dynamics equations;
- Global compensation control by uncoupling of dynamical nonlinearities interactions;
- Adaptive (model referenced) optimal control with weighted rms performance index;
- Statistical observer control with parametric (fading memory) trajectory estimator;
- Probabilistic observer control with a stochastic (Gauss-Markov) dynamics modeler.

The global compensation control is a very efficient option, bringing higher performance with some computational burden; however, practical implementations are still lacking, in reason, mainly of the

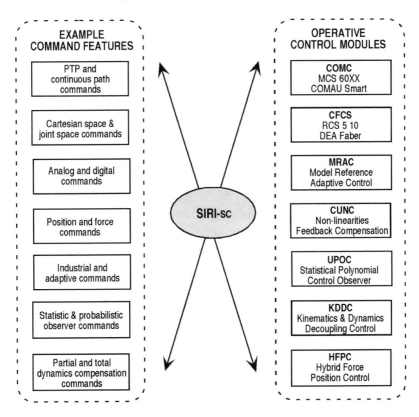

FIGURE 7.14   Features of the SIRI-SC framework.

restricted familiarity with the stability of motion of nonlinear systems. The reference to approximated models having the compensation limited to backward kinematics transforms [MPM78] is considered for the cases liable to simple use. The adaptive optimal control [AMM84a] deserves special interest for theoretic studies. The results have "robustness" limitations and generally present computation problems. The inclusion of observers with the function of (multi-variable) feed-forward compensators, is also useful. A good bargain between performance and complexity is obtained with polynomial fading-memory filtering [BMM85]. The control strategy based on the numerical inversion of the dynamics equations [Whi69a], [AMM87] is finally included in the SIRI-SC library to be employed to generate the reference trajectories for the performance estimation of the different control strategies.

The ability of changing the manipulator dynamics, by setting a command law that depends on the running behaviour, can be used to force the motion according to given requirements. Dynamics "shaping," for instance, allows the uncoupling of each joint as linear (second order) block provided that the appropriate compensation is applied. The result can be obtained by acceleration data and feedback closure or by model-based feed-forward modulation. When both options exist, the self-calibration is simply performed along with unconstrained tasks.

Specialized Options of the CAD Environment

The simulation environment SIRIxx has purposely been developed for assessing, with reliableness, the robot accuracy, dexterity, efficiency, and versatility figures. The environment has been expanded, on a modular base, for dressing the analysis abilities for innovatory robotics, into standard evaluation frames. In order to achieve additional or particular properties, as required for most advanced applications, a second group of packages has been developed aiming at more detailed goals. Example packages are:

- SIRI-AT: providing the animation of the tasks progression possibly interfaced to time-varying surroundings [CMP94]; the graphic restitution is based on standard softwares.
- SIRI-UM: showing the consequences of elastic and anelastic impacts of the end-effectors against fixed or moving obstacles [ACM96a]. Impulses are propagated along the members to the joints and the related actuators.
- SIRI-HD: generating combined position/force feed-backs, interfaced with structured or unstructured surroundings [ACC93], where redundancy improves accuracy and versatility.
- SIRI-MR: assessing the options of multi-robot equipment with co-ordination to enable recurrent job refinement [ACM91a, AMM91c]; cooperation improves dexterity and efficiency.

The list is recalled to show how diversified subjects arise during actual design cycles. As before, short comments on the typical situations faced by a designer are given.

- *Operation details analysis*—The visual restitution by means of the SIRI-AT animation is a powerful aid for robot path planning including obstacles avoidance checks. At a higher level of complexity, the assessment of the collision effects is provided by the SIRI-UM package [ACC94a]. The technical literature on the subject leaves more questions open than solved. The impacts of robot members (against rigid or compliant bodies) are usually non-central with reflected effects on both normal and tangential rebounding speed components. Fully consistent analyses do not exist (unless for the single-degree-of-freedom case) and they could lead to unnecessary complexity models since practical situations do not require control of the collision behavior but rather only to avoid unacceptable fall-offs. Then, with SIRI-UM, a simplified analysis is provided based on the introduction of suitable bouncing coefficients which supply a consistent restitution of the impact, correctly complying with energy decay and momentum balance requirements.

The analysis of details is quite often neglected, since the activity modes selected for the progression of the requested tasks should avoid the pitfall of even approaching the related risks. High performance robots, in any case, must compensate discontinuities at the engagement phase, when the unconstrained navigation phase stops giving place to the constrained work phase. The region could characterize collision effects with rebounding and bouncing phenomena, [ACC94b]. The resort to simplified identification procedures to support appropriate task setting operations looks promising, by-passing the theoretical and practical difficulties of fully developed models. Duly assessed system hypotheses help the self-calibration of the codes, to fit continuity link-up.

• *Task extension analysis*—Functional redundancy is a basic option in advanced robotics. The related technical literature is large with several suggestions [AsA88], [AsH89], [AsY89], [Ben97], [ACH86b], [AaH77], [ESG90], [KiT97], [LiY97], [Sim75]. The command redundancy is a simple option with different implementations [AMM91a], [AsI89], [CPP96], [Des96], [FFM97], [MHS97], [Pel96], [RaC85], [Whi87], [WLY96a], [YLI96]. It is studied by the SIRI-HD package by combining force and position feedbacks. Mobility redundancy is a more complex option, with several fall-offs [HuJ86], [KAG96a], [LII96], [LiA92], [MiI96], [NoH89], [UIH97], [YoZ93], [ZLY87]. It is tackled with the SIRI-MR package introducing the combined cooperation of multiple-robot fixtures. Example developments are discussed in the following paragraphs and hints are given as introductory remarks and for the re-design operations based on process-matching requests.

The ability of separately closing position force controls can be used for driving the robot to follow a trajectory, transmitting a pre-set effort law. During the work phases of the robot, independent sensing devices provide useful data for closing the appropriate feedback loops. When state expansion makes it possible to model the interfacing context, the dependence of force data and position data requires fading away of the redundancy (suppressing over-specification). Processing of the extra information is for calibration purposes.

The attribution of operation redundancy is a design trick to comply within robotics when the requested functions are not easily faced by the usual six mobilities. Addition of freedom to a single arm has anthropomorphic justifications and is a good contrivance when the end-effector, for instance, operates within a bounded work space previously reached through a narrow entry. Trajectory and control plannings usually split into sub-tasks: the approaching (or latching) and the operation (or tracking) tasks. More general setups based on multi-robot equipment with cooperation deserve special attention (Fig. 7.15). Situations leading to such directions are quite different, thus, a general purpose simulation environment is a very important design aid usefully explained with case applications.

7.3 The Design of Process-Attuned Robots

The study of specialized options, generally, profits the systematic approach built by using the general purposes-modules of SIRIxx and recording the results into the data-base of the SIRI-XE package. Still, detailed developments sometimes, require further studies when the individual applications concern, for example, subtle details or broad duty areas. In the former instances, accurate modeling of tasks progression

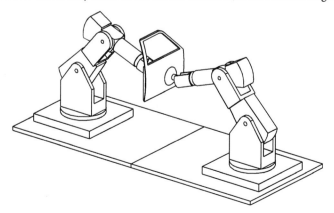

FIGURE 7.15 Robots with cooperation setup.

might result in difficulty, critically effecting the recognition of negligible effects on task planning. In the latter instances, consistent identification of unique robot setups could fail unless relevant refixturing or retro-fitting are performed to accomplish each sub-set of tasks. Operation peculiarities and job capabilities are both easily assessed when the design of the "new" fixtures follows a modularly-constrained development track.

Emphasis on modularity to expand robot's permitted activity modes has to be faced at both the CFC and the MDM frame levels. The philosophy that lies behind "earlier" robotics was that hardware fixtures should have been capable of being adapted by merely reprogramming. Selection of equipment that will best suit the needs of flexible automation is, however, a more difficult and costly exercise. The goal is simplified by joining the use of proper CAD supports (e.g., SIRIxx) with the standardization of the reference units and functions as well, so that the appropriate set-ups and fit-outs would effect outcome by combining a series of modules. The specification of the suited solution is alleviated by the previously performed analysis procedure on the standard units and functions and the synthesis procedure follows, (Fig. 7.16), according to the rule: "to determine the appropriate CFC frame, by joining the set of modules (functions and units) with the proper MDM frame, enabling the set of useful tasks with performance explicitly weighed by process specialisation." On these grounds, modularity is exploited as conditional aid of the ideation stages, (Fig. 7.2), indeed, the architecture setting has task-driven global bounds which specialize any given acceptable topology—the governing fitting has performance-driven approval tests to verify innovation appropriateness.

This "global" design cycle exploits modularity for a two-fold choice orientation:

- General purposes kinematics modeling techniques, control strategies assessments, etc., provide the classification rules for acknowledging the functional units consistency in relation to the process and for comparing their appropriateness to perform the tasks.

- Technical data, specially attributed to each unit, immediately provides the mechanical design parameters of the proposed solution (geometry, center of mass, weight, mass quadratic moments, joint stiffness, velocity/acceleration/torque limits, etc.) to help identify the expected performance figures.

Three of the four steps of the design-cycle procedure, (Fig. 7.1), are accomplished transferring the properties of the units with "global" consistency assessments. The fourth step leads to reconsidering the duty figures as actual achievements, readily starting the re-design activity with focus on the actual process where robots are introduced. The results of each new design-cycle procedure are background knowledge to start the process back-poised design of specialized equipment; in particular, by means of the already mentioned SIRI-XE package. They are stored, (Fig. 7.13), into a properly ordered database. Innovation is built step after step, acknowledging the "task requirements—functional blocks" pairs in terms of weighed performance indices (technical figures 'accuracy, dexterity, efficiency, and versatility' and economical return on investment). The designer makes inquiries from background knowledge that he can

Hardware modules adaptation (CFC frame)	
Component, Facility-configuration, and Command frame	The manipulation setup is specified as feasible topology granting the given tasks, by means of a 'global' design cycle defined by a series of transform matrices, through whom joint axes architecture, links length, mobilities spans, actuation ranges, etc., are provided.
Software modules adaptation (MDM frame)	
Monitoring, Decision-manifold, and Management frame	The functional fit-out is specified as an activity model which accomplishes the tasks with the current topology; the activities are described by the effector position and attitude in the workspace, the tip velocity/acceleration, the allowable arm deflection, the transmitted force/torque, etc.

FIGURE 7.16   Synthesis of process-attuned modular equipment.

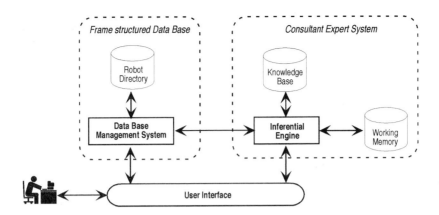

FIGURE 7.17 Decision cycle supported by the SIRI-XE package.

extract from overall formalized solutions when the process, to which the robot has to be linked, has already been studied. Most of the time, he will receive suggestions or only preliminary-by-default solutions. The logic of the SIRI-XE package, (Fig. 7.17), is built on sets of rule-based procedures that can be expanded or modified by the user to personalize, not only the reference knowledge, but the decisional patterns.

It should be noted that knowledge updating and maintenance of the expert system will become difficult to keep the reference background at satisfactory levels. Indeed, "industrial robotics" has reached the range of technological appropriateness and the number of equipment grants return-on-investment to manufacturers. In other fields, large opportunities like to be established as soon as the critical threshold of economic efficiency is achieved (this is expected to be, e.g., the case of micro-robotics).

Simultaneous Design of Robot-and-Process

The design of instrumental robots is strongly affected by the on-duty requirements concerning the facility and the related functional blocks. Actually, design does not end with the development (from conception, to construction) of the fixture. It should cover its life-long management including every action for programming and re-fixturing or for fitting and calibration. Indeed, the "quality" of the accomplished tasks and, indirectly, of the processed artifacts should be granted by restoring actions (pro-active planning), aimed at avoiding, extra costs of products approval tests and risk of regeneration jobs for the delivered services. Thereafter, return-on-investment in flexible automation is assessed by showing the advantages obtained by the setting the appropriate layout and fitting of proper governing logic. Among the advantages, process-granted "quality" is an inherent fall-out to be taken into account.

This discourse has been dealing with the design of instrumental robots, emphasizing the computer aids, prepared basically as CAD instruments for the development of the equipment for intelligent manufacturing. Industrial robots, in this context, are function-oriented equipment with duty adaptivity and 'intelligence' to that goal. For factory automation, robots have in-charge jobs (handling, inspection, assembly, etc.) required to enable unmanned running operations, in time-varying production plans and product mixes. Robotics is, thus, the reference technology for intelligent manufacturing, as opposed to fixed automation; the functional orientation is the basic design reference to comply with versatility (while productivity lays behind). Due to the variability of the production process, the action modes are large in number and tightly cross-coupled; then, computer aids, and ideation phases are relevant for on-duty iterated use, to manage hardware and software resources, and to provide helpful assistance to fit up and refit, to trim and adapt, etc. the available facilities so that the actual situations are faced with the best pre-set capabilities.

The evaluation of the return on investment appears to be a complex business since effectiveness has to be enabled along the operation life of the robot. Best performance, in fact, is achieved step by step on the condition that transparent knowledge processing is done during the robot life-cycle, from the conception along the running phases to the final breaking up, in order to make it possible

- to simulate the actual robot dynamical behavior aiming at virtual reality checks;
- to perform data fusion, with measurement restitution steered by system hypotheses;
- to promote command options, depending on control/mobility redundancies;
- to explore operation alternatives, suggested and assessed by expert modules.

Robotics is a multi-disciplinary area, obtained by merging many technologies to work out "mechatronics" issues through the cooperative effort of experts in several fields. Cooperative knowledge processing means that scientifically consistent bounds are established between engineers which concurrently work toward the solution with participated responsiveness. Experts may operate independently (at their own concern), putting in common requests and results and sharing the over-all reference knowledge. Final robots possess properties selected through the collaboration (with contribution, from designers and users), on condition that a unifying CAD environment is accessed *simultaneously* by all involved people.

This is the main idea of simultaneous engineering, with side off-springs such as

- integrating application studies with research seductive option with fruitful issues; and
- economic considerations cannot be separated from technical specifications, but need to be assessed with care, to help rank design alternatives in terms of process requests.

The criteria can be used to build an expert system, which helps to recognize good from bad fixtures. The job cannot be accomplished unless manufacturing engineers are asked to cooperate, so that no single issue dominates. The cooperation presumes structured knowledge environments and, possibly, modularly-arranged processing aids. From the collaborative effort, robots might be modified by removing or simplifying functions to agree with process fundamentals. With re-engineering, moreover, product-and-process may be changed as well, as technology-driven solutions are revised.

The Robot Setting: Equipment Modularity

In flexible automation, the main emphasis is reserved to "programmability." Robots are developed to be able to adapt themselves to any new product or process merely by reprogramming. The issue depends on (Fig. 7.18) the combined accuracy, dexterity, flexibility, and versatility figures needed by the manufacturing strategies: "versatility" means how far the task domain extends; "flexibility" specifies the on-process adaptability range; "dexterity" measures the duty level of complexity; and "accuracy" specifies the metrologic criticality of each operation. These are technical requests to be weighed against "productivity" for manufacturing applications. Robots, of course, never reach the productivity of special purpose devices used in fixed automation. Yet, single-purpose resources become useless when artifacts to be processed are modified. To preserve high productivity, while making it possible to recover the resources for different productions, 'modular robots' could be the right solution. The implementations are a typical issue of simultaneous engineering as standardization presumes the combined knowledge of robotics and the process-and-product design; the technical literature [BZL89], [CLC92], [GKY84], [HWM86], [Kan83], [KeK88], [LeR87], [MuM84], [ShS84], [SmC82], [TeB89], [Wur86] provides several useful indications.

A modular robot consists of standard units (links, joints, auxiliary rigs, etc.) which may be configured into suited arms as soon as new tasks are defined. This authorizes the exploitation of oriented devices (e.g., arms with only one or two mobilities) each time this is consistent with a given manufacturing process. Versatility is fully fixed off-process while flexibility is compressed as low as possible. The productivity can rise considerably on the current tactical horizons, due to the final fixture specialization. The stops, to implement the "new" robots of each new production program, are an unavoidable drawback to be removed

accuracy	quality obtained by combining repeatability (range of random spreads) and unbiaseness (range of systematic off-sets)
dexterity	specified in terms of tasks complexity with the related specifications on path planning and on control modulation requirements
efficiency	established by the swiftness of successfully accomplished work-cycles, with inclusion of set-up time and temptative trials
flexibility	range of the functional capabilities which assure the on-process adaptivity of the equipment for performing all the scheduled tasks
productivity	conventional measurement of the amount of items, produced according to standard schedules, over a given time span
versatility	domain of the reachable tasks, possibly, obtained, after re-fixturing and re-programming, to re-focus the robot functional orientation

FIGURE 7.18 Performance figures of instrumental robots.

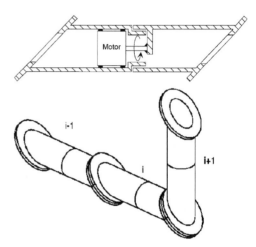

FIGURE 7.19 Example robot built with modular units.

by careful, simultaneous engineering (putting in parallel artefacts and production resources development) and standardizing previously arranged hardware and software units.

On the marketplace, (Fig. 7.19), the modular robots are offered with different levels of complexity. The assembly of small plastic artifacts, to be delivered by extended batches, is a sample case where the rig modularity has been explored to reach return on investment [ACM96e]. The development of the fixtures is, basically, concerned with an inventory of properly assorted units and by composition procedures based on pre-established patterns, such as

- Units are self-standing structures (links, joints, sensors, grippers, etc.).
- Each unit has self-contained function (e.g.: link with included actuator).
- Congruent units connect each other, through standard coupling.
- Auxiliary units (sensors, etc.) can be superposed, regardless of size, type, etc.
- Individual unit has optimized design, in terms of a selected set of charges.
- Reference setups are available, with properly tested characteristics.
- Software modularity provides fit-out schemes, with known performance.
- Similar other prescriptions automatize the access to the inventory.

The conceptual design, then, is very effectively done with help of the basic packages of the SIRIxx environment.

- SIRI-CA supports path planning and architectural analysis to minimize the number of mobilities, frozen-joints models that are used to specify actual links and the architectures are analyzed with respect to only the task-driven paths.
- SIRI-SC assures control planning and performance analysis. The structural parameters are directly transferred from the inventory and the control strategies are programed for use on related processing units.

The modularization considered in this paragraph establishes constraints mainly on the mechanical side of robotics. Sensing and computer units are already available as standard units; thus, no relevant limitation is expected. Now, robots are dextrous devices charged to accomplish, with autonomy, given tasks; being allowed to use their *intelligence* of the world they are interacting with according to this definition, the modular fixtures happen to be classified as 'robots,' depending on the amount of 'intelligence' they are using, properly equipped for on-process duties. Overall flexibility is, in any case, the winning option to working out the appropriate setup by the off-line investigation and to transfer the efficient supervision at the level of on-line operativity. In summary, the design of process-attuned robots, by integrating series of properly standardized units, is aimed at timely equipment re-setting (CFC frame) while performing the overall re-engineering of product-and-process.

The Robot Fitting: Versatility by Process Back-Poising

Reference to standard units only is often too restrictive resulting in final configurations too complex or not properly balanced and, in general, poorly optimized. A different kind of modularity has been discussed in the chapter based on the idea of process back-poised standardization. At this stage, the innovation could bring unsuspected options and few hints are given. Instrumental robots are operation-oriented devices with programmable functions related to allocated tasks and adaptivity depending on the autonomy latitude and performance ranges. Their development requires the previous acknowledgement of the on-duty behavior to be established within actual operation conditions with model-based computer simulation (starting at earlier conceptualization, during the design steps, and covering the life-long task programming to manage the on-duty fitting). The process-attuned standardization simply means to fix a set of procedures (rules or algorithms) in order to systematically combine the "activity modes" (of production agendas) and the "equipment set-ups" (of processing resources).

Then, upgrading in instrumental robotics is explored by mapping functions (MDM frame) into equipments (CFC frame) which share, as a standard feature, the knowledge of the tasks to be performed. Once the resources are detailed, the planning is the critical request and the effectiveness endowment is dramatically dependent on the capability (based on system hypotheses) of continuously assessing the task progression, aiming at adapting the operation sequences in such a way that disturbances are filtered and offsetting influences are avoided. The issue is reached (Fig. 7.20) by exploiting dated process information and the conditioning relational contexts and is based on

- The availability of models with manipulation dynamics so that robot behavior is predicted with the requested approximation;
- The inclusion of standard sensing devices, to provide directly or indirectly, visibility on every quantity which may affect robot performance;
- The ranking of the feasible redundant setups in order to supply control fits in and/or mobility options for higher robot effectiveness;
- The access to common decision aids with incorporated 'expertise' to simplify the robot setup and to improve its operation efficiency.

With process-attuned standards, an instrumental equipment is directly related to the tasks and, in a moderate manner, affected by methods or procedures that man has discovered to obtain, results. This becomes the starting point to innovation. Actually, robot potentialities are considered more and more to supply effective solutions to the many manipulation tasks that actually are out of man's possibilities.

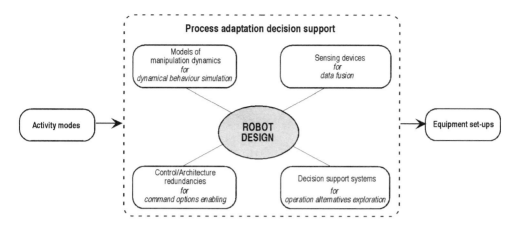

FIGURE 7.20 Basic requirements for robot's design effectiveness.

medical apparatus and services	prosthetic devices with constraints on weight, on articulation and function range, on mutual stability of operative sequences, etc.
surgical equipment	catheter-based micro-manipulators for repairing blood vessels, automatic micro-lancets for remote operations, etc.
instruments for scientific research	micro-actuators for force-balance sensors, micro-optical benches, micro-handling of biological materials, etc.
manufacturing equipment	components for aligning and assembly of optical fibers, active probes for micro-electronic applications, etc.
clean room instrumentation	the down-sizing reduces costs of keeping the production facilities of drug and biochemical industries, etc. free of particulate.
advanced consumer products	printer mechanisms made smaller, micro-sized servo actuators with multiple force-path constraints, etc.

FIGURE 7.21 Example applications in micro-robotics.

Between the new fields of instrumental robotics, the area of *micro-dynamical systems*, for instance, deserves attention and several example applications, (Fig. 7.21), already cover specialized market areas. These new applications of instrumental robotics are characterized by the need of exploiting to the highest degree the amount of information available, since the success critically depends on the ability of acknowledging the task progression with continuity. This acknowledgement is based on the availability of appropriate models of the controlled manipulation dynamics; extended (direct or indirect) visibility on the process variables; and convenient redundancy of the controlling capabilities. In other words, we seek a robot with versatility capable of being properly balanced to the process requirements. As a particular instance, micro-robotics is recognized as requiring sophisticated model-based computer-simulators, observers-driven sensors-data fusion, and dynamics-shaping control capabilities. These research efforts are, however, common trends for instrumental robotics, for developing specialized and effective devices with the low-cost desirability which assures market rentability.

The equipment resetting and retro-fitting are accomplishments that introduce the shift from a "special" technology, such as *robotics*, to a broader engineering context, namely, the *design activity*. At the end of the evolution, robots might not preserve a technological independence any more (adding multi-disciplinary knowledge); by now, innovation is issued with aids which help solve an engineering design problem. Micro-robotics has been used to exemplify the trend: designers look at "on-duty" (process-attuned) prerequisites as a principal concern of a re-engineering business; not as advanced achievement of robot technology.

Robot Dynamics with Constrained Motion Duties

Aimed at process tuning, the robot's manueuvers will principally be concerned with constrained motion models, (Fig. 7.8). In Section 7·2, the connection between a robot and the environment has already been given by means of the "impedance" G(s); binding the incremental deviation of the coupling force (and torque) δF_E with the position (and attitude) infinitesimal off-sets of the end-effector from the equilibrium location, the linearized model follows:

$$\delta \tilde{F}_E = G(s)\delta \tilde{x} \quad \text{with: } G(s) = Ms^2 + Hs + K \qquad \text{(repeated)} \quad (7.7)$$

When the coupled work path starts, the model (7.4) describing the dynamical behaviour of the robot can be linearized to become:

$$\delta Q_A = A_q^* \delta \ddot{q} + B_q^* \delta q + D_q^* \delta \dot{q} + J^T \delta F_E \qquad (7.8)$$

where δQ_A is the incremental actuation torque; the A_q^* (reduced inertial terms); the B_q^* (reduced stiffness) heavily depend on the configuration, the reduced damping D_q^* is less sensitive, and the gyroscopic term, $D(q, \dot{q})$, since the second order in $\dot{q}$ becomes negligible after linearization.

To reduce off-set errors, the industrial robot controllers resort to joint commands (7.5) with, possibly, an integrative term,

$$\delta Q_m = E_1 \int (q_d - q)\mathrm{dt} + E_p(q_d - q) - E_v(q_d - q) \qquad (7.9)$$

δQ_m computed at the joint axes (the desired acceleration may not appear, being seldom available). Therefore, if the behavior of each actuator is approximated inside the motors linear range (before saturation), by the "equivalent" inertia and damping terms:

$$\delta Q_m - (\eta^2 J_m \ddot{q} + \eta^2 h_m \dot{q}) = \delta Q_A \qquad (7.10)$$

(where η is the matrix of gear ratios from motors to joint axes), then the closed loop dynamics is stated, with respect to the linear approximation (7.8), provided that the external inputs δx or δF_E and the actuation commands δQ_A are bounded both in magnitude and in frequency.

The effectiveness of instrumental robots by process-data acknowledgement shall not divert attention from their inborn faculties, namely, the behavioral adaptivity. This, rather than abstract qualification, corresponds to the ability of figuring out the dynamics by means of models by granting quantitative prediction of the motion. The aspects are considered to suggest a patterned approach for feeding in the process "knowledge" (possibly, by modularly arranged CAD supports) and to introduce concepts quite general in respect to the example applications of the following sections. In fact, a model remains a representation of the actual behavior for a certain duty range. The deviation is called the model (structural) uncertainty and the interlacing between deep knowledge frames (e.g., the above stated equations (7.7)–(7.10)) and shallow knowledge links (e.g., the rules of qualitative reasoning) should be explored to reduce this uncertainty [MCR96].

According to the said goal, the show of useful process data is enabled, recognized, and exploited, during the duty progression, mainly, for two reasons:

- For instrumental robots, as outcome of task departure monitoring; and
- For autonomous robots, as instance of task-planning success/failure.

In the first case, robot effectiveness is improved by means of the recalled options, (e.g., dynamics shaping to compensate the inertial coupling between joints) and redundant control (to have independent modulation of the effector position and force). In the second case, the availability of structured knowledge frames of the arm behavior can help modify task-planning, with relevant advantages for highly sensorized devices.

The concept of dynamics shaping has been defined as the capability of modifying the steering planning by combined adaptive feedbacks [Yos93]. The compensation of the reflected inertia dynamical coupling is quite an obvious modulation to counteract the internally generated inertia load. Up-grading is apparent for rigid arms required to track high speed navigation paths. Moving to the robot back-loads generated by external couplings, earlier studies, before any snags due to limited stiffness, considered strategies based on separate loops for position or force commands. The extension is justified by considering how redundancy is used by people to preserve the stability of motion, e.g., during everyday walking, or for expanding dexterity and versatility, for complex exercises. The model-based connection of force and position feedbacks is a primary goal for development in robotics. Most of the recalled micro-manipulation tasks, (Fig. 7.21), needing the graduation of the impressed force jointly with the effector displacement steering, require control redundancy. In fact, due to dynamics nonlinearity, any changes of accuracy or efficiency figures will also require task-driven adjustments, making control planning necessary in addition to path planning. Simple example developments are discussed in Section 7.4, for explanatory purposes.

One question is, moreover, the fact that robots could be required to operate with totally or partially unstructured constraints. As a general rule, the request is faced by "intelligence" of the outside world mastered by learning schemes to expand the range of successful duties. The practical evaluation of the effectiveness figures of abstractly defined duty ranges, however, worsen due to the arbitrariness of the reference standards and to difficulties of establishing consistent gauges. Instrumental robotics of recent years has preferred the gathering of "intelligence" of the outside world expanding experimental information and profitably doing data-fusion by combining sensor measurements and system hypotheses; whereas autonomous robots may struggle against vagueness another way, by learning cycles built, e.g., on qualitative reasoning. Computer aids, with knowledge frames similarly based on the said four logic steps, (Fig. 7.1), can be used for exploring the feasibility of prospective tasks when open-duty activity modes are addressed for goal-oriented planning performed into unknown surroundings and unpredictable disturbances. Steering self-adaptability, learning ability, and recovery options are evaluated by fully autonomous agents with virtual reality experiments, supplying a rival show of closed-duty applications of the fixed automation. Iteration of functional design cycles, then, operatively provides a decision pattern for training procedures, granting self-learning abilities.

The integration of on-line measurements is the last, but not the least, of the problems related to process-tuning. Sensing devices are extensively used to measure internal coordinates e.g., encoders for joint angles. The addition of angular accelerometers makes it possible to obtain signals for the compensation, as shown in Fig. 7.7, of the nonlinear inertial couplings during the unconstrained navigation phases. Image analysis and optical scanning are useful means to derive surveillance functions for presetting the engagement phases; sensing devices at end-effector or wrist provide data for tactile recognition and hectic reactions and the incorporation into system knowledge profits by identifying a structured relationship of interactions at the robot/surroundings interface. Availability of reliable, low-cost devices supports the trend [ArM83], [FLG97], [DeS98], [HiH83], [Hil85], [LiG93], [RoM66], toward inclusion of new measurements. Quality and effectiveness require accurate calibration procedures and the related cost and time should be compressed and accounted for. The answer is "intelligent instrumentation," having standard self-calibration capabilities. The robot is specified through model-based control accounting for actual nonlinear dynamics; and, to conclude calibration, is endowed by a duly modeled interface. The data, collected by specified duty sequences, are processed and compared with "virtual" measurements for automatic calibration of the sensing and restitution devices.

The discussion on the on-process opportunities for expanding current information by means of learning capabilities and artificial reasoning or of measurement devices and data fusion, however, should not divert attention from pertinent models, such as the one represented by the Eqs. (7.7) through (7.10), which help describe the manipulator dynamics within given application ranges. This assumes a "nearly" valid prediction of the process evolution to be compared with measurements to have an insight on whether external disturbances would superimpose.

A Challenging Option: Robots with Cooperation

Process conditioning pops up as brain wave and multi-robot systems for task parallelism appear as innovation aiming at better productivity and/or effectiveness. The subject has already been concerned in several studies and sample applications exist with loose cooperation figures between units. To assess these figures, the specification of the multi-functional framework is requested, explicitly defining the relational structures of the task/performance cross-dependence and for the job-flow/resources concurrence. The co-operation problem is stated, at this stage, as distinguishing control loops (closure of physical feedback) and decision schemes (closure of logic nets), and join the efficiency of the in-line command operation with the flexibility of the in-process adaptivity whenever requested by the application. For duty-specification, the aspects to investigate include:

- Functional description of the job to assess the advantages of robot cooperation.
- Executional constraints with specification of task programming requirements.
- Govern and information fit-out to select control and communication setup.

Schedule meshing analysis, (Fig. 7.22), is done at first, to recognise if duties are closely bounded, sequentially related, or mainly self-sufficient. Cooperation, in fact, increases plant productivity as robots share portions of the job and are able to perform a large variety of actions. Therefore, task complexity, (Fig. 7.23), is analyzed to set the handling architecture and to fix the govern level hierarchy, namely:

- Logic sequencing, at lower scheduler level, to comply with the nesting of (off-process specified) tasks "closed-duty" agendas, accomplishing in parallel independent actions, to improve productivity.
- Communicate synchronized coordination, at intermediate planner level, to obtain the task-coordination by means of "sync-duty" operations, respecting the sequentiality of actions with priority constraints.
- Decisional mechanisms activation, at upper controller level, for matching tasks and "open-duty" environment in order to fulfilled jobs actually requiring collaborative effort to grant reliable results.

A third issue, data sharing requests are considered, to be satisfied at

- Operation range: scheduling/sequencing; devising/planning; observing/controlling;
- Govern range: centralized (controller level) or decentralized (scheduler level) policy.

The design of efficient multi-robot equipment depends on the application. It can be viewed as the most satisfying setup between conflicting goals such as: duty flexibility vs. setting quickness; task versatility vs. plant productivity; and job autonomy vs. quality assurance. Due to the complexity of the contrasts, choice of solutions needs to be explored since conceptualization, (referring to actual running conditions to check functional and decisional options exactly in the duty specification frame of the particular case). To limit development costs, general purpose CAD packages are a convenient means. A number of alternatives can be explored by comparing charges and benefits and contrasting functional and decisional options with quantitative figures of the robots' performance.

Handling and governing structures are the central issue in developing the overall fixture. The first characterizes defining the functional components: end-effectors, joints, kinematic chains, actuators, sensors, etc., and needs to be adapted to the manipulation surroundings (i.e. workspace, job requests, tasks agenda, information interfaces, control operations, etc.). Dexterity and accuracy push toward integrated sensing/command blocks and hybrid position/force control loops for the arm-coordination. The study is, accordingly, carried on by appropriate functional models: first, for the preliminary command-setting and path-planning with simple kinematic models; then, with full dynamic models that should include a library of control strategies, for operation checking and performances evaluation. The governing structure, (Fig. 7.24), has to continuously adapt actions to current situations related to on-going

CLOSED-DUTY	
the agendas are carried out simply managing the job parallelism	
Several robots may operate in a given workspace, supervised by schedulers, with 'passive' constraints (for collision avoidance, etc.).	• the decisional schemes are moved off-process • the command logic is pre-set, with reference to the executional stages (ruled by a scheduler)
SYNC-DUTY	
the agendas are implemented exploiting appropriate functional sequencing	
The planners govern the robots, with 'active' constraints on the job to be performed in parallel and/or in sequence	• the operation characterization is detailed within the set of '*a priori*' system-hypotheses • the task co-ordination follows a logic, previously assigned by (fixed) procedural knowledge
OPEN-DUTY	
the agendas are built with procedural knowledge, shared by the decentralized control units	
The job progression is ruled by controllers that schedule, with embedded decisional aids, the operations concurrence, based on updated information on the actual state	• the functional characterization of robots is given, with the class of authorized tasks • the coordination is adapted with the on-process knowledge

FIGURE 7.22　Specification of the schedules by "duty modes."

MANDATORY TASK COOPERATION	
two or more robots, simultaneously or jointly, perform the job, with links on individual tasks such as	
joint operated tasks	- the robots are doing a part of or the total job, for the fulfillment of which coupled cooperation is required
simultaneous tasks	- the operations require more than one robot, e.g., one robot serves as a programmable fixture for other robots
CONCURRENT TASK COOPERATION	
two or more robots carry out, in parallel, portions of the same job, having independent charges such as	
joint parallel tasks	- the robots work together on different facets of the same job, decreasing the total cycle time
split parallel tasks	- the robot diverse capabilities are exploited for specialized operations, e.g.: positioning, precision assembly, etc.
OPTIONAL TASK COOPERATION	
any one of several robots can fulfill the job and only one is required since the cooperation is based on	
interchangeable tasks	- the responsibilities can dynamically be assigned among the robots and the job accomplishment is covered with failure backup

FIGURE 7.23　Cooperation classing by "task modes."

Structures of the decision logic	• *hierarchic information tree-structure*: the cooperation among the robots is assured by a centralized control, under an explicitly established supervisor • *parallel-distributed information network*: cooperation exists in a multi-agent cluster of units (sharing common interest data) interfaced through an intelligent layer
Modes of the decision support	• to fulfill pre-scheduled steady operations, after command decentralization • to perform job planning, resetting the programmed-mode conditions • to recover the on-line control of the multi-robot facility, at emergencies
Outputs of the govern module	• at the executional level, for enabling the operations of each individual robot • at the coordination level, for controlling the cooperation between robots • at the organizational level, for acknowledging the programmed tasks

FIGURE 7.24　Decision-and-govern modes of multi-robots.

job progression. The scheduler activates the tasks parallelism and, once verified, the planned job sequencing. The controller, congruent with flexible surroundings, requires full visibility of tasks progression to exploit the updated knowledge on current situations; to modify the state depending on the scheduled duties; and eventually, to adapt the robot behavior in relation to the situational changes. The context brings a hierarchic knowledge reference framework to distinguish the "external" from the "internal" structural conditions and to prepare solving procedures consistent with the acknowledged relational schemes.

The preparation of the activity modes can be separated from their execution by the tasks given to the individual arm and the trajectories (with the related motion-wait conditions). Job fulfillment is programmed (planner level) off-line and synchronisation only is enabled during implementation. To govern cooperating robots, thus, requires a communication structure between units assuring

- At the scheduler level, the monitoring of closed-duty agendas;
- At the planner level, the sequencing of sync-duty agendas;
- At the controller level, the coordination of open-duty agendas.

7.4   Modulated-Control Example Developments

Improvement of robot potential has been related to the ability of modulating the behavior to reach accuracy, dexterity, flexibility, and versatility so that the specified handling tasks (even out of man's capability) are performed. The challenge, characterized by the use of dynamics shaping and/or force modulation, to subdue unwanted effects on the manipulator behavior, is by adapting the control to the on-going duties. Dynamics shaping corresponds, in fact [Yos93], to compensate systematic offsets or drifts which may arise due to: actuation nonlinearities, mobilities inertial couplings, transmission compliance, actuation backlash, sensors' bias, or the like, using error signals, measured or computed with respect to pre-set dynamics, in the joint-frame. Conversely, aiming at the force modulation, typical studies [AnH89] have considered strategies based on two commands (position or force), conveniently switched to drive the arm as the duty is modified to constrained motion manueuvers. A simpler setting lies in the impedance control ([Dra77], [Sal80], [Hog80], [Hog81], [Hog85], [KKN95], [Mil96], [Pel96], [CaB97]) which enables a force feedback mapped from position data on the condition that the coupling stiffness matrices are known. Both approaches could be explored to supply task-orientation aimed at position and/or force combined control. This strategy is suggested by the observation of how the redundancy is exploited by (trained) living subjects to preserve stability of motion and to improve dexterity and versatility (even the running of a man on discontinuous ground requires multifarious adaptivity, to select a complex balance of position and force reactions and to carry on with a stable gait). The availability of redundant information enables sets of (hybrid) options for governing the robots to stick to the planned task, even when biasing effects arise.

The above considerations show that to enable control adaptation means, at least: redundancy, as far as actuation actions, and transparency, as far as dynamics effects. The redundancy is investigated by models properly extended to cover (with the robot dynamics) the effects of the surroundings. The visibility is assured either

1. By a state observer, generated according to the *a priori* knowledge of the process. This will generate the robot behavior by means of the model. The basic control strategies to uncouple the workspace errors for the actuation feedback requires explicit shaping of the dynamics as internal reference knowledge; or
2. By a self-sufficient observation scheme based on sensors and processing units. State coordinates (in the workspace and in the jointspace) are measured to assure tasks execution (in workspace) with up-dated commands (in jointspace) and proper control of both, tip position and transmitted force.

The two options can be exploited simultaneously, building combined solutions:

- The internal model is important for the on-line processing of the uncoupling feedback based, for instance, on the "computed" torque method; and
- The sensors provide further data, useful to adapt, via supervisory mode, in-progress schedules and co-ordination requirements.

To improve robot performance, the reflected loads shall be "accommodated" rather than resisted. Earlier studies aimed at compensating the effects in a non-conflicting way. The manueuvers are recognized by the way the position controls unconstrained motion degrees-of-freedom (force control for constrained motion). Example cases are further discussed to show the relevance of the control planning, with respect to more traditional approaches (confined at the stage-of-path planning, only); in particular, for situations that might lead to an erratic robot behaviour, suppressing wavering through errors compensation, and improving accuracy and dexterity without penalizing versatility and efficiency; namely,

- An example deals with inconsistencies of the control capabilities that could appear depending on the reflected-inertial dynamic coupling.
- The second, aiming at measurement robots gives hints to design a high performance device when handling effectiveness ought be joined with steady accuracy.
- The third introduces position/force control problems showing the coupled stiffness effects on the path planning repeatability.
- The last considers haptic manipulation using touch information to close position-and-force feedback to keep 'sufficient' stability margins to the robot motion while the tasks progress.

The Process-Adapted Control Planning Setup

Robots are non-linear mechanical systems. The handling dynamics shows undesirable behavior such as: joints cross-coupling due to the reflected inertial terms; driving misfits produced by backlash, stiction, saturation, etc.; vibrations and accuracy losses rising from lumped and distributed compliance; and so on. From the actuation stand, each motor has to drive a load which depends on varying mechanical parameters with modulating inertial, compliance, and damping couplings effects. The locally linearized model with inclusion of motor and transmission effects reduced to the robot joint axes combines Eqs. (7.8) and (7.10), to obtain:

$$\delta \tilde{Q}_m = [(\eta^2 J_m + A_q^*)s^2 + (\eta^2 h_m + D_q^*)s + (k_m^* + B_q^*)]\delta \tilde{q} + J^T \delta \tilde{F}_E \frac{1}{k_m^*} = \frac{1}{\eta^2 k_m} + \frac{1}{k_A}$$

$$(7.11)$$

where motors' effects are scaled by the gear ratio η and the reflected dynamics adds to give: $J_T^* = (\eta^2 J_m + A_q^*)$, equivalent inertia; $h_T^* = (\eta^2 h_m + D_q^*)$; equivalent damping; and $k_T^* = (k_m^* + B_q^*)$, equivalent stiffness. The terms A_q^* and B_q^*, as said, heavily depend on the configuration, but are somewhat equalized when high gear ratios are used.

The model approximation is apparent each time the architecture deviations are not negligible; the motion does not progress slowly and without impacts, and one of the following facts holds:

- The motors are not geared through a speed reducer, i.e., direct drive actuation is used;
- Quick changes are tracked in the joint space (the 'equivalent' parameters vary rapidly as no steady contribution dominates), even if the end-effector moves slowly;
- High accuracy is required for given tasks, such as at contact-transient for quick plough during assembly operations.

The back effects of the constrained motion appear in the work space and have to be transformed in the jointspace leading to:

$$\delta \tilde{Q}_m = [J_T^* s^2 + h_T^* s + k_T^*]\delta \tilde{q} + J^T G(s)J\delta \tilde{q} \quad \text{since: } \delta \tilde{F}_E = G(s)\delta \tilde{x} = G(s)J\delta \tilde{q} \tag{7.12}$$

Then the closure of feedback loops by with the usual error signals (7.5) or (7.9). The real behavior differs from the model (12) due to the set of discrepancies, to be evaluated in the practice, resorting to harmonic describing functions. These are defined for fixed classes of bounded magnitude and frequency [GrM61]; thereafter, letting $K_q^F(j\omega; \cdot ;;\ )$ represent the true manipulator dynamics at the selected configuration and for a driving harmonic wave of given amplitude, Fig. 25, the local deviation is expressed by:

$$K_q^F(j\omega; \omega_j, \Delta \omega, \Delta \delta q) - \{(h_T^* j\omega + k_T^* - J_T^* \omega^2) + J^T G(j\omega)J\} = e_q^E(j\omega; \omega_j, \Delta \omega, \Delta \delta q) \tag{7.13}$$

which confines the identified model, with respect to the true open-loop behavior, into specified bounds, separately dealing with:

- Manipulation uncertainty which includes the local linearization of the dynamics and the additional non-linearities on gearing stiction and stiffness; and
- Operating uncertainty which corresponds to insufficient knowledge on the coupled impedance $G(j\omega) = [(K - \omega^2 M) + j\omega H]$ parametrization.

The closed-loop behavior, (7.5) or (7.9), deals with further non-linearities such as motor saturation, which can be modeled as open-loop gain reduction for each given frequencies band.

For developing high-performance robots, models shall be "nearly" valid within each considered operation range. In this practice, the designer is mainly concerned with the two-duty conditions: free motion

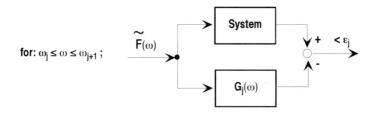

FIGURE 7.25   The identification rule for the set of describing models.

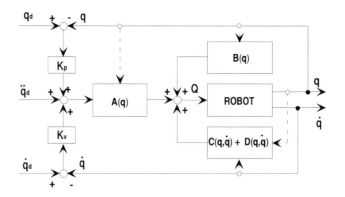

FIGURE 7.26   Computed-torque control.

and constrained motion. In any case, the relevance of the non-linear uncoupling and the perspectives (that *dynamics shaping* opens), are well explained by examples. The approaches proposed to figure out controls able to cope with the non-linear effects of the dynamics [ACM93], [AAC94] have as common an idea to change the actuation law in order to suppress the coupling so that the robot is forced to keep a linear behaviour while performing the assigned tasks. Coupled motion is, instead, controlled looking for "accommodation" attempts ruled by devising mechanical impedance figures in the frequency domain and by designing proper compensators by means of the (linearized) harmonic describing functions.

The effectiveness of the compensation strategy, with the first example, is proved looking at an existing robot: its dynamics is, first, verified by simulation with either the existing command or via the compensation, (Fig. 7.5), based on dynamics shaping aiming at fully suppressing the handling behavior inconsistencies by means of the "computed torque" method. The results show the usefulness of simulation for the transparent assessment of the task progression and for helping the control planning. The ability of expanding versatility and dexterity, moreover, is related to the possibility of including touch data on the end-effector path tracking. The exploitation of haptic artificial reality, however, cannot be effectively enabled, unless the composition laws of position and force data are related to the robot desired behavior, for performing the dynamics shaping. The basic comments are introduced by the second example.

Command Planning: Tip Wavering Under Inertial Coupling

The idea behind the "computed" torque method is that once gravity unbalances and transport terms are compensated, the forcing contribution Q^* related to the inertial coupling, should be superposed to cancel out undesirable effects and to assure that globally each robot mobility will behave as a perturbed double integrator with pre-established stability and robustness margin. Choice of the uncoupling compensation can be done by optimizing a performance index purposely defined for the set of the prescribed task paths (for improving steady efficiency), or according to general stability criteria in front of sharp disturbances (for a better transient behavior).

To consider this second option, Fig. 26, the following compensation:

$$Q_c = \Delta Q + Q^* = A(q)[E_p(q_d - q) + E_v(\dot{q}_d - \dot{q}) + \ddot{q}_d] + B(q) + C(q, \dot{q}) + D(q, \dot{q))$$

$$(7.14)$$

will force the joint actuation, after uncoupling, to behave as a linear second-order block with critical damping. Downstream of the inertial compensation module, the model includes the additional non-linearity due to motors' saturation. With higher gains, the control reacts promptly as long as joint motors remain in the proportional range (before saturation). This may reduce the loop safety margin, yielding to intolerable oscillations, unless the appropriate feed-forward compensation is computed, by modifying the command gains when the driving setting exceeds the saturation threshold. In such a case, the compensation of the reflected coupling inertial terms can only be approximated (at least for the individual feedback of the saturated joint actuator).

A quite instructive example problem was obtained during the investigations carried out with an industrial robot from a leading enterprise of the field. In order to reach a very accurate description of the robot actual dynamics, the setting of the centroids and of the mass quadratic moments of each individual member were obtained with a 3D solid modeler interfaced to the SIRIxx package. The original control modules provided by the robot manufacturer were analyzed and conveniently modeled. Then, simulation was carried out according to the testing procedures programmed with the standard operation prescriptions, in order to have the set of validation references for trimming of the SIRIxx code (see Fig. 2.27, for a typical test sheet), before carrying on comparative studies with more complex control strategies.

A particular trial is represented by the synchronous steering commands: a joint-interpolated trajectory is assigned between two points close to work-space boundaries, (Fig. 7.28), with the joint actuators that simultaneously start and stop and only exploit the convenient speed modulation for controlling the end-effector position and attitude. Even if the driving command settings do not exceed the saturation threshold of the motors, an undesirable wavering appears on the fifth joint of the robot (and of the model simulated with the original control setup). This causes current oscillations in the related motor, as can be seen in Fig. 7.29. The hindrance completely disappears when the compensation of the reflected inertial

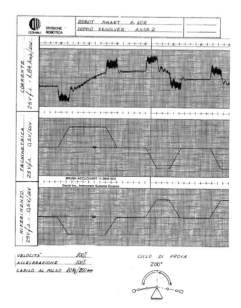

FIGURE 7.27   Test sheet for the simulation trial (courtesy COMAU).

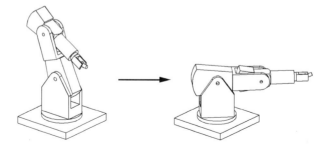

FIGURE 7.28   The synchronous steering commands test.

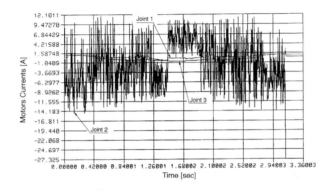

FIGURE 7.29   Disturbing effects of the non-linear reflected inertia couplings.

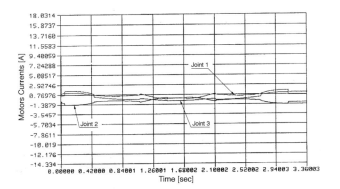

FIGURE 7.30   Manipulator's behavior with dynamics compensation control.

terms is introduced, (Fig. 7.30), in the model. The coupling dynamic modulation is very slight during usual operation tasks and is, in most of the cases, practically negligible on the end-effector when measured as out-of-all robot performance. The robot manufacturer was aware of this "anomalous" joint behaviour that could be avoided only by modifying the "admissible" task. Instead of looking at the inertial coupling effects, he was wondering about compliances in joint transmissions, aimed at increased stiffness without removing the dithering effects.

Explanatory evidence is easily reached by simulation once the robot behavior is modeled by means of a convenient code, (such as SIRIxx) making explicit reference to real architectures and contrasting the accuracy and dexterity limitations, and experimenting on facilities having a control strategy based on dynamics shaping. These kinds of results are, in the practice, quite often disregarded by robot manufacturers, as the "anomalies" appear (normally) for quite "pathological" manueuvers and are completely absent for rather extended duty ranges.

Measurement Robot Based on Controlled Laser Scanning

Measurement robots [CDM96] offer the cue for applications in control planning once the handling set-ups are properly selected. The form features restitution by contact or proximity sensors and needs free-motion high speed manueuvers and wide work-spaces. Laser scanning is a different option, assuring the detection of 3-D contours for remote shape recognition, with accuracy depending on the correct aware-ness of the sweeping path. In every situation, the balanced aims of low price and high performance require the careful design of the equipment to improve the effectiveness while keeping the mechanism to simplest dispositions. An example case is, for instance, the recognition of the cutting edge contour of tools to assess the wear-out degree and to verify the fitness-with purposes condition of on-going machin-ing tasks. High contrast is mandatory and structured light is good by projecting light stripes onto the work surface or by layered illumination with laser beams. The design of the rig, accurately performing the mechanical scanning at high speed, shall certainly have to resort to uncoupling the actuation of each arm mobility. Then the fit of the sweeping path has to face abrupt swerves with reversal motors motion. The selection of the structural elements rizes according to critical changes. Hints about appropriate solutions are given hereafter, with purposely focused concern on the conception of a wrist with high manueuvers repeatability.

The measurement setting basically comprises the carrier of the optical beam source for back-lighting of the selected tool and the CCD camera. The image processor is run on a PC (by Speroni Power Vision software) to achieve contour resolution up to the requested detail level. Spacing and alignment are trimmed at fitting out of the carrying yoke; this shall approach sequentially the tools during their idle periods, properly fixing the position and the attitude of the yoke, to accomplish the complete analysis of the 3D cutting edge with reference to the given form features. The instrumented end-effector bears inherent complexity to allow setting and fitting operations; additional requisites concern its smooth driving, without quivers and jerks.

Some monitoring tasks are better performed by front illumination so an optional episcopic vision kit, with adjustable light intensity, can be used for the set of tasks that are properly carried out this way (Fig. 31). For that purpose, two cameras with different fields of vision can be used according to the present needs.

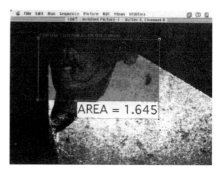

FIGURE 7.31   Shots of sound (a) and broken (b) tools illuminated by the episcopic vision kit. (Courtesy Speroni S.P.A.)

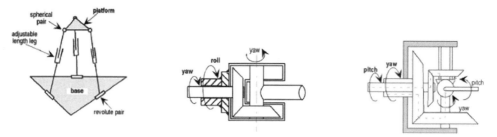

FIGURE 7.32   Concept solutions for alternative wrist settings.

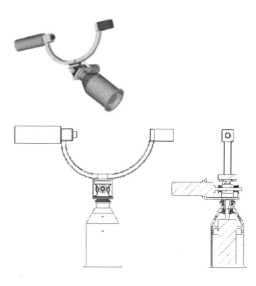

FIGURE 7.33   Assembly view (a) and design table (b) of the chosen wrist.

The appropriate equipment characterizes as a five-mobilities arm, with three links assuring the navigation path, ending with a high accuracy wrist. This needs the positioning assured in elevation of the sight line during arm-navigation. Then, a roll and yaw motion is performed to gather information on the cutting edge wear. Different concepts have been considered, (Fig. 7.32), including an in-parallel actuated arrangement [FaH97]; basic design specifications shall consider

- Backlash rejection to keep high accuracy during the laser beam scanning;
- Design compactness to increase dexterity and avoid impediments when carrying on data acquisition;
- Low weight to reduce dynamics back-effects during the work-cycle;
- Cost-effective design to increase product competitiveness.

The final solution, (Fig. 7.33), uses a direct drive of the roll and yaw axes, even if this implies higher loads to be carried by the roll motor and a bulkier structure. It seems to be the best solution, granting the requested high accuracy for camera's motion. Driving is realized by means of two AC brushless servomotors with harmonic drive reducers plus resolvers and holding brakes.

The solution, quite simple and composed of few parts, dramatically reduces backlash effects. The presence of a motor, directly mounted on the yaw axis, requires minimizing the masses to reduce the inertial coupling, while preserving high dexterity to the wrist motion. The development takes advantage from parametrical CAD tools for both the structural and dynamical analyses. The outcoming lay-out, to satisfy the functional requirements, brings to a set of technical features (such as: close together axes, rugged housing, camera nearby axes, etc.) that provide large stiffness and increase the driving efficiency, allowing high accuracy and repeatability. A simple PD controller is finally used, with the tuning of proportional and derivative gains obtained testing the robot (by extended simulation) during the execution of real operative tasks. This kinematical setting is uncoupled which makes it easier to design the control with the exception, of only a singularity at the center of the working space, when the roll axis is aligned with the yoke plate axis.

Modulated Command Options: Position/Force Feedback

Sometimes information redundancy is needed to perform complex tasks especially when the robot interacts with poorly structured surroundings or the task itself requires control of the contact forces (like: precision assembly, parts finishing, etc.). In these cases; *position/force* methods with distinct feedback loops might be used to be able to trim the value of the applied force *and* the attained position along the given axes of the (possibly) compliant engagement. The evaluation of the interaction force between robot and environment is sometimes directly measured and sometimes indirectly assessed with the *impedance control* that simply aims at monitoring the tip displacement and at making use of the contact path of the end-effector with the environment, to obtain the information.

The SIRI-HD package, [AMM91a], [ACC93], (as already pointed out in Fig. 7.9), has been based on the redundant control options, and independent feedback loops are closed for position-attitude and for force-torque errors. Joint rates are monitored while force rates can either be measured by a wrist sensor or deduced through a model of the coupling. The resulting control forcing terms F_c are finally transformed into the appropriate actuation signals as usual, by means of the transposed Jacobian matrix (7.6). In general, the control logic is switched by convenient $[S]$ and $[\bar{S}]$ matrices that select the combination of "position" or "force" commands separately for unconstrained or constrained maneuvers. The choice depends on the task progression. Compliance and dynamics of the manipulated object can be considered explicitly, during SIRI-HD programming. The overall scheme, finally, allows:

- The setting of joint-space commands for controlling the interfacing force;
- The closure of force loops by respect of sensors at the robot wrist;
- The closure of position loops by respect of sensors in the work space.

With the *impedance control*, when the tip is in contact with its environment and a new reference location is commanded, the resulting interactions are under control. The fixture accepts position-attitude set points and reflecting force-torque outputs which depend on the (assumed) interfacing impedance. The method easily applies to the usual position-controlled robots, reprogramming the feedback gains by means of the force mapped signals. On the contrary, the retrofit of position-control fixtures requires (costly) force sensors and the balancing of the two separate feedbacks on the transmitted force and on the tip location (with respect to an absolute reference frame). Many times, however, contact force control is sufficient to successfully accomplish manipulation tasks as the tip constrained motion provides path steering with "convenient" tolerances. These situations are, thus, consistent to the "computed" torque method on the condition, of course, that, throughout, checks based on the actual running requirements are performed.

An example simulational testing program, carried out to assess the actual force control behavior, has been widely done with employed industrial robots, (Fig. 7.34). The equipment was forced to follow different trajectories and its controlled behaviour did successfully perform the given tasks, always applying the prescribed normal pressures even when tracking very complex lines needing the simultaneous activation of the six axes. The results of (Fig. 7.35) are related to the tracking of a rectilinear path with time-varying reference force, exerted on a 100 kN/m stiffness environment. It is shown that the transmission of a sinusoidal forcer causes a noisy response in the position-commanded mobilities due to the engagement coupling, but the amount of the errors is quite small all along the duty range. (Besides that, after an initial transient, the force signal is tracked and preserved with nice approximation.)

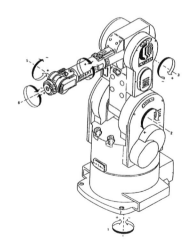

FIGURE 8.34 The Robot COMAU Smart 6.10R (courtesy COMAU).

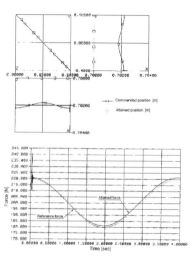

FIGURE 7.35 Reference and attained path for the simulation task (a) and for behavior of the manipulator (b).

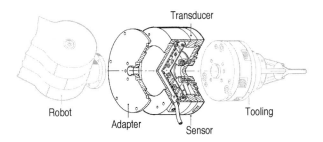

Transducer

Robot

Adapter

Sensor

Tooling

FIGURE 7.36 Schematic of a force/torque transducer.

The force feedback is, for the said reasons, employed for assembly and machining operations requiring comparative accurate skill. The simulated trials show that, for a given robot and equal applied forces, the performances slightly decrease as the stiffness function (13) of the robot-environment interaction increases. Force errors, however, stay within the range of the hundredths of the programmed set-points unless a rigid-wall engagement is approached. Beyond some given stiffness thresholds, in fact, the end-effector presents the wavering behavior with downgrading of tip attitude. As a general rule, for a stiffness ranging between 10 and 1000 kN/m and for the considered quite severe tasks, tests have confirmed the good performances that the existing robotic equipment could realize if provided with a control strategy that duly combines the position/force feedbacks during the work cycle with dynamic compensation along the navigation phase. Dynamics shaping assures high manipulation performance for very fast approaching paths. The force feedback grants adaptive end-effector operativity for a comparatively wide stiffness range. In fact, during constrained manueuvers, compliant arms operate on interfaced objects supported by compliant jigs, etc. The "computed" torque method deals with joint impedance effects requiring that the desired contact force is a function of the tip current position. When this function is poorly known, force and position are also, poorly assessed. The "measured" torque method by-passes such uncertainty, on condition, to put force sensors at the end-effectors or at the supporting jigs.

Expert Steering Commands: Compliant Assembly by Force Control

The introduction of a force/torque transducer at the connecting wrist, (Fig. 7.36), is still unusual for industrial applications due to extra costs of the setup and also in terms of the current programming software. However, the independent measurement of the force/torque components exerted by the robot tip is being viewed with increasing interest since efficient and low cost instrumented wrists started to become available. Haptic manipulating feel, indeed, is being considered for instrumental robotics as an additional opportunity to expand versatility and dexterity up to (and above) the range of human potentialities. This is a critical requirement; for instance, to technically and commercially assess the appropriateness of the devices of the emerging fields of micro-machines and of micro-dynamics. Touch information is, in fact, useful for modulating the feedback in order to keep "sufficient" stability margins to the robot motion, while the task progresses.

It has been demonstrated, [AnH89], that independent position-and-force controls may lead to instability. Unless the overall dynamics is taken into consideration, to reset the feedback gains by writing the manipulation laws directly in terms of the effector frame (which does not correspond with the work frame anymore) with joint weighing the 'robot-arm and coupled-environment' process. Due to the variety of real occurrences, however, the models for combining position and force controls should be assessed only about actual task situations. These models usually consider a global work-space frame $G\{y\}$ separate from the tip frame $P\{x\}$, which follows the strain of the interfacing medium. Unless for very high stiffness, the relations (7.7) and (7.12) have to be modified to include the transforms between these two frames:

$$y = Tx \quad F = G(s)x \quad \text{with: } G(s) = Ms^2 + Hs + k \tag{7.15}$$

$$F^* = T^{-1}F \quad F^* = G^*(s)y \quad \text{with: } G^*(s) = M^*s^2 + H^*s + K^* \tag{7.16}$$

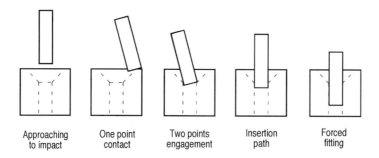

| Approaching to impact | One point contact | Two points engagement | Insertion path | Forced fitting |

FIGURE 7.37   Assembly with reactive steering.

The CAD facilities are useful to specialize the knowledge setup required in order to establish the programs for tasks scheduling; to fix the agendas for job planning; and to design the decisional schemes for operation testing. Expert supervisors follow joining continuous controllers for the work phases with the logic steering of the task depending on the acknowledgement of given thresholds. A typical study deals with plugging a part in a (prepared in advance) seat (slot, hole, etc.). The reference phases would cover: part picking up; approaching motion; and insertion with check of operation soundness. This last conveniently splits into sub-phases aimed at the simple "peg-in-hole" forced fitting, (Fig. 7.37). Several situations may arise and condition monitoring by the wrist transducer provides the visibility on the process, to grant the result.

The situational analysis is performed by the '*expert*' supervisor, recognizing:

- The peg collision feature at the end of the navigation path, with a preliminary guess about the engagement line slant;
- The peg tilting feature at single point bump, with identification of topology constraints due to the front seat shape;
- The peg engagement feature at two zones contact, with assessment of tip maneuvers to recover the proper attitude;
- The peg insertion feature at the forced operation, with path control to keep zero offset for the wrist torque;
- The peg fitting feature, at the plugging end, with stroke settling as for specifications and actual part mating tolerances.

Monitoring provides information about incorrect settings, such as jammed or loose fit; the force-torque sensor, (in addition to the steering commands), makes it possible to acquire the data for on-process quality checks.

The assembly task, according to the said description, is an occurrence-driven process with relevant advantages supplied by an "expert" supervisor typically incorporating a heuristic decision logic process. Instrumental robotics will possibly consider special applications such as those of the micro-dynamical systems where the option would bring noteworthy advantages, thus, leading to fixtures with return-on-investment. The domain of sophisticated fixturing is further discussed in the next section with focus on the possibility of functional redundancy in addition to the command redundancy.

7.5   Redundant Mobility Robots with Cooperation

Operation-driven design is powerful help for setting robot's effectiveness, provided that functional models are stated detailing manipulation dynamics up to the certainty ranges of the needed performance figures. The field of instrumental robotics is fitted up by talented solutions for factory automation. Industrial manipulators support most of the work cycles and only loosely assessed manufacturing processes meet drawbacks, such as surface deburr. Most of the time, the enterprises in these cases have resorted to hand interventions when geometric constraints, edging compliance, shape variability, and tolerated span represent a highly demanding mix of requirements despite unsteady machining patterns. Automated deburring, in fact, now runs into deficiencies when modeling the process and this prevents accurate and efficient issues. The fixtures have perhaps been developed by miming too much manual habits, even if no reason at all exists that task-oriented solutions should profit by anthropocentric rules. By this conventional approach, a deburring robot presents a performant manipulating arm, with the finishing tool at the tip, conveniently actuated and extensively sensorized. It's functionality shows limitations, that actually, a skillful and trained operator overcomes with craft and ingenuity, adapting the operation modes to the task progression.

The switching to robotic equipment for precision deburr has to be reached by looking over the process again, to establish a setup aiming at smooth engagement; position-and-force control; steady repetitiveness; and, in general, highly adaptive fit-outs based on skillful survey of the work progress to restore correctness. By robotizing, once accuracy figures are achieved, productivity and tolerances are preserved according to total quality conditions, therefore, assuring improved product finishing as compared to manual operation. Robotic equipment, with cooperation, is an important alternative to be considered. The example case addresses this target. The deburring tool is operated by a six degrees-of-freedom arm. The work-piece is borne by a similar six degrees-of-freedom rig whose mobilities are controlled to interact with the machining end-effector. The rig, in this context, reduces to a platform whose position and attitude are driven by task-oriented requests. Functional innovation, "cooperation" task setting, is related to the ability of establishing work sequences that depend on the deburr cycles to be executed. The dynamics of the bearing rig and of the operating arm need to be programmed concurrently.

To that purpose, the analysis of the cooperation opportunities is the preliminary step to correctly and efficiently integrating work and handling facilities. This can be done with the already mentioned package SIRI-MR which combines a series of blocks; that is, (SIRI-CA) for configuration analyses; SIRI-AD for dynamics generation; and SIRI-SC for control strategy choice. With the SIRI-MR package, a virtual reality duplication of the multi-agent 'environment' is provided to characterize robots cooperation. A design sequence with SIRI-MR presents as follows:

1. *Specification of functional cooperation figures*: primary goal is paths selection and tasks timing. The SIRI-MR package is employed as "planner-frame" (essentially based on the SIRI-CA block) and between "feasible" paths of the individual arms, the 'job-consistent' trajectories for the multirobot system are singled out.
2. *Specification of operational coordination figures*: this second design step is aimed at trajectories setting and control trimming; to change "feasible" dynamics into "tasks-consistent" dynamics; and the SIRI-MR package is employed as 'controller-frame' based on the SIRI-SC block with the included SIRI-AD block options.
3. *Specification of multirobot performances*: efficiency testing and accuracy check are performed on job-consistent (according to given functional cooperation figures) paths and with tasks-consistent (according to the chosen duty coordination figures) dynamics. The SIRI-MR package is fully enabled as "cooperation-frame" according to the selected duty (Fig. 7.21), task (Fig. 22), and govern (Fig. 7.23) modes.

The following presentation refers to the above ideas to exemplify how cooperating fixtures are selected to supply process attuned solutions. The deburr process is reviewed first; then discussion is turned to a powered platform purposely built as a cooperating rig when provided by position/attitude commands with control of the interfacing force components.

Process Conditioning Environments: Deburr Operations

Machining of work-pieces commonly results in burrs left on material bodies. These burrs have to be removed, due to piece safe handling, fitting, or assembling, because of functional requests on the surface shape (e.g., for fluid flows mating) on the body properties (e.g., stress intensity concentrators relief).

Today, burr removal is still mostly carried out by hand, which means a time-consuming and boring job. Finishing moreover, highly depends on skill and mastery of trained workers. Since labor becomes more and more expensive, artifact's cost is influenced considerably by this process. In some cases, the deburr process causes 35% of the final price [KBK86]. In addition, manual burr removal does not allow persistency of tolerated figures and replication of exactly defined chamfer profiles which is critical for a variety of artefacts. This so-called precision deburring is, however, requested to reach total quality, particularly in the production of diversified turbomachinery blades and nozzles (to reduce turbulent flows); the manufacturing of gear, shaft, or cranks (to relieve local stress); the assembly of high speed rotors (to keep dynamic balance); or etc. By robotic deburring, moreover, the reject rate of products would be far lower. In the near future, automatic burr removal processes could aim at zero-defects production and wider exploitation of the equipment, today developed as technology driven contrivance might become a market driven option for factory automation.

For robot execution, the process shall first be quantitatively modeled; starting with a proper estimation of burr size and shape [HoG87] , [KWB88], [AsT96]. To satisfy finishing results, the fixtured unit should be able to avoid the so-called "worst case burr"; namely, a maximum size burr occasionally occurring when the machining forces do not concentrate according to given geometries. The amount of material to be removed per unit time, Fig. 7.38, the so-called "material removal rate" (MRR), can be expressed by a simple balance:

$$MRR = (A_B + A_C)v_T = A_C(R_M + 1)v_T \quad R_M = A_B/A_C \qquad (7.17)$$

where A_B is the burr's cross sectional area, A_C is the chamfer's cross sectional area, and v_T is the tool speed along the edge to be deburred, (Fig. 7.38). Of course, the mentioned areas can be expressed in function of other parameters such as contact forces, strength of the material, etc., to connect geometries, strains, stresses, and machining operations.

The tangential area ratio: $R_M = A_B/A_C$, typically varies between 0 and 2 depending on burr size, where the value 2 refers to "worst case burr." In practice, large variations causing diversified situations need to be faced. Process variability ranges need to be further analyzed to fix standard reference figures. During deburring, the normal F_n and the tangential F_t components stress, (both robot and piece). The current cross sectional area of burr and chamfer, therefore, depend on both the normal and tangential projections. Then, the variation of normal and tangential components (ΔF_n and ΔF_t) is related to the respective projections of the cross sectional areas of burr and chamfer, that is:

$$\Delta F_n = f_n(\Delta A_B/\Delta A_c)_n \quad \Delta F_t = f_t(\Delta A_B/\Delta A_C)_t \qquad (7.18)$$

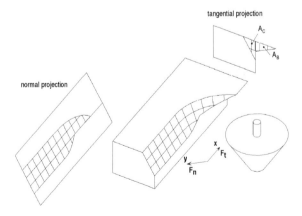

FIGURE 7. 38 Details of the burrs removal process.

Of course, the chamfer normal and tangential projections do not depend on the burr size. On the contrary, the burr size modifies its cross sectional area projections. Burr size, more precisely, greatly affects the tangential force F_t, whereas it has considerably smaller affect on the normal force F_n. Moreover, a constant chamfer surface quality is required regardless of burr size. This means that the material removal rate (MRR) shall remain steady; otherwise, the dressing process with small burrs suddenly changes to rough-machining process, yielding the "worst case burr" occurrence with bad surface finishing. Since R_M varies between 0 and 2, the control system must monitor the variations of F_t and F_n and modify the tool speed to keep MRR constant. The option requires control of the mutual positions of the tool cutter and the piece surface attitude with closure of independent force/torque feedbacks. The analysis of the deburr process [PeV96], [RLC97], [KBK86] shows that the proper robotic solution should carry on, in parallel, the position/attitude command and the force/torque modulation at the engagement boundary between cutting tool and compliant surface.

A six mobilities arm, without force feedback, suffers from considerable drawbacks, represented by three-dimensional vibrations that upset the chamfer path. This results in unsatisfactory surface edge quality which is important in precision deburring. The wavering behavior depends on the discontinuities of the exchanged machining forces and on the compliance of piece supports and tool drivers. An attempt at preserving the finished surface quality has been sought through stochastic control [Pek64] or by means of adaptive end-effectors. Redundant mobilities can be added at the deburring front-end, in the way that the conventional 6 d.o.f. serial arm makes the main engagement (e.g., force control setting) and the extra mobilities carry out secondary compensation (e.g., position trimming). The set-up could be explored through passive adaptation. As usual, the behavior of the fixture is described by a mass-spring-damper model with normal direction mechanical impedance (ratio of contact force to end-effector deflection, as a frequency function) given by:

$$F_n(j\omega) = G_n(j\omega)x_n(j\omega) \quad \text{with: } G_n(j\omega) = (K_n - \omega^2 M) + j\omega H_n \qquad (7.19)$$

A large normal impedance causes the end-point to balance grinder forces remaining close to the pre-set trajectory. Given the volume of metal to be removed, the desired tolerance in the normal direction prescribes that the value of this impedance shall not exceed conditions yielding burr excitation resonance. At the same time, it is necessary not to produce high tangential contact forces since tool stall (or even breakage) may occur with dangerous normal skips. It follows, Eqs. (7.18) and (7.19), that the end-effector needs to operate all along with bounded interaction forces which implies small tangential impedance. On the other hand, uncertainties in the end-effector position are smoothed by a large compliance in the normal direction, at least up to the robot resonance range. All in all, the end-effector shall show the following behavior in the normal direction:

$= \; >|G_n (j\omega)|$, large for all ω in the ω_R band; $|G_n (j\omega)|$, small for all ω in the ω_B band;

$=> \omega_R < \sqrt{K_n / M} < \omega_B$ where: ω_B, burr resonance range; ω_R, robot resonance range.

The Automation of Precision Deburr Operations

It is possible to design a passive end-effector with such dynamic characteristics but it would be impossible to let it also meet the condition on the tangential direction (large compliance). Because of the role played by the constant mass of the grinder, making equal the dynamic behavior of the end-effector in both directions at high frequencies, when a large normal stiffness is chosen to improve the quality of the surface finish, and then the end-effector will not be compliant enough to compensate for robot oscillations. This is why an active system is required to optimize the process parameters and to compensate for robot oscillations while showing large stiffness in the normal direction [BEL91], [HeK91], [YOY94], [KIK90], [KuW92], [StS90], [VaP96], [WET90], [WhT92], [WKT90]. Active dynamical systems can either operate by control redundancy or by functional cooperation. In the first case, distinct position/attitude and force/torque sensors are used to accomplish redundant tool-tip control. The setup still suffers deficiencies as the uncoupling of normal and tangential behavior is hindered by inertial effects of the equal massic terms. As for functional cooperation, redundant mobilities are required, namely:

- Addition of independently actuated members to the arm (serial d.o.f.) [YHM94];
- Inclusion of an actuated rig for holding the piece to be deburred (parallel d.o.f.).

The first solution suffers from given snags: low stiffness of the open chain with critical control setups constraints requiring nasty trimming and arduous presetting operations; and band limitation in particular with variable operation ranges depending on extended mixes of pieces to be deburred.

The second solution offers several advantages:

- The redundant mobilities extend versatility and dexterity enhancing the robot accessibility along the surface edge to be deburred. The rig d.o.f. can be used to hold the piece in a pose that favors the robot end-effectors work-trajectories;
- The adaptivity can be upgraded by intersecting paths operation modes with efficient sweep of the workspace and exploration of task planning which avoids collisions at engagements or undue penetration during deburr;
- The efficiency can be improved: the execution times can be reduced with low absolute speeds of each cooperating robots, but high relative speeds of the work-tip;
- The same accuracy can cover the full workspace; robot position/attitude along the main movements may be compensated by the position/attitude tracking of the cooperative rig, which is responsible for the servoed movements;
- Critical tasks can be faced with repetitiveness: sharp corners, for instance, are tracked without considerable speed reduction by obtaining smooth paths by split tracks (this is important for precision deburr, when corner-rounding is not admissible);
- Closed kinematic chain allows a lighter rig design which results in lower mass inertia and better dynamic behavior of the cooperating equipment used to get rid of the partner robot oscillations, as is the case with precision deburr;
- Further quality and efficiency betterments are obtained by adaptive job planning aiming at preventing worst case burr or, at least, avoiding uttermost courses within the variability range of the removed burrs.

Obviously, there are certain drawbacks, too. Two robots, instead of one, result in higher costs. Coupled motion requires more sophisticated control which further increases costs. Programming is more time-consuming compared to a single robot, particularly in case of adaptive path planning and tightly bound dynamics. These handicaps are reduced when proper standardization is reached by the cooperating rig, the control architecture, and the programming aid. In the first instance, the in-parallel actuated platform offers quite an effective option. It is close-packed, easily powered, and suited for position/attitude tracking. Control and programming burdens are drastically reduced by referring to CAD packages, such as the SIRI families of codes, and using them for the design, development, setting, and fitting operations in virtual reality surroundings, all along the robots life-cycle.

A Cooperative Fixture for Work-Parameters Adaptation

The cooperating engagement of the piece-supporting rig needs proper performance in terms of position tracking ability, reactive stiffness, attitude controllability, etc., in a way to upgrade arm's accuracy, dexterity, efficiency, and versatility according to requests. The mechanical architecture of this rig is based on an in-parallel actuated platform (originally designed at the Polytechnic of Turin [RoS92], to support assembly operations). The development has also been studied [ACC94c], in view of micro-robotics applications [ACM95]. A solid model of the fixture provided the principal features of the rig. It consists of a platform, actuated, in parallel, by three driving blocks, each one displacing a vertex of an equilateral triangle which specifies attitude and position of the reference plane. Each driving block is obtained by

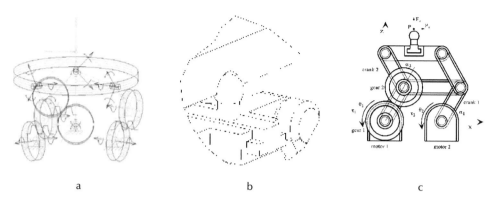

FIGURE 7.39   The powered co-operating rig example: (a) multibody model; (b) linking of table and upper parallelogram (particular); and (c) side view of one platform's driver.

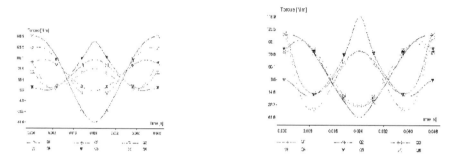

FIGURE 7.40   Inverse dynamics: plot of required torques [N m] (a): full payload (5 kg) – (b): no payload.

two superposed planar parallelograms moving vertically. The platform is fixed to the upper parallelograms. The linking is assured by spherical bolts that are impinged at its lower side and can also slide along guideways, fixed to the top beam of the upper parallelogram. The guiding slots, placed orthogonally to their carrying beams, form an angle of 120° between each other; the coupled parallelograms have beams linked by ball bearings to reduce friction. They are driven by a pair of DC motors, solid with the rig base to reduce the inertial effects. For position accuracy, the upper four bars, Fig. 7.39, are moved by a backlash-free gear train, not linked to the bottom cranks; the lower four bars are directly driven by the twin motor.

The setup repeats three times and the final the assembly with the plate results in a system with six degrees of freedom; thus, the rig has six servo-motors to be controlled. The rig dynamics has been analyzed [MAC97], by assessing

- The actuation kinematics using the geometrical constraints to model the forward and backward mappings, which link workspace and platform control coordinates; and
- The open-loop dynamics combining inertial terms and constrained motion to generate the reflected loads on the driving commands.

The modeled platform was used for the functional validation of the cooperating rig. The prototype weighs about 5 kg and is supposed to be able to carry a same amount as pay-load. The working space is small with path continuity hindrances at the out-boards but the platform is requested to accomplish only small oscillations around its "central" positions. Both direct and inverse dynamics simulations have been performed: the imposed trajectories of inverse simulations have been chosen to be straight lines (in the working space) tracked with sinusoidal time laws. The period of the sinusoids has been chosen so that maximum accelerations around 1 g are obtained, except the few cases in which high acceleration motions (10 g) have been considered. Fig. 7.40, for instance, shows the torques needed to track an oblique

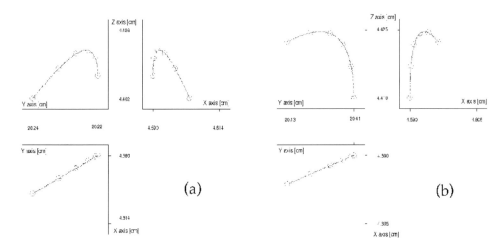

FIGURE 7.41   Direct dynamics: orthographic projection of platform's free motion (initial vertical velocity of 0.1 m/s): (a): full payload (5 kg) (b): no payload.

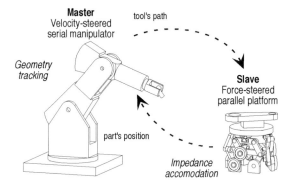

FIGURE 7.42   Block-schema of a deburr stand with cooperating robots.

straight line of 3 mm stroke with change of attitude about a "horizontal" configuration, with and without the maximum allowed payload. Figure 7. 41, instead, shows the three orthographic projections of platform's free motion paths, when an initial vertical velocity of 0.1 m/s is assigned.

Comparison of results obtained by several test cases provides the main characteristics of the equipment.

- The global dynamical behavior of the system is mainly affected by the dynamics of the actuation system, as the contribution of the payload/table group is quite negligible.
- The motion is more easily obtained when it is parallel to one actuation side rather than normal to it.
- As already pointed out by other researchers [SoC93], the working space is considerably restricted.
- Compared to the small-scale realizations [ACM95], the equipment is moderately sensitive to the influence of the gravitational field.

This kind of fixture deserves particular interest for its ability of accurate tracking, in position and attitude, any three dimensional surface. The rig, Fig. 42, simultaneously controlled with the arm equipped by deburr chamfer gives rise to a redundant mobilities setup. Joint force-and-displacement governing strategies can be enabled to reach the very high versatility and dexterity figures of human operators while improving the efficiency achievements with operation continuity and the steady accuracy of the surface finishing by an impedance control, with the displacement term represented by the relative motion between the deburring robot and the supporting platform.

The Impedance Control of the Cooperating Fixture

As noted on pages 7–31 through 7–33, the starting point of impedance control design is the choice of two stiffness K_P and damping H_P matrices to obtain desired coupling effects at the tip or, which is the same, to have interaction forces at the interface given by

$$F_E = K_P(x - x_d) + H_P(\dot{x} - \dot{x}_d) \tag{7.20}$$

Such external forces are, of course, independent variables and the motor torques can be computed so that the actual tip displacements are related to the developed forces (7.20), for instance (for gravity compensation) by applying

$$Q_P = B(q_p) - J^T(q_p)[K_P(x - x_d) + H_P(\dot{x} - \dot{x}_d)] \tag{7.21}$$

This approach, thus, consists of monitoring the dynamic relationship between force and position, rather than separately measuring the two quantities. It must be noted that, by varying the K_P and H_P matrices, either a force control or a trajectory control is obtained. In fact, by increasing the values of the stiffness elements in the K_P matrix, the control system tends to keep the end-effector closer to the assigned path; while a decrease of such values ends up with a more compliant end-effector. Commonly the H_P matrix is chosen to reach critical damping along the trajectory controlled directions.

Stiffness and damping matrices are usually expressed in a local work-frame {L}, attached at the piece in the contact point with the tool and with x and y axes parallel to the tangential and normal directions. Therefore, calling K'_P and H'_P these local matrices, a time-varying mapping with the global frame {G} is needed. That is why, also in case of constant process requirements, (i.e., fixed stiffness and damping values for the various directions) actual K_P and H_P matrices change during normal contouring operations.

$$K_P(q_P) = {}^L_G[R(q_P)]^T K'_P \, {}^L_G[R(q_P)] \quad H_P(q_P) = {}^L_G[R(q_P)]^T H'_P \, {}^L_G[R(q_P)] \tag{7.22}$$

where ${}^L_G[R(q_P)]$ is the rotation matrix between global and local frames.

The stiffness matrix K'_P can be selected diagonal with principal elements chosen to grant the desired compliant task; namely, the terms related to tangential translations have low values (i.e., the interaction is characterized by low stiffness) for the direction along which force must be limited; the terms for the directions along which trajectory has to be controlled, i.e., and the other two directions have large values (only limited by the available control bandwidth). The matrix H'_P is, moreover, chosen to be diagonal and composed of the desired damping coefficients in each direction with critical damping selected along trajectory controlled directions. As for rotations, the requirement is to follow the assigned attitude as close as possible. The related (high) stiffness, accordingly, will be isotropic in the working space. In this case, best choice seems to be the use of the equivalent angle-axis representation that expresses the rotation between reference and actual tool frames giving the axis r along which rotation occurred and the related angle ϑ . Thereafter, the relation (7.20) is preserved for the impedance control of the linear motions, while for the rotational ones, the following equivalent formula is used:

$$M_E = k_p \vartheta r + h_p \omega \tag{7.23}$$

The setup is completed by the compensation of gravity terms via feed-forward cancellation of the related contributions. Exact compensation can be computationally heavy as the full forward kinematics, which are rather complex [ACC94c], shall be evaluated on-line. A good compromise is the off-line evaluation of the gravity terms corresponding to the assigned path, with their on-line updating, Fig. 7.43, at lower rates with respect to the inner control loop. Computation of the trimming terms and reflection to the motors are easily performed once the dynamics is known [ACM95], [MAC97].

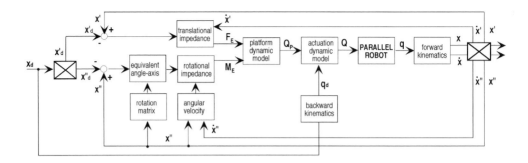

FIGURE 7.43 Scheme of the impedance control system (prime symbols are related to translations and double prime to rotations).

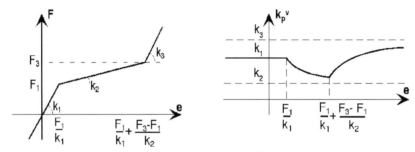

FIGURE 7.44 Variable structure stiffness (a) and virtual equivalent stiffness (b).

A further improvement can be obtained, for the present application, by adopting a non-linear law for the stiffness along the tangentia direction, Fig. 7.44. The "optimal" value (k_2) can be used for working conditions near the reference state while the compliance is stiffened if the tool is working outside the standard range. By this way, even for difficult tasks, the need of resetting the cooperative fixtures (serial robot and platform) is almost avoided. In fact, if the platform is going out of the working space, an increased stiffness brings the reset to standard working conditions. To be able to keep the usual matrix forms, a virtual stiffness k_s^v is introduced with the behavior shown in the Fig. 7.44, plotted against the position error e in the tangential direction. The related damping factor k_s^v must be accommodated accordingly, within the same working range.

The whole system has been studied by computer simulation with Pro/MECHANICA (by Parametric Co.). It is a complex multibody package that has been used to solve the complex DAE model of platform dynamics and to test the proposed control system. Several tricks have been used to simplify the model while preserving the correctness of the dynamical behaviour and finally, Fig. 7.39, a fixture with 10 parts and 9 kinematic pairs has been worked out.

To perform a few simulation trials, the deburring process model was also needed. With reference to the above considerations, this has been particularized for the case of deburring of aluminium aeronautical components, for which many experimental data were available [Hic85], [KiH86]. Therefore, the tangential and normal forces between tool and component are synthetically expressed [JKL97], [Jok97] by experimental relations:

$$F_t = F_t(V_{tool}, \omega_{burr}, \dot{x}, x, t); \quad F_n = F_n(V_{tool}, \omega_{burr}, \dot{x}, xt) \tag{7.24}$$

with t, time, x, $\dot{x}$ displacement and velocity of platform in tangentialdirection, V_{tool} velocity of the tool, and ω_{burr}, frequency of the surface grooves.

The presence of the motors modifies the system's response and the described model needs to be augmented with the addition of a (first order) dynamic block for each torque motor. As expected, the overall response is affected by higher damping. Then, to finish a surface with a steady undulation (left by previous

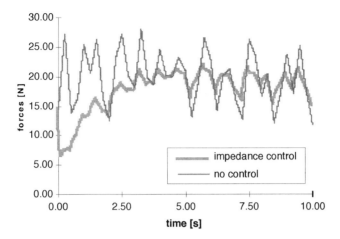

FIGURE 7.45 Simulation output: deburr forces with and without compliant fixture.

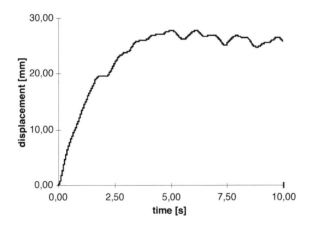

FIGURE 7.46 Typical results: platform tangential displacements.

machining), the transient trend, Fig. 7.45, shows the slight accomodation of the sinusoidal interaction force and features the relevant path stability, Fig. 7.46. Once the system that is charged by the sudden force and torque step comes back home, the surface finishing, it shall be pointed out, does not depend on global displacements (rather only on the relative displacements).

This control, as expected, characterizes, by simplicity and robustness, parametric uncertainty even if with limited dynamic performance. Indeed, it is not required to explicitly solve the manipulator inverse kinematics, since the actuation law is given in terms of work-space errors; moreover, it does not require measurement the interaction forces or to explicitly assess the environment stiffness.

The Multi-Robot Assembly of Compliant Components

The domain of robots with cooperation opens several other possibilities as in the case, for instance, aiming at improving assembly effectiveness. During the joining tasks, some components may characterize by large compliance and settling cannot neglect the mutual deflections during processing. An automotive body, for instance, is composed of different bent sheet metal pieces; these are positioned by clamping rigs to be spot welded into parts, further handled to be joined together to form structural bodies (passenger compartment, engine box, rear trunk, wheel shields, etc.). To achieve proper dimension tolerances, the shaping accuracy around some 0.5 mm is needed, regardless of sheet warping, by quick and reliable part positioning. Valuable aid is supplied by multi-robot assembly, based, e.g., on a position-controlled master with a force-controlled slave.

In front of large compliances, a better set-up would address a coordinated control, with a supervisor steering two robots, Fig. 7.47, each holding a deformable payload, (Fig. 7.15), to be positioned and joined together, within tolerated figures. The analysis develops by modeling the components of known compliance, so that the navigation paths bring the pieces with due assessment of their perturbed geometry, Fig. 7.48. For assembly, (e.g., the (outer) forged sheet-steel and the (inner) pressed trimming face) to obtain a car door, some simplifying assumptions are, generally, say,

- The manipulator links and transmission compliances are neglected;
- The effects of the gravitational potential energy are omitted;
- The grip zones hold both pieces without local energy storage build-up;
- The back-coupling of the pieces strain conditions on the arm is ignored;
- The contact mechanics assumes the central impact between matching shapes;
- The pieces joining is fulfilled by a single stroke with damping out of the efforts;

and other similar hypotheses, to bound the overall degrees-of-freedom in handling and assembly.

The dynamics of the cooperating robots will, finally, be described by using the model (7.5) for the free motion phase, followed by the linearized approximation (7.8) as soon as the joining operation starts. The interfacing force δF_E is given by assessing the contact model between the cooperating robots grip points with interposed compliant payload. The description consists of separate coordinate frames for each tip and

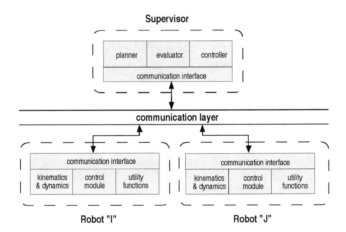

FIGURE 7.47 The coordinated control architecture of multi-robot fixtures.

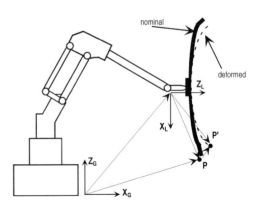

FIGURE 7.48 Handling model for the carried compliant shells.

for the workspace using the transforms (7.15) and (7.16). The last could possibly be omitted, by enabling a steering logic for bringing the joining movement to follow a normal direction with respect to the clinching surface. Thereafter the contact model is simply expressed by the reaction force at any of the grip points:

$$F^* = G^*(s)y \tag{7.20}$$

where y-relative displacement vector between tips.

The identification of the generalized mass, damping, and stiffness can be obtained by means of a finite-elements code, directly used to study the strain behaviour of the pieces once the contact is established. A detailed investigation is given in [MiI96], with rather extended simulation results. All things considered, an effective assembly stand can use standard cylindrical arms to accomplish the picking and approaching maneuvers; the wrists shall properly allow small accomodations, say:

- Angular shifts:—pan, around a vertical axis;—tilt, around a transverse horizontal axis;—yaw, around a normal horizontal axis; and
- Linear shifts: vertical and lateral moves aside;

while the linear move for clinch fastening is, as said, impressed through the arms by an impedance control.

The set-up leads to effective solutions on condition of slightly modifying the pieces (for safe feeding and handling). The transposition of similar arrangements to micro-manipulation tasks deserves special attention especially in front of duty sequences in the microscopic range out of many possibilities.

7.6 Conclusions

The development of instrumental robots has been addressed in this chapter, mainly as a "technical" problem looking for "highly effective" rigs according to function-oriented requirements. The goal is basically related to the capability of modeling the devices' behaviour. Several texts, [Ard87], [AsS86], [Cra89], [Koi89], [Pau81], [Riv88], [VuK89], [FGL87], [McK91], show how to obtain the manipulators dynamics, based on the Lagrange's approach or on the Newton-Euler equations. Suitable computer codes are, as well, already available to provide solutions at different levels of accuracy. The issues, however, could suffer 'economical' drawbacks when the robot abilities happen to be overemphasised as compared to the duty actually required to achieve the instrumental scopes. Therefore, "leanness" is not stressed enough to address artifact-and-process re-design and all business re-engineering, so that the finally chosen instrumental robot will perform the assigned tasks with no function, duty, or resource redundancy.

On such a preamble, the design activity will outgrow the possibilities of most teams and only iterative attempts might try to approach "balanced" solutions, on condition of being able to assess the fixtures actual behavior, within real operation conditions. The "obvious" idea of experimenting on prototypes is not considered while rising costs and the interacting surroundings will sometimes supply incomplete, incorrect or improbable settings. This is the main reason for developing the CAD series of packages, such as the SIRIxx environment, Fig. 7.49, to have an integrated reference for assessing the controlled dynamics of task-oriented high-performances manipulators by virtual reality testing.

The virtual reality simulation is used as CAD support in the ideation and technical specification phases of the equipment and helps, as off-process reference, for tasks programming and control tuning, each time the actual use of the equipment needs to be modified. The main feature of simulation is that it allows different types of analyses (so that it covers most advanced performance robotics requirements) by integrated knowledge frames (in order not to lose the specialization effectiveness, aiming at each sectorial field application). An illustration of the capabilities has been given, emphasizing the ability of shaping the dynamics and of managing the redundancy.

The design of an instrumental robot starts by acknowledging sets of competing task-driven solutions, supporting the setup of highly effective operation modes. Structured functional models must be established,

SIRI-AD [AMM84b]	The nonlinear manipulation dynamics is generated, for open-chain arms with any number of mobilities and joint connections; the interfacing with 32 modelers helps the computation of mass quadratic moments and the location of mass centers.
SIRI-CA [ACM86]	The trajectory planning is provided for 32 families of open chain manipulation set-ups, by means of algorithmic models describing both forward and backward kinematical transforms; parametrized standardizations are available
SIRI-CL [ACM96c]	The kinematics and the dynamics of 2 families of closed chain manipulation set ups are established with parametrical models; topologic analysis criteria are given for the closed chain set-ups commercially available
SIRI-SC [AMM87]	The control planning is established for open-chain arms with any number of mobilities and joint connections, referring to an expandable library of strategies; the dynamics shaping with nonlinearities compensation is considered
SIRI-AT [CMP94]	The manipulators postures/shapes are given with virtual reality restitution, to investigate the work-space singularities (collision avoidance, etc.); *by-default* blocks automatically generate solid members, once lengths are available
SIRI-MR [AMM91b]	The robotic systems with cooperation are provided, for performing trajectory and control planning for master/slave setups, for supervised coordinated robots, for parallely operated fixtures, etc. when operation redundancy is used
SIRI-HD [AMM91a]	The command redundancy with force and/or position feedback is provided, for adaptive task planning (and dynamics shaping), with and without state expansion by modeling robot-surroundings interactions
SIRI-UM [ACC94b]	The *engagement* phase (between the *navigation* and the *work* phases) is given with different collision and bouncing conditions; the description is parametrized for including the results from experimental investigations

FIGURE 7.49 Synoptic presentation of the main SIRIxx packages.

so that the dynamics of each robotic equipment can be generated throughout assessment of the accuracy, dexterity, efficiency, and versatility figures achieved by each particular solution.

Once the functional models are available, the design procedure extensively exploits CAD-based tests for virtual reality experimentation before actually building prototypal devices that might fail to reach the requested technical and economical effectiveness. The approach success, however, depends on the appropriateness of the model. In fact, whether reductive equivalencies (to lower the degrees of freedom), approximations (to suppress nonlinearities), motion constraints (to simplify cross-coupling effects), etc. are not properly stated, the generated dynamics does not provide correct reference to assess the robot performance, in terms of accuracy, dexterity efficiency, and versatility.

The engineering practice suggests different ways for "proper" modeling and we have already pointed out how developments in instrumental robotics will share the methods of the integrated design activities. As a general rule, it is worth distinguishing

- The manipulation dynamics: the forced motion of hinged solid bodies in the joint-space is the basic model with the unavoidable transport effects of inertial terms and Coriolis acceleration. The refining might cover: joints and links compliance; transmission and actuation effects; etc.

- The interfaced surroundings coupling: the constrained motion of the robot tip, active in an independently defined workspace, is the basic model with (possibly) reduction to joint-space driven-commands through "impedance" control state extension with account of the measured variables.

- The logic steering govern: the programming abilities are the basic means to deal with task variability and with occurrence uncertainty. Shallow knowledge models are used to expand the relational frames with "expert" modules, having overseeing and decision support functions.

- The monitoring of actual returns: value cycle models are defined to assess the real cost of the innovation in terms of beneficial fallout on the products once potentialities are fully explored (including on-process exploitation of quality data), so that checks are run on the economical side.

The chapter concern is, mainly, to design instrumental robots whose performance is driven by the capability of accomplishing a given set of tasks. Then attention is focused on the deep knowledge needed to describe their dynamics. The addition of nonexploited abilities results in unacceptable costs; thus, the reasonable connection between tasks domain and functions assignments has to be done from the ideation steps to understand the technical appropriateness of each prospected solution in terms of actual returns. To that purpose, the SIRIxx environment provides the conditioning references, in terms of structured (deep knowledge) constraints. Technically "advanced" options have, in particular, to be explored before implementing real facilities. Aiming at that, the study profits by moving along the design cycle with standard steps.

The basic development stages for the design of instrumental robots, Fig. 7.11, require clear visibility on both the material resources (CFC frame) and on the logic resources (MDM frame) properties. On these premises, effectiveness can be reached by iterating the design cycle, Fig. 7.1, with due account of a few simple suggestions. Dynamics shaping and control planning are steps directly faced to achieve high robot performance by conventional setups. The availability of a library of control modules makes it easy to characterize the dynamic behavior in competing running conditions. The library SIRIxx includes common and sophisticated schemes; it can be generally used for control planning operations and as specialized aid for performing dynamics shaping. Of course the dynamic modelization implies knowledge about all links' centroids and mass quadratic moments; these parameters might be measured or experimentally identified, when actual robots are already available; alternatively, they are evaluated with the help of a solid modeler (interfaced to the SIRIxx package), at the earlier robot design or development stages. The inertial coupling is seldom considered by existing manufacturers of robotic equipment; such effects may be liable, however, of serious consequences on actual performance, since they introduce task modulation on the feedback gains; dynamics shaping, thus, is expected to become an important feature for advanced robotics, as dexterity and accuracy must be joined to high speed requests.

The function modulation aspects are commented in the Section 7.2 and expounded in Section 7.4, considering example cases: dynamics inconsistencies induced by coupling inertial effects; accuracy upgrading by the redesign of a fixtured wrist; dexterity improvement by redundant force control; versatility expansion by expert steering. The latter cases introduce the opportunities given by further sophistication; this is a step ahead in terms of sophistication and the option needs to be carefully evaluated to assess the return on investment. In the Section 7.3, concepts are reviewed by addressing the operation redundancy as a re-engineering issue to obtain process-attuned robots. The job can be grounded on standard rules by exploiting modularity, against the proper classification of the activity modes and the distinct presetting of the functional units. When the application area grants return (e.g., in micro-dynamics), the sophistication leads to mobility redundancy (cooperating robots) and command redundancy (position and force control).

Section 7.5 is devoted to these advanced developments of robotics with attention on manufacturing applications. The basic motivations of using robots with cooperation for automatic precision deburring is discussed, with hints on the machining process to show how the 'external' conditions are faced by manual workers and how a cooperating fixture might automatize the operations (with steady quality issue). The functional redundancy appears as a worthy opportunity, also, for assembly tasks, particularly to join highly compliant pieces or in front of micro-handling cases. All these situations require properly sophisticated models of the rigs dynamics, so that the design choices might be, step by step, validated with virtual reality simulation, before moving to the implementation of real fixtures.

Acknowledgments

We gratefully acknowledge the financial support of CNR (Italian Research Council) for the basic developments of these studies leading to the implementation of series of SIRIXX packages, under the frame of the project PFR (Progetto Finalizzato Robotica). We also thank the manufacturing companies that cooperated with us for the different developments, particularly: COMAU Robotica (Beinasco, Torino) and Speroni S.p.A. (Spessa, Pavia).

References

[AAC94] G.M. Acaccia, C. Aiachini, M. Callegari, R.C. Michelini, R.M. Molfino. 1994. Dynamics shaping for the control of instrumental robots, *Proc. 10th ISPE/IFAC Intl. Conf. on CAD/CAM, Robotics and Factories of the Future: CARs & FOF '94,* Ottawa, Aug. 21–24. 514–520.

[ACC93] G.M. Acaccia, M. Callegari, R. Caracciolo, R.C. Michelini, R.M. Molfino, M. Torbidoni. 1993. Redundant position/force control for advanced robot applications, *Advances in Computer Cybernetics and Information Engineering,* George E. Lasker, Ed. (The International Institute for Advanced Studies in Systems Research and Cybernetics Publ., Windsor, CA). 63–72.

[ACC94a] G.M. Acaccia, P.C. Cagetti, M. Callegari, R.C. Michelini, R.M. Molfino. 1994. Contact mechanics description of robot engagement tasks, *Proc. IASTED Intl. Conf. Applied Modelling and Simulation: AMS '94,* Lugano, Jun. 22–24. 32–35.

[ACC94b] G.M. Acaccia, P.C. Cagetti, M. Callegari, R.C. Michelini, R.M. Molfino. 1994. Modelling the impact dynamics of robotic manipulators, *Preprints 4th IFAC Symp. on Robot Control: SY.RO.CO. '94,* Capri, Sept. 19–21. 559–564.

[ACC94c] G.M. Acaccia, P.C. Cagetti, M. Callegari, R.C. Michelini, R.M. Molfino. 1994. Dynamic model of a robotised fixture for co-operative deburring operations, *Proc. 2nd European Solid Mechanics Conference: EUROMECH' 94,* Genova, Sept. 12–16.

[ACC95] G.M. Acaccia, M. Callegari, L. Consano, R.C. Michelini, R.M. Molfino, S. Pampagnin, R.P. Razzoli. 1995. Universal master for remote micro-manipulation, *Proc. First ECPD Int. Conf. on Advanced Robotics and Intelligent Automation,* Athens, Sept. 6–8. 499–504.

[ACH95] G.M. Acaccia, M. Callegari, D. Hagemann, R.C. Michelini, R.M. Molfino, S. Pampagnin, R. Razzoli, H. Schwenke. 1995. Robotic fixture for experimenting antropomorphic vision, *Proc. 7th Intl. Conf. on Advanced Robotics: ICAR '95,* Sant Feliu de Guìxols, Sept. 20–22. 237–244.

[ACM86] G.M. Acaccia, M. Callegari, R.C. Michelini, R.M. Molfino. 1986. Architectural analysis of robotic industrial manipulators, *Proc. AFCET/IASTED Intl. Symp. on Robotics and Artificial Intelligence,* Toulouse, Jun. 18–20. 579–598.

[ACM88] G.M. Acaccia, M. Callegari, R.C. Michelini, R.M. Molfino, P.A. Piaggio. 1988. X-ARS: a consultation program for selecting the industrial robot architectures, *Artificial Intelligence in Engineering: Robotics and Processes,* J.S. Gero Ed. (Elsevier Publ., Bath, UK). 35–58.

[ACM91a] G.M. Acaccia, M. Callegari, R.C. Michelini, R.M. Molfino, M. Recine. 1991. Functional coordination of multirobot equipment, *Proc. 14th IASTED Intl. Symp. Manufacturing and Robotics,* Lugano, Jun. 25–27. 122–125.

[ACM91b] G.M. Acaccia, M. Callegari, R.C. Michelini, R.M. Molfino, M. Recine. 1991. SIRIxx: a simulation programming environment for the design and the management of industrial robots, *Atti Convegno ANIPLA,* Milano, 29–30 ott. 461–474.

[ACM93] G.M. Acaccia, M. Callegari, R.C. Michelini, R.M. Molfino. 1993. Dynamic control of robots, *Proc. 3rd Intl. Symp. on Measurement and Control in Robotics: ISMCR '93,* Turin, Sept. 21–24, 1993. Bs.I-13:Bs.I–18.

[ACM95] G.M. Acaccia, M. Callegari, R.C. Michelini, R.M. Molfino, R. Razzoli. 1995. Dynamics of a multi-powered platform for task-steered instrumental robots, *Proc. IX World Congress on the Theory of Machines and Mechanisms,* Milano, Aug. 30–31/Sept. 1–2. 1816–1820.

[ACM96a] G.M. Acaccia, M. Callegari, R.C. Michelini, R.M. Molfino. 1996. The impact dynamics of robotic arms, *Proc. 2nd ECPD Int. Conf. on Advanced Robotics, Intelligent Automation and Active Systems,* Vienna, Sept. 26–28. 313–320.

[ACM96b] G.M. Acaccia, M. Callegari, R.C. Michelini, R.M. Molfino. 1996. Innovations in instrumental robotics: concurrency operations and co-operation, redundancy modulation and control, *Proc. 27th Intl. Symp. on Industrial Robots: ISIR '96,* Milano, Oct. 6–8. 37–42.

[ACM96c] G.M. Acaccia, M. Callegari, R.C. Michelini, R.M. Molfino, R.P. Razzoli. 1996. Assessing the dynamics of articulated manipulators with closed kinematic chains, *Proc. 27th Intl. Symp. on Industrial Robots: ISIR '96,* Milano, Oct. 6–8, 1996. 575–580.

[ACM96d] G.M. Acaccia, M. Callegari, R.C. Michelini, R.M. Molfino, R.P. Razzoli. 1996. The design of robotic equipment for flexible automation, *Proc. 2nd ECPD Int. Conf. on Advanced Robotics, Intelligent Automation and Active Systems,* Vienna, Austria, Sept. 26–28. 554–559.

[ACM96e] G.M. Acaccia, M. Callegari, R.C. Michelini, R.M. Molfino. Simulational assessment of a modular assembly facility. 1996. *Proc. Intl. Conf. on Concurrent Engineering and Electronic Design Automation: CEE '96,* Robinson College, Cambridge, Apr. 10–12. 37–41.

[ADA89] E.W. Aboaf, S.M. Drucker, C.G. Atkeson. 1989. Task-level robot learning; juggling a tennis ball more accurately, *Proc. IEEE Int. Conf. on Robotics and Automation,*. XXX. 1290–1295.

[Alg97] E.A. AlGallaf. 1997. Manipulation of ill-conditioned configurations by a robot hand: employment at local and global dexterities, *Mechatronics,* 7(5): 479–503.

[AMM84a] G.M. Acaccia, R.C. Michelini, R.M. Molfino, P.A. Piaggio. 1984. Simulation of adaptive controlling strategies for industrial robots, *Proc. Intl. Symp. Applied Modelling and Simulation,* Nice, 19–21 June.

[AMM84b] G.M. Acaccia, R.C. Michelini, R.M. Molfino. 1984. Computer-aided design procedures for the design of industrial robots: on-line generation of the dynamics equations, *Proc. Intl. Symp. Computer Aided Design,* Nice, Jun. 19–21. 183–187.

[AMM87] G.M. Acaccia, R.C. Michelini, R.M. Molfino. 1987. Development of CAD codes for the job integration of industrial robots, *Intl. J. Robotics,* 3(3/4): 371–388.

[AMM90] G.M. Acaccia, R.C. Michelini, R.M. Molfino, M.A. Recine. 1990. Simulational programming environment for the development of industrial multirobot systems, *Proc. ISCIE-ASME Symp. on Flexible Automation,* Kyoto, Jul. 9–12. 849–855.

[AMM91a] G.M. Acaccia, R.C. Michelini, R.M. Molfino, M.A. Recine. 1991. Assessment of position/force dynamic control performances for advanced robotics, *Proc. 5th Intl. Conf Advanced Robotics,* Pisa, Jun. 19–22. 1465–1468.

[AMM91b] G.M. Acaccia, R.C. Michelini, R.M. Molfino, M.A. Recine. 1991. Information reference setup for the development of industrial multirobot systems, *Intl. J. Computer Applications in Technology,* 4(3): 137–148.

[AMM91c] G.M. Acaccia, R.C. Michelini, R.M. Molfino, M.A. Recine. 1991. Modeling the coordination of multi-robot equipment, *Proc. 6th Intl. Conf. CAD/CAM, Robotics and Factories of the Future,* London, Aug. 19–22. 870–876.

[AMP89] H. Asada, Z.D. Ma, J.H. Park. 1989. Inverse dynamics of flexible robot arms: feasible solutions and arm design guidelines, *Proc. ASME Winter Meting, Robotics Research,* ASME DSC-Vol. 14. 279–287.

[AnH89] C.H. An, J.M. Hollerbach. 1989. The role of dynamic models in cartesian force control of manipulators, *Intl. J. of Robotics Research ,* 8(4). 51–72.

[Ara83] S. Aramaki. 1983. Flexible playback control of an artificial hand, *Trans. Society of Instrument and Control Engineer.* 19 (6).

[Ard87] D.D. Ardayfio. 1987. *Fundamentals of Robotics,* Marcel Dekker Inc., New York, 1987.

[ArM83] S. Arimoto, F. Miyazaki. 1983. Stability and robustness of P.I.D. feedback control for robot manipulators of sensory capability, *Proc. 1st Int. Symp. Robotics Research.*

[AsA88] H. Asada, Y. Asari. 1988. The direct teaching of tool manipulation skills via impedance identification of human motions, *Proc. IEEE Int. Conf. on Robotics and Automation.* 1269–1274.

[Asd83] H. Asada. 1983. A geometrical representation of manipulator dynamics and its application to arm design, *ASME J. Dynamic Systems, Measurement and Control,* 105. 131–1335.

[AsH79] H. Asada, H. Hanafusa. 1979. Playback control of force teached robots, *Trans. Society of Instrument and Control Engineers.* 15(3).

[AsH89] H. Asada, S. Hirai. 1989. Towards a symbolic-level force feedback recognition of assembly process states, *Proc. 5th Int. Symp. of Robotic Research,* Tokyo.

[AsI89] H. Asada, H. Izumi. 1989. Automatic program generation from teaching data for the hybrid control of robots, *IEEE Trans. on Robotics and Automation,* 5(2): 163–173.

[AsS86] H. Asada, J.J.E. Slotine. 1986. *Robotic Analysis and Control,* John Wiley Inc., New York, NY.

[AsT96] N. Asakawa, Y. Takeuch. 1996. Automatic deburring of cast iron workpiece: removal of projection on a convex sculptured surface, *Proc. 3rd Japan-France Congress on Mechatronics*, Besancon. 686–690.

[AsY87] H. Asada, K. Youcef-Toumi. 1987. *Direct-Drive Robots: Theory and Practice*, The MIT Press, Mass.

[AsY89] H. Asada, H., B.-H. Yang. 1989. Skill acquisition from human expert through pattern processing of teaching data, *Proc. IEEE Int. Conf. on Robotics and Automation*. 1302–1307.

[BAL91] D.F. Baldwin, T.E. Abell, M.C. Lui, T.L. De Fazio, D.E. Whitney. 1991. An integrated computer aid for generating and evaluating assembly sequences for mechanical products, *IEEE J. Robotics and Automation,* 7(1): 78–94.

[BCQ86] M.A. Bronez, M.M. Clark, R. Quinn. 1986. Requirements development for a free-flying robot: the ROBIN, *Proc. IEEE Int. Conf. on Robotics and Automation*. 667–672.

[BEL91] G.M. Bone, M.A. Elbestawi, R. Lingarkar, L. Liu. 1991. Force control for robotic deburring, *ASME J. Dynamic Systems, Meas. and Control.* 113 (3): 395–400.

[Ben97] D. Benarieh. 1997. Task management in a multi-robot environment, *Computer Integrated Manufacturing Systems.* 10(2): 123–131.

[BeP90] W.K. Belvin, K.C. Park. 1990. Structural tailoring and feedback control synthesis: an interdisciplinary approach, *J. of Guidance, Control, and Dynamics.* 13(3): 424–429.

[BeP97] N.P. Belfiore, E. Pennestri. 1997. An atlas of linkage-type robotic grippers, *Mechanism and Machine Design.* 32(7): 811–833.

[BMM85] F. Bonsignorio, R.C. Michelini, R.M. Molfino, P.A. Piaggio. 1985. Polynomial control for assembly robots, *Proc. VII Intl. Symp. Robotics & Automation*, Lugano, Jun. 24–26.

[BoJ85] D.S. Bodden, J.L. Junkins. 1985. Eigenvalue optimization algorithms for structure/controller design iterations, *J. of Guidance, Control, and Dynamics.* 8(6): 697–706.

[BZL89] B. Benhabib, G. Zak, M.G. Lipton. 1989. A generalized kinematic modeling method for modular robots, *J. of Robotic Systems.* 6(5): 545–571.

[CaB97] C. Canudas DeWit, B. Brogliato. 1997. Direct adaptive impedance control, *Automatica.* 33(4): 643–654.

[CBZ90] R. Cohen, B. Benhabib, G. Zak. 1990. Kinematic modeling of modular robots with non-parallel and near-parallel axes units, *Proc. ASME Mechanisms Conference*, DE-Vol. 25, Chicago, Sept. 147–152.

[CCS91] P. Chiacchio, S. Chiaverini, L. Sciavicco, B. Siciliano. 1991. Task space dynamic analysis of multiarm system configurations. *Intl. J. of Robotics Research,* 10(6): 708–715.

[CDM96] M. Callegari, F. Drago, R.M. Molfino, F. Principe, D. Speroni. 1996. High repeatability active vision wrist for 3D shapes measurements, *Proc. 27th Intl. Symp. on Industrial Robots*, Milano, Oct. 6–8. 347–352.

[ChL97] J.H. Chin, S.T. Lin. 1997. The path pre-compensation method for flexible arm robot, *J. Robotics and Computer Integrated Manufacturing.* 13(3): 203–215.

[CLD92] R. Cohen, M.G. Lipton, M.Q. Dai, B. Benhabib. 1992. Conceptual design of a modular robot, *ASME J. Dynamic Systems, Measurement and Control.* 114. 117–125.

[CMP94] P. Cagetti, R.C. Michelini, F. Pampagnin, R. Razzoli. 1994. SIRIAT: an animation module for virtual reality simulation of robotic manipulators, *Proc. 27th ISATA on Mechatronics*, Aachen, Oct. 31–Nov. 4. 609–616.

[CPP96] P.H. Chang, B.S. Park, K.C. Park. 1996. An Experimental Study on Improving Hybrid Position/Force Control of a Robot Using Time Delay Control. *Mechatronics.* 6(8): 915–931.

[Cra89] J.J. Craig. 1989. *Introduction to Robotics: Mechanics & Control*, Addison Wesley, Reading.

[DeL87] J. DeSchutter, J. Leysen. 1987. Tracking in compliant motion automatic generation of the task frame trajectory based on observation natural constraints, *Proc. 4th Int. of Robotics Research.* 215–22.

[DeS89] C.W. DeSilva. 1989. *Control Sensors and Actuators*, Prentice Hall Inc., New Yersey.

[Des96] R.M. DeSantis. 1996. Motion/Force Control of Robotic Manipulators. *ASME J. Dynamic Systems, Measurement and Control,* 118(2): 386–389.

[Dra77] S.H. Drake. 1977. Using compliance in lieu of sensory feedback for automatic assembly, *Proc. IFAC Symp. on Information and Control Problems in Manufacturing Technology,* Tokyo.

[ESG90] S.D. Eppinger, R.P. Smith, D.A. Gebala, D.E. Whitney. 1990. Organizing the tasks in complex design projects, *Proc. ASME Design Automation Conference: Design Theory and Methodology,* Vol. DE–27, Chicago, Sept. 39–46.

[FaH97] Y. Fang, Z. Huang. 1997. Kinematics of a three-degrees-of-freedom in-parallel actuated manipulator mechanism, *Mechanism and Machine Design.* **32**(7): 789–796.

[Fer66] W.R. Ferrell. 1996. Delayed force feedback, *Human Factors.* 449–455.

[FFM97] L. Ferrarini, G. Ferretti, C. Maffezzoni, G. Magnani. 1997. Hybrid Modeling and Simulation for the Design of an Advanced Industrial Robot Controller. *IEEE Robotics & Automation Magazine.* **4**(2): 45–51.

[FGL87] K.S. Lu, R.C. Gonzales, C.S.G. Lee. 1987. *Robotics: Control, Sensing, Vision and Intelligence.* McGraw-Hill, New York, NY.

[FiM92] W.D. Fisher, M.S. Mujtaba. 1992. Hybrid position/force control: a correct formulation. *Intl. J. of Robotics Research,* **11**(4): 299–311.

[FWY86] S. Fortune, G. Wilfgong, C. Yap. 1986. Coordinated motion of two robot arms, *Proc. IEEE Int. Conf. on Robotics and Automation,* San Francisco, Apr. 7–10. 1216–1223.

[GHW83] R.E. Gustavson, M.J. Hennessey, D.E. Whitney. 1983. Designing chamfers, *Robotics Research.* **2** (4): 3–18.

[GKY84] S.D.V. Gruzdev, O.B. Korytko, E.I. Yurevich. 1984. Modular electro-mechanical industrial robots, *Elektroteknika.* **55** (4): 4–7.

[GrM61] D. Graham, D. McRuer. 1961. *Analysis of Non-Linear Control Systems,* John Wiley, New York.

[GrR88] S.C. Graves, C.H. Redfield. 1988. Equipment selection and task assignment for multiproduct assembly system design, *Int. J. of Flexible Mfr. Sys.* **1**: 31–50.

[Gus88] R.E. Gustavson. 1988. Design of cost-effective assembly systems, *Proc. Successfully Planning and Implementation of Flexible Assembly Systems,* SME, Mar., Ann Arbor, MI.

[HaA77] H. Hanafusa, H. Asada. 1977. A robotic hand with elastic finger and its application to assembly processes, *Proc. IFAC Symp. on Information Control Problems in Production Engineering,* 127–138.

[HaN89] A.M.A. Hamdan, A.H. Nayfeh. 1989. Measure of modal controllability and observability for first- and second-order linear systems, *J. of Guidance, Control, and Dynamics.* **12**(3): 421–428.

[HeK91] M.G. Herr, H. Kazerooni. 1991. Automated robotic deburring of parts using compliance control, *ASME J. Dynamic Systems, Meas. and Control.* **113**(1): 60–66.

[Hic85] P.K. Hickman. 1985. An Analysis of Burrs and Burr Removal on Aircraft Engine Parts. SB Thesis. MIT.

[HiH83] G. Hirzinger, J. Heindl. 1983. Sensor programming: a new way for teaching robot parts and forces/torques simultaneously, *Proc. 3rd Int. Conf. on Robot Vision and Sensory Controls.* 549–558.

[HiL85] G. Hirzinger, K. Landzettel. 1985. Sensory feedback structures for robots with supervised learning, *Proc. IEEE Int. Conf. on Robotics and Automation.* 627–635.

[HKS92] T. Hamilton, A. Kondoleon, D. Seltzer. 1992. Automation of inertial instruments, C.S. Draper Lab report P-3190, *Presented at Joint Services Data Exchange for GN&C,* Palm Springs, Oct.

[Hog79] N. Hogan. 1979. Adaptive stiffness control in human movement, *ASME J. Advances in Bioengineering,* 53–54.

[Hog80] N. Hogan. 1990. Control of mechanical impedance of prosthetic arms, *Proc. JACC.*

[Hog81] N. Hogan. 1981. Impedance control of a robotic manipulator, *Winter Annual Meeting of the ASME,* Washington.

[Hog85] N. Hogan. 1985. Impedance control: an approach to manipulation, Part I-III, *ASME J. Dynamic Systems, Measurement, and Control.* **107**(1): 1–23.

[HoG87] R. Hollowell, R. Guile. 1987. An analysis of robotic chamfering and deburring, *ASME Winter Annual Meeting: Robotics Theory and Applications.*

[HuJ86] S.S. Hussaini, D.E. Jakopac. 1986. Multiple manipulators and robotic workcell coordination, *Proc. IEEE Int. Conf. on Robotics and Automation,* San Francisco, Apr. 7–10. 1236–1241.

[HWM86] R. Harrison, R.H. Weston, P.R. Moore, T.W. Thatcher. 1986. Industrial applications of pneumatic servo-controlled modular robots, *Proc. 1st National Conf. on Production research (U.K.)*. 229–236.

[IOY94] M. Ichinohe, K. Ohara, K. Yamaguchi, K. Maeda. 1994. Development of deburring robot for cast iron with vision and force sensing, *Proc. 24th Intl. Symp. on Industrial Robots*. 49–54.

[IYI96] S. Ito, H. Yuasa, K. Ito, M. Ito. Energy-based pattern transition in quadrupedal locomotion with oscillator and mechanical model, *Proc. Intl. IEEE Conf. Man and Cybernetics*, Beijing. 2321–2326.

[Jen86] L.M. Jenkins. 1986. Telerobotic work system: space robotics applications, *Proc. IEEE Int. Conf. on Robotics and Automation*. 804–806.

[JKL97] H.E. Jenkins, T.R. Kurfess, S.J. Ludwick. 1997. Determination of a dynamic grinding model. *ASME J. Dynamic Systems, Measurement and Control*, **119**(2): 289–293.

[Jok97] T.A.E. Jo Ko. 1997. A dynamic surface roughness model for face milling, *Precision Engineering*, **20**(3): 171–178.

[KAG96] M. Katayama, K. Asada, X.Z. Zheng, M. Yamakita, K. Ito. 1996. Self-organisation of a task oriented visuo-motor map for a redundant arm, *Proc. IEEE Conf. Emerging Technologies and Factory Automation*, Hawaii. 302–308.

[Kam83] L.J. Kamm. 1983. Recent applications of modular technology robots, *Proc.,13th Int. Symp. on Industrial Robots*. 11.66–11.74.

[Kaz87] H. Kazerooni. 1987. Automated robotic deburring using electronic compliance impedance control, *Proc. IEEE Intl. Conf. on Robotics and Automation*, Raleigh, USA, Mar. 31-Apr. 3. 1025–1032.

[Kaz89] H. Kazerooni. 1989. On the Robot Compliant Motion Control. *ASME J. Dynamic Systems, Measurement and Control*, **111**(3): 416–425.

[KBK86] H. Kazerooni, J.J. Bausch, B.M. Kramer. 1986. An approach to automated deburring by robot manipulators, *ASME J. Dynamic Systems, Meas. and Control*. **108**(4): 354–359.

[KeK88] L. Kelmar, P.K. Khosla. 1988. Automatic generation of kinematics for a reconfigurable modular manipulator system, *IEEE Proc. Int. Conf. on Robotics and Automation*. 663–668.

[KHB92] W.S. Kim, B. Hannaford, A. Bejczy. 1992. Force reflection and shared compliant control in operating telemanipulators with time delay, *IEEE Trans. Robotics & Automation*, **8**(2): 176–185.

[KiH86] R. King, R. Hahn. 1986. *Handbook of Modern Grinding Technology*. 34–38.

[KIK90] O. Kashiwagi, K. Ono, E. Izumi, T. Kurenuwa, K. Yamada. 1990. Force-controlled robot for grinding, *IEEE Int. Workshop on Intelligent Robots and Systems*. 1001–1006.

[KiT97] S. Kirk, E. Tebaldi. 1997. Design of robotic facilities for agile automobile manufacturing. *Industrial Robot*. **24**(1): 72–77.

[KKM90] R.L. Kosut, G.M. Kabuli, S. Morrison, Y.P. Harn. 1990. Simultaneous control and structure design for large space structures, *Proc. American Control Conference*. 860–865.

[KKN95] A. Kato, N. Kondo, N. Narita, K. Ito, Z.W. Luo. 1995. Compliance control of direct drive manipulator using ultrasonic motor, *Theory and Practice of Robots and Manipulators*. 125–130.

[Koi89] A.J. Koivo. 1989. *Fundamentals for Control of Robotic Manipulators*, John Wiley, New York, NY.

[Kov97] J. Kovecses. 1997. Joint motion dynamics and reaction forces in flexible link robotic mechanism, *Mechanism and Machine Design*. **32**(7): 869–880.

[KuW92] T.R. Kurfess, D.E. Whitney. 1992. Predictive control of a robotic grinding system. *ASME J. Dynamic Systems, Measurement and Control*, **114**(4): 412–420.

[KWB88] T.R. Kurfess, D.E. Whitney, M.L. Brown. 1988. Verification of a dynamic grinding model, *ASME J. Dynamic Systems, Meas. and Control*. **110**(4): 403–409.

[LAM88] H.G. Lee, S. Arimoto, F. Miyazaki. 1988. Liapunov stability analysis for PDS control of flexible multi–link manipulators, *Proc. 27th Conf. on Decision and Control*. 75–80.

[LBH89] D.K. Lindner, J. Babendreier, A.M.A. Hamdan. 1989. Measure of controllability, observability and residues, *IEEE Trans. Automatic Control*, **34**(6): 648–650.

[LeR87] G. Legnani, R. Riva. 1987. Kinematics of modular robots, *Proc. World Congress on Mechanisms and Machine Theory*, Spain. 1159–1162.

[Li97] Y. Li. 1997. Hybrid control approach to the peg-in-hole problem. *IEEE Robotics & Automation Magazine.* **4**(2): 52–60.

[LiA92] S. Liu, H. Asada. 1992. Transferring manipulative skills to robots: representation and acquisition of tool manipulative skills using a process dynamics model. *ASME J. Dynamic Systems, Measurement and Control,* **114**(2): 220–228.

[LiG93] K.B. Lim, W. Gawronski. 1993. Actuator and sensor placement for control of flexible structures, *Control and Dynamic Systems: Advances in Theory and Applications,* C.T. Leondes Ed., Academic Press.

[LII96a] Z.W. Luo, K. Ito, M. Ito, A. Kato. 1996. Dynamic co-operative manipulation of flexible objects, *Japan-USA Symp. Flexible Automation,* Boston. 229–232.

[LII96b] B.L. Lu, K. Ito, M. Ito. 1996. Solving inverse kinematics problems of redundant manipulators in an environment with obstacles, using separable nonlinear programming, *Proc. Japan-USA Symp. Flexible Automation,* Boston. 79–82.

[LII96c] Z.W. Luo, K. Ito, M. Ito, A. Kato. 1996. On co-operative manipulation of dynamic objects, *J. Advanced Robotics.* **10**(6): 621–636.

[LiJ89] K.B. Lim, J.L. Junkins. 1989. Robust optimization of structural and controller parameters, *J. of Guidance, Control, and Dynamics.* **12**(1): 89–96.

[LIK96] Z.W. Luo, M. Ito, A. Kato, K. Ito. 1996. Nonlinear robust control for compliant manipulation on dynamic environment, *J. Advanced Robotics.* **10**(2): 213–227.

[LuI96] B.L. Lu, K. Ito. 1996. A parallel and modular multi-sieving neural network architecture with multiple control networks, *Proc. Intl. IEEE Conf. Man and Cybernetics,* Beijing. 1303–1308.

[LWP80] J.Y.S. Luh, M.W. Walker, P.R. Paul. 1980. On-line computational scheme for mechanical manipulators, *ASME J. of Dynamic Systems, Measurement and Control.* **102**: 69–76.

[LYK97] S.H. Lee, B.J. Yi, Y.K. Kwak. 1997. Optimal kinematic design of an anthropomorphic robot module with redundant actuators, *Mechatronics.* **7**(5): 443–464.

[MAC93] R.C. Michelini, G.M. Acaccia, M. Callegari, R.M. Molfino. 1993. Virtual reality technique for the development and integration of robotic manipulators, *Proc. 9th Intl. Conf. on CAD/CAM, Robotics and Factories of the Future,* St. Petersburg, May 17–20. 425–430.

[MaC96] L. Markov, R.M.H. Cheng. 1996. Conceptual design of robotic filament winding complexes. *Mechatronics.* **6**(8): 881–896.

[MAC97] R.C. Michelini, G.M. Acaccia, M. Callegari, R.M. Molfino, R.P. Razzoli. 1997. Dynamics of a cooperating robotic fixture for supporting automatic deburring tasks, *Proc. Intl. Conf. Informatics and Control,* St. Petersburg, Jun. 9–13, 1244–1254.

[MAC98] R.C. Michelini, G.M. Acaccia, M. Callegari, R.M. Molfino, R.P. Razzoli. 1998. Techniques in computer integrated assembly for cost effective developments, in *Computer Aided and Integrated Manufacturing Systems Techniques and Applications,* Cornelius T. Leondes, Ed., Gordon & Breach Publ., Newark, NJ, 1998.

[Mak80] H. Makino. 1980. Research and development of the SCARA robot, *Proc. 4th Intl. Conf. on Production Engineering,* Tokyo, Japan Society of Precision Engineering. 885–890.

[MaR89] T. Marilier, J.A. Richard. 1989. Non-linear mechanic and electronic behavior of a robot axis with a harmonic drive gear, *J. of Robotics and Integrated Manufacturing.* **213**(5): 129–136.

[McK91] P.J. McKerrow. 1991. *Introduction to Robotics.* Addison-Wesley, Sydney.

[MCR96] R.C. Michelini, M. Callegari, G.B. Rossi. Robots with uncertainty and intelligent automation, *Proc. 2nd ECPD Int. Conf. on Advanced Robotics, Intelligent Automation and Active Systems,* Vienna, Sept. 26–28. 31–39.

[MHS97] B.J. McCarragher, G. Hovland, P. Sikka, P. Aigner, D. Austin. 1997. Hybrid dynamic modeling and control of constrained manipulation systems. *IEEE Robotics & Automation Magazine.* **4**(2): 27–44.

[Mic92] R.C. Michelini. 1992. Decision anthropocentric manifold in robotic manufacturing, *Proc. 4th ASME Intl. Symp. on Flexible Automation,* San Francisco, USA, Jul. 12–15. 467–474.

[MiI96] J.K. Mills, J.G.-L. Ing. 1996. Dynamic modeling and control of a multi-robot system for assembly of flexible payloads with applications to automotive body assembly. *J. Robotic Systems.* **13**(12): 817–836.

[Mil96] J.K. Mills. 1996. Simultaneous control of robot manipulator impedance and generalized force and position. *Mechanisms and Machine Theory.* **31**(8): 1069–1080.

[MiS87] D.F. Miller, J. Shim. 1987. Gradient-based combined structural and control optimization, *J. of Guidance, Control, and Dynamics,* **10**(3)

[MMA83] R.C. Michelini, R.M. Molfino, G.M. Acaccia. 1983. The development of modular simulation procedures for the design of task-dependant industrial robots, *Proc. Intl. Symp. Robotics & Automation,* Lugano, Jun. 22–24.

[MPM78] R.C. Michelini, P.L. Polledro, C. Marcantoni Taddei. 1978. Position steering of industrial robots by statistical controllers, *Proc. 8th Intl. Symp. on Industrial Robots* Stuttgart, May 31–Jun. 1.

[MuM84] R. Muck, J.A.G. Mammern. 1984. Modular mechanical engineering, IFS, *Proc. Int. Conf. on Advances in Manufacturing.* 271–282.

[MuP97] P. Muraca, P. Pugliese. 1997. A variable structure regulator for robotics, *Automatica,* **33**(7): 1423–1426.

[NeW78] J.L. Nevins, D.E. Whitney. 1978. Computer controlled assembly, *Scientific American,* **238**(2): 62–74.

[Nil69] N. Nilsson. 1969. A mobile automation: an application of artificial intelligence techniques, *Proc. Int. Joint Conf. on Artificial Intelligence.* 509–520.

[NoH89] S.Y. Nof, D. Hanna. 1989. Operational characteristics of multi-robot systems with cooperation, *Intl. J. Production Researches.* **27**(3)

[NWD89] J.L. Nevins, D.E. Whitney, T.L. De Fazio, R.E. Gustavson, A.C. Edsall, R.W. Metzinger, W.A. Dvorak. 1989. A strategy for the next generation in *Concurrent Design of Products and Processes in Manufacturing,* McGraw-Hill, New York, NY.

[OKA97] F. Ozturk, N. Kaya, O.B. Alankus, S. Sevinc. 1997. Machining features and algorithms for set-up planning and fixture design, *Computer-Integrated Manufacturing Systems.* **9**(4): 207–216.

[PaA91] J.H. Park, H. Asada. 1991. Dynamic analysis of noncollocated flexible arms and design of torque transmission mechanisms, *Proc. American Control Conference.* 1885–1890.

[PaA94] J.H. Park, H. Asada. 1994. Concurrent design optimisation of mechanical structure and control for high speed robots, *ASME Trans. J. Dynamic Systems, Measurement and Control.* **116**: 244–256.

[Pau8] R.P. Paul. 1981. *Robot Manipulator: Mathematics, Programming, and Control,* The MIT Press, Mass.

[Pek64] J. Peklenik. 1964. Contributions to the theory of surface characterisation, *CIRP Annals.* **12**: 173–178.

[Pel96] M. Pelletier. 1996. Synthesis of hybrid impedance control strategies for robot manipulators. *ASME J. Dynamic Systems, Measurement and Control,* **118**(3): 566–571.

[PeV96] W. Persoons, P. Vanherck. 1996. A process model for robotics cup grinding, *CIRP Annals.* **45**(1): 319–325.

[RaC81] M.H. Raibert, J.J. Craig. 1981. Hybrid position/force control of manipulators, *ASME, J. of Dynamic Systems, Measurement and Control,* **102**(2): 126–133.

[Riv88] E.I. Rivin. 1988. *Mechanical Design of Robotics,* McGraw Hill, New York.

[RLC97] W.B. Rowe, Y. Li, X. Chen, B. Mills. 1997. Case-based reasoning for selection of grinding conditions, *Computer-Integrated Manufacturing Systems.* **9**(4): 197–205.

[ROB97] S. Reignier, F.B. Ouezdou, P. Bidaud. 1997. Distributed method for inverse kinematics of all-serial manipulators, *Mechanism and Machine Design.* **32**(7): 855–867.

[RoM66] R.A. Rothchild, R.W. Mann. 1966. An EMG controlled force sensing proportional rate elbow prosthesis, *Proc. Symp. on Biomedical Engineering,* Milwaukee.

[RoS92] A. Romiti, M. Sorli. 1992. A parallel 6 d.o.f. manipulator for cooperative work between robots in deburring, *Proc. 23rd Intl. Symp. on Industrial Robotics:*Barcelona. 437–442.

[SaK97] V. Santibanez, R. Kelly. 1997. Strict Lyapunov functions for control robotic manipulators, *Automatica*. **33**(4): 675–682.

[Sal80] J.K. Salisbury. 1980. Active stiffness control of a manipulator in cartesian coordinates, *Proc. 19th IEEE Conf. on Decision and Control*. Albuquerque

[She86] T. Sheridan. 1986. Merging mind and machine, *Technol. Rev*. **23**(7): 32–40.

[ShS84] V.I. Shub, M.K. Selder. 1984. Pneumatic modular industrial robots, *Elektrotekhnika*. **55**(4): 7–9.

[Sim75] S.N. Simunovic. 1975. Force information in assembly processes, *Proc. 5th Intal. Symposium on Industrial Robots*, Chicago, IL

[SmC82] R.C. Smith, K. Cazes. 1982. Modularity in robotics: technical aspects and applications, *IFS Proc., Int. Conf. on Robotics in the Automotive Industry (UK)*. 115–122.

[SoC93] M. Sorli, M. Ceccarelli. 1993. On the workspace of a 6 d.o.f. platform with three articulated double-parallelograms. *Proc. Intl. Conf. on Advanced Robotics*: Tokyo. 147–152.

[SpF91] V.A. Spector, H. Flashner. 1991. Modeling and design implications of noncollocated control in flexible systems, *ASME J. Dynamic Systems, Measurement and Control*. **112**: 186–193.

[SSG87] T.M. Stepien, L. Sweet, M. Good, M. Tomizuka. 1987. Control of tool-workpiece contact force with application to robotic deburring, *IEEE Trans. on Robotics and Automation*. **3**.

[Ste81] D.V. Steward. 1981. The design structure matrix, *IEEE Trans. Eng. Mgt.*, **28**(3): 71–74.

[StS90] D.M. Stokic, D. Surdilovic. 1990. Simulation and control of robotic deburring, *Intl. J. of Robotics and Automation*. **5**(3): 107–114.

[TeB89] D. Tesar, M.S. Butler. 1989. A generalized modular architecture for robot structures, *ASME J. of Manufacturing Review*. **2**(2): 91–117.

[UIH87] M. Uchiyama, N. Iwasawa, K. Hakomori. 1987. Hybrid position/force control for the coordination of two-arms robot, *IEEE Conf. Robotics and Automation*, Raleigh, USA

[VaP96] H. VanBrussels, W. Persoons. 1996. Robotic deburring of small series of castings, *CIRP Annals*, **45**(1): 405–410.

[Ver83] S.A. Vere. 1983. Planning in time: windows and durations for activities and goals, *IEEE Trans. on Pattern Analysis and Machine Intelligence*. **5**(3): 246–266.

[VuK89] M. Vukobratovic, N. Kirkanski. 1989. *Real-Time Dynamics of Manipulation Robots*, Springer Verlag, New York, NY.

[WaD75] P.C. Watson, S.H. Drake. 1975. Pedestal and wrist force sensors for industrial assembly, *Proc. 5th Intl. Symposium on Industrial Robots*, Chicago, IL.

[Wat76] P.C. Watson. 1976. A multidimensional system analysis of the assembly process as performed by a manipulator, *Proc. 1st North American Robot Conference*, Chicago, IL.

[WET90] D.E. Whitney, A.C. Edsall, A.B. Todtenkopf, T.R. Kurfess, A.R. Tate. 1990. Development and control of an automated robotic weld bead grinding system, *ASME J. Dynamic Systems, Meas. and Control*. **112**(2): 166–176.

[Whi69a] D.E. Whitney. 1969. Resolved motion rate control of manipulators and human prostheses, *IEEE Trans. Man-Machine Systems*. **10**(2): 47–53.

[Whi69b] D.E. Whitney. 1969. State space models of remote manipulation tasks, *IEEE Trans. Automatic Control*. **14**(6): 617–623.

[Whi72] D.E. Whitney. 1972. Mathematics of coordinated control of prosthetic arms and remote manipulators, *ASME J. Dynamic Systems, Measurement and Control*. **93**(4): 303–309.

[Whi77] D.E. Whitney. 1977. Force feedback control of manipulator fine motions, *ASME Journal of Dynamic Systems, Measurement and Control*. **99**(2): 91–97.

[Whi82] D.E. Whitney. 1982. Quasi-static assembly of compliantly supported rigid parts, *ASME J. Dynamic Systems, Measurement and Control*. **104**: 65–77.

[Whi93] D.E. Whitney. 1993. From robots to design, *ASME Trans. J. Dynamic Systems, Measurement and Control*. **115**: 262–270.

[WhR86] D.E. Whitney, J.M. Rourke. 1986. Mechanical behavior and design equations for elastomer shear pad remote center compliances, *ASME Journal of Dynamic Systems, Measurement and Control*. **108**(3): 223–232.

[WhT92] D.E. Whitney, E.D. Tung. 1992. Robot grinding and finishing of cast iron stamping dies, *ASME J. Dynamic Systems, Measurement and Control.* **114**: 132–140.

[WKT90] D.E. Whitney, T.R. Kurfess, A.B. Todtenkopf, M.L. Brown, A.C. Edsall. 1990. Development and control of an automated robotic weld bead grinding system, *ASME J. Dynamic Systems, Measurement and Control.* **112**(2): 166–176.

[WLY96a] J.Q. Wu, Z.W. Luo, M. Yamakita, K. Ito. 1996. Adaptive hybrid control of manipulators on uncertain flexible objects, *J. Advanced Robotics.* **10**(5): 469–485.

[WLY96b] J.Q. Wu, Z.W. Luo, M. Yamakita, K. Ito. 1996. Gain scheduled control of robot manipulators for contct tasks on uncertain flexible objects, *Proc. Intl. IEEE Conf. Man and Cybernetics,* Beijing. 41–46.

[WLY96c] J.Q. Wu, Z.W. Luo, M. Yamakita, K. Ito. 1996. Adaptive hybrid control for a robot interacting with uncertain flexible environments, *13th IFAC World Congress,* San Francisco. 235–240.

[Wur86] K.H. Wurst. 1986. The conception and construction of a modular robot system, *IFS, Proc. Int. Symp. on Industrial Robotics,* Belgium. 37–44.

[YHM94] T. Yoshikawa, K. Harada, H. Murakami. 1994. Dynamic hybrid position/force control of flexible arms by macro-micro manipulator systems, *Proc. ISCIE-ASME Intl. Conf. Flexible Automation,* Jul. 11–18. 65–72.

[YLI96] M. Yamakita, Z.W. Luo, K. Ito. 1996. Potential field representation of environment model and its application to robot's force/position hybrid control, *Proc. Intl. IEEE Conf. Emerging Technologies and Factory Automation,* Hawaii. 316–321.

[Yos93] T. Yoshikawa. 1993. Dynamics shaping in robot force control and artificial reality, *Proc. Intl. Conf. on Advanced Robotics,* Tokyo. 3–8.

[YoZ93] T. Yoshikawa, X.-Z. Zheng. 1993. Coordinated dynamic hybrid position/force control for multi-robot manipulators handling one constrained object. *Intl. J. Robotics Research,* **12**(3): 219–230.

[ZLJ87] Y.F. Zheng, J.Y.S. Luh, P.F. Jia. 1987. A real time distributed computer system for coordinated motion control of two industrial robots, *IEEE Conf. Robotics and Automation,* Raleigh, USA.

[ZOY96a] X.Z. Zheng, K. Ono, M. Yamakita, M. Katayama, K. Ito. 1996. Trajectory planning and control for robotic batting/catching tasks, *Proc. Japan-USA Symp. Flexible Automation,* Boston. 17–23.

[ZOY96b] X.Z. Zheng, K. Ono, M. Yamakita, M. Katayama, K. Ito. 1996. A control structure for robotic dynamic manipulation, *Intl. IEEE Conf. Man and Cybernetics,* Beijing. 1489–1494.

[ZOY96c] X.Z. Zheng, K. Ono, M. Yamakita, M. Katayama, K. Ito. 1996. A robotic dynamic manipulation system with trajectory planning and control, *Intl. IEEE Conf. Emerging Technologies and Factory Automation,* Hawaii. 309–315.

8

Object-Oriented Techniques and Automated Methods for Robotic Assembly in Manufacturing Systems

Samuel Pierre
École Polytechnique Montréal

Monjy Rabemanantsoa
École Polytechnique Montréal

Wilfried G. Probst
Université du Québec

8.1 Introduction

Object-oriented techniques and automated methods are used in various situations for improving characterization and solutions of certain types of engineering problems related to manufacturing systems. Conceptual assembly, particularly robotic assembly, are fields where these techniques and methods are widely applied [8, 11].

Conceptual assembly is usually a complex task involving geometric and physical constraints between components which requires a large amount of information and computation. A mechanical assembly is a set of interconnected parts representing a stable unit in which each part is a solid object. Surface contacts between parts reduce the degrees of freedom for relative motion. A subassembly is a non-empty subset of these parts, having one or more elements in which every part has at least one surface contact [24, 25].

In most industrial organizations, the construction of a mechanical assembly is typically achieved by a series of assembly operations (i.e., the insertion of a bolt into a hole). The first stage in programming an assembly system is to identify the operations necessary to manufacture the given assembly and to specify the sequence in which they are to be performed. The generation of such an ordered sequence of operations is called the *assembly planning problem* [20].

The problem of robotic assembly planning can be viewed as a problem of generating assembly sequences dedicated to computer-integrated manufacturing. In fact, most computer-integrated manufacturing systems develop structures based on technical criteria for acquiring and processing data and for determining how well those systems support design documentation requirements. A flexible robotic assembly is often achieved by integrating both data processing and knowledge processing for use by manufacturing engineers in offline programming from an assembly plan, which is an ordered sequence of operations that constructs the product from its component parts [25]. The determination of a plan involves searching in a state space where a feasible course of action would transfer the parts from their initial states to goal states.

There exists in the literature various approaches to performing assembly planning [27, 28]. Some of these generate only monotone and linear plans while others generate monotone and non-linear plans. Other plans deal with non-monotone and non-linear assembly planning using the technique of *ray tracing* in which parts are represented by the Constructive Solid Geometry (CSG) model [15, 30]. A plan is linear if each operation performed in it moves only one part at a time. In cases where groups of parts can be moved together, plans are considered to be non-linear and are referred to as *subassemblies.*

Considering certain combinatorial aspects related to plan generation, and hence to robotic assembly, many researchers have proposed the use of heuristic-search techniques and intelligent-automated methods to deal with robotic assembly in manufacturing systems. This chapter follows this orientation and presents various object-oriented techniques, which perform robotic assembly in manufacturing systems. It is organized as follows: Section 8.2 summarizes the fundamentals of robotic assembly and discusses some robotic systems. Section 8.3 expounds on the basic principles of object-oriented techniques. Section 8.4 analyzes certain robotic and automated assembly architectures. Finally, Section 8.5 describes some automated assembly applications.

8.2   Intelligent Assembly Planning and Knowledge Representation

Planning is a process that searches for a sequence of actions (or operations) in order to achieve a goal statement. This original objective is normally too complex or too abstract to be accomplished through a single operation, so it is often necessary to break it down into smaller, simpler subgoals. In an ideal situation, a subplan for each subgoal can be formulated independently and a complete plan can be derived by simply combining the subplans. However, in practice, subplans often interact (i.e., the achievement of certain goals may actually prevent the accomplishment of others). This conflict problem presents major difficulties in a planning process but Artificial Intelligence techniques may be helpful in addressing such difficulties [5, 16, 29].

Structure of an Intelligent Assembly Planning System

An intelligent assembly planning system can be structured as a typical knowledge-based system [3]. As shown in Fig. 8.1, it contains a knowledge base, a control structure, and a blackboard. The knowledge base records information about the assembly problem domain and the expertise of the assembly planning. It includes workpiece structures, assembly operations, and assembly principles.

The workpiece structures describe the relationships among workpiece components and their associated properties; whereas, the assembly operations are descriptions of robot actions that can be used to assemble workpieces. The assembly principles can be viewed as "rules of thumb," (i.e., empirical guidelines of assembly experts), including the methods of ordering the component assembly sequence and posting constraints should certain situations arise [5]. The control structure utilizes knowledge-base information to generate assembly plans and integrates a structure analyzer and a plan generator.

As shown in Fig. 8.2, the planning process consists of two phases: structure analysis and plan generation [5]. During structure analysis, the system analyzes the workpiece structures according to the guidelines of assembly principles. This includes: refining an abstract workpiece (one that can be disassembled into

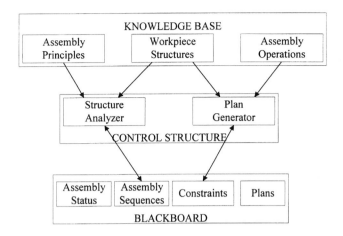

FIGURE 8.1　Components of a Mechanical Assembly Planning System.

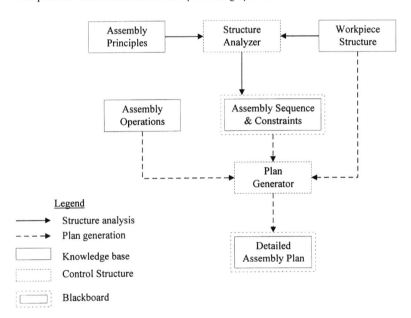

FIGURE 8.2　Different Phases of a Planning Process.

more than one part) into sub-components; deciding on the subcomponent assembly sequence; and posting assembly operation constraints for each subcomponent. A plan generator subsequently follows analysis information in order to select assembly operations. The status of the assembly domain, the properties of the components, and the constraints determine the selection of operations.

Finally, the blackboard records intermediate hypotheses and decisions, which the system manipulates. This includes: the current state of the assembly or assembly status; components being assembled or assembly sequences; and constraints and plans.

Knowledge Representation in Intelligent Robotic Systems

The area of robotics and automation systems has been undergoing an expanded revolution over the past three decades. At first, computing and sensory capabilities were not present in robotic systems. The first generation evolved into systems with limited computational and feedback capabilities. The second generation began including systems with multi-sensory and decision-making capabilities. The third generation robotic systems, called intelligent robotic systems, are equipped with a diverse set of visual and non-visual sensors and are being designed to adapt to changes within their workspace environment [23].

Regarding intelligent robotic systems, their modeling requires utilization and implementation of concepts and ideas drawn from various fields. In point of fact, Artificial Intelligence and, more precisely, knowledge-based systems as well as Operations Research and Control System Theory, constitute the background required to tackle the multiple challenges of such a modeling. Industrial automation systems, Flexible Manufacturing Systems (FMS), Computer Integrated Manufacturing (CIM) systems in general as well as telerobotic systems, are among the more frequent applications [24].

In intelligent robotic systems, knowledge may be represented either symbolically or numerically. Symbolic knowledge representation includes: rule-based, frame-based, associative networks, logic-based, and object-oriented systems. Numerical representation refers to mathematical models and is contingent upon the specific application domain.

For intelligent robotic and automation systems, time may or may not be critical in knowledge representation techniques depending on the level of knowledge processing. For example, the time factor is critical in knowledge-based systems for control applications as most of these have been developed to enhance, monitor, or modify online the dynamic system operation. Generally, an offline knowledge-based system may be developed for high level planning and decision making and a real-time system for online information processing. However, overall timing constraints may have to be taken into account regarding the specific application under consideration [32].

At least two criteria are frequently used to compare and judge knowledge representation schemes for the particular types of systems [14]:

1. Expressive power of the representation in comparison with other representation schemes and the range of control problems for which the representation is suitable; and
2. Computational complexity of reasoning, using particular representations, in comparison with other representations and the limitations of its applicability imposed by issues of complexity.

Recently, the topic of intelligent robots has attracted a great deal of attention among researchers. Considering the social situation of the lack of highly skilled, experienced technicians and workers, it would seem logical to automate the manufacturing and fabricating process. Proceeding from this idea, Kamrani et al. [13] have proposed an intelligent knowledge-based system as a solution to the problem of selecting an optimal robot for cell design. New tasks are being defined for robots in order to meet the challenges of flexible manufacturing systems. Associated with this is an increasing variety of robots from which to choose. The selection of an optimal robot for a particular task constitutes a major problem.

Various parameters must be considered and the user should choose an industrial robot whose characteristics satisfy the requirements of the intended tasks. Massay et al. [17] have proposed a hierarchical design approach, which can serve as a basis for the off-line development of effective robotic systems.

On the other hand, a hardware and software methodology leading to a knowledge-based system has been introduced in [32] for use in intelligent robotic systems. This knowledge-based system has been derived for the organizational level of such a system and is being used to develop off-line plan scenarios to execute a user-requested job. The knowledge base contains all the information related to the class of problems that the system has been designed to solve, while the inference engine operates upon the knowledge base. Although separate entities, the knowledge base and inference engine have been so designed as to enable them to operate closely together as a unit.

Similarly, an expert support system has been proposed to provide flexible manufacturing systems (FMS) with the capability of adjusting in real-time to changes in the manufacturing environment. The key component of this support system is the "information cell," which is controlled by the flow of information between the cell and its environment.

The concept and prototype of a hierarchically structured knowledge-based system for coordinated control of a welding robot and a positioning table is presented in [31]. The knowledge-based system determines appropriate weld parameters (voltage, speed, feedrate) on the basis of the job description. It is also capable of planning the optimum table orientation and robot trajectory and controlling the welding process [19].

An integrated environment for intelligent manufacturing automation has been proposed in [3]. The knowledge-based integrated environment is based on the following components: interface to external environment, meta-knowledge base, database, inference mechanism, static blackboard, and interface to other subsystems. This approach has been applied to the product design in a manufacturing process.

Planning a sequence of robot actions is especially difficult when the outcome of actions is uncertain, as is inevitable when interacting with the physical environment. Christiansen and Golbert [6] have compared two algorithms for automatic planning by robots in stochastic environments: an exponential-time algorithm maximizing probability and a polynomial-time algorithm maximizing a lower bound on the probability. They have considered the case of finite state and action spaces where actions can be modeled as Markov transitions. As these algorithms trade off plan time for plan quality, their performance is compared to a mechanical system for orienting parts. This leads to two properties of stochastic actions which can be used to choose between these planning algorithms for other applications.

Regardless of specific applications, the knowledge-based systems described above have been developed to enhance the capabilities and flexibility of robotic systems dedicated to industrial automation. Their common limitation is that they are not easily modifiable to accommodate different versions or potential deviations from the initial class of problems for which they have been designed [32]. Object-oriented techniques integrating concepts, such as class, instance, inheritance, and genericity, can deal with such a limitation.

8.3 Principles of Object-Oriented Techniques

Object-Oriented (OO) technology constitutes one of the most important software evolutions of the 1990s. It refers to object-oriented programming, design, analysis, and databases, and covers methods for either design or system analysis. An overview of such a technology is presented in this section.

Basic Concepts and Definitions

In the field of object-oriented techniques, designers think in terms of objects: they create objects; add behaviour to them; make them interact; and observe the results. From a programming point of view, an *object* represents anything real or abstract (whose attributes are represented by data types and behaviors) which are controlled by operations. An object is defined by a list of abstract attributes often called *instances* or *class variables*. Communication between objects is done across well-defined *interfaces*.

Figure 8.3 illustrates an object, whereas Fig. 8.4 gives an object-oriented example of an "Employee." Moreover, *objects* can be categorized into object types, which is specified during object-oriented analysis. *Classes* refer to the software implementation of such object types. Details of classes are determined in object-oriented design.

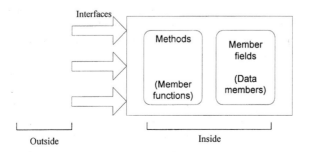

FIGURE 8.3 Concept of an Object.

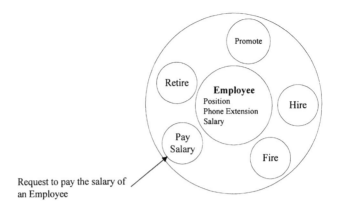

FIGURE 8.4   An Example of an Object.

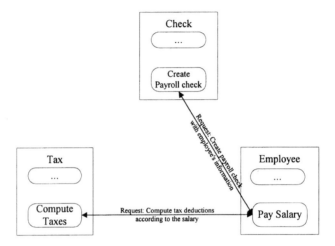

FIGURE 8.5   Messages Passing among Classes.

In order to interact with an object, a request is sent to it, causing an operation to be triggered. Such operations are called *member functions* or *methods* in object-oriented programming and are elicited from responses to messages transmitted to objects. Therefore, a method constitutes a procedural specification that alters the state of an object or determines the message to be successfully processed by such an object.

Fig. 8.5 illustrates the message passing between three classes of objects. Concretely, a class represents a collection of objects sharing attributes and methods.

Encapsulation refers to the practice of including everything within an object that it requires, in such a way, that no other object ever needs to be aware of its internal structure which resulted in packaging together its data and operations. Consequently, details of its implementation are hidden from its user. This is referred to as *information hiding*, in that the object conceals its data from other objects and allows the data to be accessed via its own operations.

An *abstract data type* (ADT) is an abstraction similar to a class that describes a set of objects in terms of an encapsulated or hidden data structure. All interfacing occurs through operations defined within the ADT, providing a well-defined means for accessing objects of a data type and running the appropriate method. Therefore, although objects know how to communicate with one another across well-defined interfaces, they are not normally allowed to know how other objects are implemented, thus protecting their data from arbitrary and unintended use. To summarize, the ADT gives objects a public interface through its permitted operations. Furthermore, all operations that apply to a particular ADT also apply to the instances of that ADT.

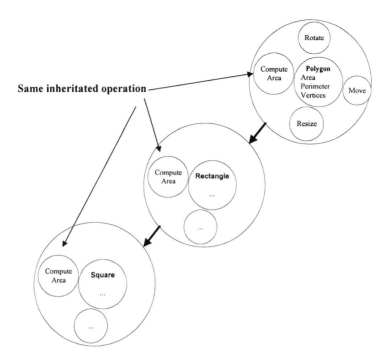

FIGURE 8.6 Example of Inheritance.

Furthermore, deriving new classes from existing classes is called *class inheritance* (or simply *inheritance*) and constitutes an implementation of generalization. The latter states that the properties of an object type apply to its subtypes which have the derived classes. With single inheritance, a class is derived from a one-base class.

In multiple inheritance, a class can inherit data structures and operations from more than one superclass. Fig. 8.6 illustrates a simple inheritance, where the class *Rectangle* inherits several operations from *Polygon* and *Square* inherits certain operations from *Rectangle*. The real strength of inheritance comes from the ability to make data structures and operations of a class physically available for reuse by its subclasses.

Polymorphism refers to the ability of objects, belonging to different classes but related by inheritance, to respond dissimilarly to the same member function call. It enables the writing of programs in a general fashion in order to handle a wide variety of existing and specified related classes. One strength of polymorphism is that a request for an operation can be made without knowing which method should be invoked. It is particularly effective for implementing layered software systems. For instance, in operating systems, each type of physical device may operate quite differently from the others, whereas commands to read or write data from and to those devices can have a measure of uniformity.

Therefore, object-oriented techniques provide facilities for representing two impressive concepts: abstraction and inheritance. They also change the way designers or developers think about systems. In fact, this way of thinking is more natural for most people than the techniques of structured analysis and design. As well, all entities consist of objects that have certain types of behavior. Certain objects may be quite different despite sharing many of the same attributes and exhibiting similar behaviors.

Special Features

In the domain of object-oriented techniques, analysts, designers, and programmers use the same conceptual models for representing object types and their behaviors. They all draw hierarchies of such object types where the subtypes share the properties of their parent. Also, they consider objects as being composed of other objects, use generalization or encapsulation, and think about events changing the states of objects and triggering certain operations.

Program complexity is a measure of program understandability. Avoiding such complexity improves program reliability and reduces the effort needed for its development and maintenance. Object-oriented techniques provide two important ways to divide complex software into simple procedures. First, methods result in a state change of an object which is usually simple and easy to implement. Secondly, each operation is isolated from cause and effect. In fact, after being triggered by a number of events, an operation executes the associated method to change the state of an object but does not recognize its cause and effect. That is to say, the operation has no knowledge of what event caused it to happen, nor why, and is unaware of what operations will be set off as a consequence of its events. Therefore, reduction in complexity is partly due to the fact that object-oriented classes can be self-contained and divided into methods.

Moreover, systems can be built from existing objects since inheriting operations from a superclass enables code sharing and structure reuse among classes. When creating a new class, instead of completely rewriting data members and member functions, the programmer can designate the new class to inherit them from a previously defined class. This leads to a high degree of reusability, which saves money, shortens development time, and increases system reliability.

Today, software factories manage well as possible libraries of reusable classes, to maximize reusability and minimize maintenance costs. Developers who create new classes are evaluated according to the re-use of such classes. On the other hand, designers who utilize existing classes are evaluated regarding the way they re-use those classes. In other words, developers or designers should strive to increase the functionality of the classes they use by building new objects from existing objects. Moreover, object types can be designed for customization according to the needs of various systems, resulting in a method that can be reused on different software platforms and on multiple interacting processors.

Object-oriented techniques allow for the building of complex and reliable systems in a simple way. The code is directly related to defined objects and to methods that manipulate those objects. Each method is relatively simple; easy to change; and can be quickly created using pictures, tables, declarative statements, equations, database access, report generation, rules, and nonprocedural techniques. As systems become increasingly complex, new functionality can be directly added to existing classes for building these powerful systems. Therefore, by using good class libraries, a designer can assemble complex software in a fluid way. There is no need to be concerned with the internal workings of a class. If an object type functions well, it can be treated as a black box whose interior never necessitates scrutiny. Furthermore, created objects can be quickly and repeatedly modified.

Obviously, object-oriented techniques result in numerous advantages. First, each object in a system turns out to be relatively small, self-contained, and manageable. This reduces the complexity of systems development and may lead to higher quality systems, which are less expensive to build and maintain. In addition, once an object is defined, implemented, and tested, it can be reused in other systems. Indeed, reusability can greatly increase productivity since the reused objects are generally proven products. Finally, an object-oriented system can be modified or enhanced very easily by changing some types of objects or by adding new types of objects without interfering with the rest of the system. Those potential benefits constitute the driving force behind the object-oriented revolution.

Methodology and Models

Object-oriented techniques enable the building of real-life models that include two important features: a representation of the object types with their structures and a representation of the object behaviors. A set of diagrams should be created for each of these aspects: object-structure diagrams showing the objects and their interrelationships, and event diagrams showing what happens to the objects. The first aspect concerns object types, relationships between objects, and inheritance. It refers to the Object Structure Analysis (OSA) and corresponds to the class structure design. The second view concerns the behavior of objects and what happens to them over time, which refers to the Object Behaviour Analysis (OBA) and method design.

The OSA defines the object types and the way in which they are associated. This leads to the following questions:

- What types of objects exist? What are their functions and how are they related? What are the useful subtypes and supertypes?
- Is a certain kind of object composed of other objects?
- Such questions allow for the identification of the classes and the definition of superclasses, subclasses, their inheritance relationships and the methods to be used, which result in the detailed design of the data structure.

The OBA relates to the following questions:

- Which states can the object types be in? What types of events change these states?
- What succession of events occurs?
- What operations result in these events and how are they triggered?
- Which rules govern the actions taken when events occur?

These questions lead to the detailed design of methods, using either procedural or nonprocedural techniques. Consequently, the input to the code generator is developed, screen design is done, dialogues between objects are designed and generated, and prototypes are built.

Events cause an object-oriented system to take various actions. In order to describe processes in terms of events, triggers, conditions, and operations, event diagrams constitute a primary means of communicating object-oriented behavior, showing events and the operations set in motion by the methods. They express processes in a more rigorous fashion which makes the code easy to be generated using a sequence of operations (drawn with rounded-cornered boxes and easily understood). Such diagrams must be precise (with an engineering-like precision), which can speed up work, improve the results, enhance creativity, and simplify maintenance. For strategic-level planning, an Object-Flow Diagram (OFD) is useful for indicating the constructed objects and the activities that produce and exchange them. A three-dimensional box is used to represent real-life objects that flow between activities.

In an object-oriented context, the data structure of an object type can be manipulated only through the methods of the object class. As for changing the state of an object, requests must be sent in order to activate the associated methods. Each state change is usually simple to program, which leads to the division of programming into relatively uncomplicated parts. Also, each object performs a specific function independently of the others, and responds to requests neither knowing the reasons for such requests nor the consequences of their actions. As a result, classes can be largely changed independent of other classes, which renders them relatively easy to test and to modify.

8.4 Applications to Robotic and Automated Assembly

In a robotic and assembly manufacturing environment, the available flexibility introduces another degree of complexity in decision making. The assembly operations are descriptions of robot actions that can be used to match components and subcomponents. It is important that robot planners not be obliged to analyze an assembly but rather to use a path-planning generated by the assembly in order to structure the issuing of task level commands. To achieve this, the architecture must be designed to retrieve data through contents and not through a fixed predefined structure. "Which components have a champfix and five edges?" is a type of question which requires an answer that object-oriented databases can satisfactorily provide.

Object-Oriented Modeling

The way object-oriented databases reason about geometry is by recognizing certain geometric features of objects [18]. To this end, representations of designed objects in terms of features must be developed. As a result, an important use of the database is the design of a knowledge-based system. Learning techniques can also be used to obtain relevant information automatically and to guide the system in search of good planning solutions [2].

As presented earlier, object-oriented concepts are based on fundamental principles of complexity management which are very helpful for complex object modeling and data manipulation related to complex entities such as computer-aided design (CAD) or computer-aided engineering (CAE) system environment [1, 10]. In the assembly domain, the objects are typically composites or aggregates of components. The term "composite object" is commonly used to denote a layered abstraction model. The use of abstract data types enhance program modularity since modifying the object structure does not affect the manipulation of external objects. Abstract data type can be implemented as an object collection with the same structure and representing a class.

Each object is characterized by a specific identifier OID (object identifier) [25]. However, two similar objects with the same dimensions, but a dissimilar OID, are said to be different. Thus, through the development phases of composite objects, the OID provides a natural paradigm for maintaining the uniqueness of objects independent of structure or content.

Object identity allows for direct graph modeling and objects are characterized by properties. A property may be an object characteristic such as an attribute, a function, or a subobject component. For example, the object *circle* can have the following properties:

- Simple attribute: radius (R);
- Composite attribute: center (x, y, z);
- Function: surface (πR^2).

A class may be defined as a description of the behavior of a collection of objects in a modular way (e.g., circles with the same radius, polygons wth a common shape). The concept of "class hierarchy" eliminates the possibility of specifying and storing redundant information. It integrates the previous notions of "superclass" as a higher level mode; "subclass" as a lower level mode; and "methods" as operations that can either retrieve or update the state of an object.

For assembly process planning, methods are very useful in an inheritance hierarchy (a method defined for a class is inherited by its subclasses). Data integrity is ensured by a procedure called "demon," a triggered operation associated with an object when a particular condition occurs. In this context, geometrical and topological information is not sufficient for establishing assembly planning design rules. Geometric interference, physical constraints, tolerances, and kinematic relationships are all important issues to be addressed in assembly planning design.

Geometry helps to make a representation scheme for parts and products, and it refers to the places where the objects are located. Topology defines the connectivity between geometric elements and is in charge of processing the rules for connecting elements of geometry to produce a part. It also refers to the reasons why objects are located in some specific places. Tolerances can be defined as acceptable variations between design and manufacturing processes that allow for the various components to be assembled into a product. As a result, the use of an object-oriented approach provides the tool, without any interactive means, for modeling entities and properties, as well as for managing design information effectively. Fig. 8.7 shows an object-oriented representation of a superclass SOLID-3D.

Knowledge Representation

Various approaches have been proposed in assembly-planning literature [1, 4, 7, 9, 28]. Rabemanantsoa and Pierre [21–23] have also proposed an approach based on the idea of viewing the planning with intermediate states and considering the whole assembly as a hierarchy of structures. Therefore, the knowledge base integrated in our system consists of three parts: component structures, intermediate states, and primitive structures. These parts contain the physical information related to knowledge of the design intent such as geometry topology, features, and tolerances. Each component is represented by a frame, which consists of a name and a number of properties. For example, Fig. 8.8 shows the panel frame of a mechanical part: number of faces, number of flat surface, composite surface, etc.

Class: Solid_3D
Inherit: Object
Class structure:

SuperClass	:	(Sphere, Ellipsoid, Paraboloid, Cone, Cylinder, Hyperboloid)
Class	:	(Subsphere, Subellipsoid, Subparaboloid, SubCone, Subcylinder, Subhyperboloid)
SubClass	:	(Arcs, Edges, Vertices, Points)

Structure:

Attribute, Transformation

FIGURE 8.7   Object-Oriented Representation of a Complex Object.

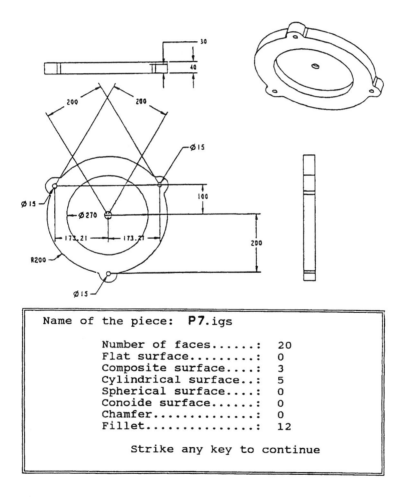

FIGURE 8.8   Panel of Mechanical Parts.

To characterize the intermediate states, the input scheme is represented by a set of assembly operations along with their states. In this context, the state of a component is defined as either the initial component or another position induced by a set of predefined operations [20]. Given a product p with "n" components, we have $P = (p_1, p_2, \ldots, p_n)$, and the set of states of each component is $S = (s_1, s_2, \ldots, s_k)$ for "k" maximum states.

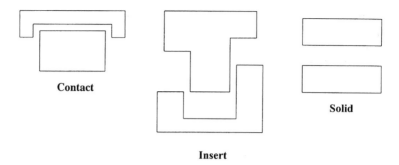

FIGURE 8.9   Primitive structures.

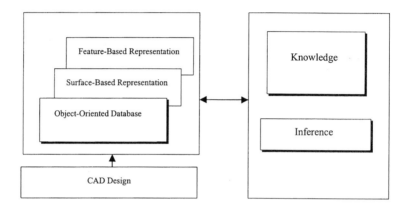

FIGURE 8.10   Database Structure.

Let the set of predefined assembly trajectories for each component in P be $T = (t_1, t_2, \ldots, t_j)$ where j is the total number of trajectories. The knowledge contains a change of state along a fixed trajectory, that is, for component $p_i \in P$, an assembly operation is denoted as a triple:

$$A = \{P, S, T\}$$

The primitive structures are necessary to ensure the completeness of the knowledge. They involve a small number of parts or components which interact positionally with one another in simple ways [21]. The three types of primitive structures currently being considered are shown in Fig. 8.9:

1. Contact: in this structure; two components interact in such a way that they are in contact and each component can be moved into the structure in any direction through an infinite half-space.
2. Insert: in this structure, two components interact in such a way that each one has to mate in a specific direction if the other is already present.
3. Solid: this structure is similar to (1), except for the two components not being in contact with each other.

Another way to consider the knowledge representation is using a CAD database. This architecture provides different levels of abstraction for modeling the parts through the object-oriented database illustrated in Fig. 8.10.

The link structure of the system between the database and the CAD design uses the method developed by Shah and Bhatnagar [27]: intrinsic and explicit characteristics (type 1); intrinsic and implicit characteristics (type 2); and extrinsic characteristics (type 3). For example, a hole should be described in terms of diameter, length, and an orientation vector; these parameters are type 1 because they may be available in the database itself. The radius of a hole is type 2 because it may be derived from the center coordinates and the diameter. However, if the angle between the axes of two holes of equal diameter is needed, this type 3 information is not available in the database itself, but would depend on the concept utilized in the model.

The Artificial Intelligence (AI) techniques relating to *decision tree* and *production rule* provide better understanding of the surfaces. For example; if two surfaces have the same ordinate, the AI techniques make it possible to check whether these surfaces are adjacent or in contact. The positions of the surfaces depend on the spatial relationships between each pair of components by using the coordinate frame (body axis system) attached to the basic component. The knowledge is expressed in terms of mobility (*M*) and contact functions for each pair of parts. The two functions *C* and *M* are then used by an expert system called XGEN to generate assembly sequences [25].

As defined earlier, a mechanical assembly is a composition of interconnected parts forming a stable unit. Interconnection implies that one or more surfaces are in contact. The body axis system has six directions attached to each basic component. Contact and mobility of one component with respect to another are then evaluated for each of the *n* components within the six degrees of freedom in a spatial relationship. This results in a combination of two into *n*.

As illustrated in Fig. 8.11, the contact C of component "*b*" with respect to component "*a*" is a function $C(a, b) = (V_1, V_2, V_3, V_4, V_5, V_6)$, and the relative mobility *M* of component "*b*" with respect to component "*a*" is a function $M(a, b) = (V_1, V_2, V_3, V_4, V_5, V_6)$. The part union angle θ is used to express the contact angle, while the rotation ϕ is used to express the relative mobility angle. A part mating along an assembly axis having $\theta_i = 0$ and $\phi_i = 0$ is defined as an orthogonal assembly trajectory; when $\theta_i \neq 0$ and $\phi_i \neq 0$, we have a non-orthogonal assembly trajectory. An attachment by means of a screw is defined as a circular assembly trajectory. Hence, contact and mobility functions are expressed as follows:

$$C(a,b) = (V_1 : \theta_1, \ V_2 : \theta_2, V_3 : \theta_3, \ V_4, V_5, V_6) \text{ and}$$

$$M(a,b) = (V_1 : \phi_x, \ V_2 : \phi_y, V_3 : \phi_z, \ V_4, V_5, V_6)$$

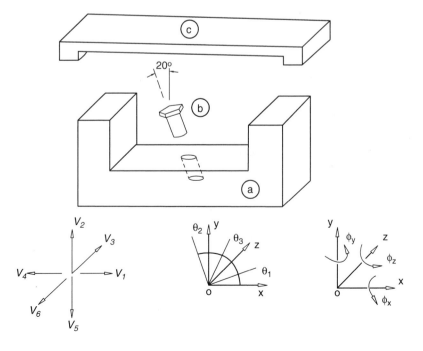

FIGURE 8.11 Example of a Non-Orthogonal Assembly Trajectory.

TABLE 8.1   Knowledge Representation

C: Contact	M: Mobility
$C(a,b) = (2,1{:}20,2,2,1,2)$	$M(a,b) = (0,4,1{:}{-}20,0,0,0)$
$C(a,c) = (2,0,0,2,1,0)$	$M(a,c) = (0,1,1,0,0,1)$
$C(b,c) = (0,0,0,0,0,0)$	$M(b,c) = (1,1,1,1,0,1)$
$\theta_1 = 0;\ \theta_2 = 20°;\ \theta_3 = 0$	$\phi_y = 0;\ \phi_z = -20°$
Note:	Note:
0 = no contact	0 = no relative mobility
1 = presence of contact	1 = possibility of moboility
2 = lateral contact	3 = circular trajectory
	4 = non-orthogonal trajectory

Figure 8.11 is an example of a non-orthogonal assembly trajectory developed with this system. For the set of parts composed of three components, there are 3! = 6 possible solutions. To overcome the *combinatorial explosion* risk, our approach uses V_k ($k = 1,6$) to represent the six degrees of freedom. The spatial relationship of one component, with respect to another one, considers one pair of components, that is, the combination of 2 into n, where $n = 3$. The referential {R2} provides an "explicit relative mobility condition." Both referentials are attached to the lower component when considering each pair of components. Furthermore, from the semantic data modeling point of view, each V_k ($k = 1,6$) is coded.

In the contact function C, we have $V_k = 0,1,2$ representing the absence of contact, presence of contact, and presence of lateral contact respectively. When we take into account these mating conditions, the contact function of component "*b*" in relation to component "*a*" (Fig. 8.11) is expressed as follows: $C(a,b) = (2,1{:}20,2,2,1,2)$, in which $V_1 = V_3, = V_4 = V_6 = 2$, indicating the presence of lateral contact along these directions. ($V_2{:}\theta_2$) = (1: 20) indicates the contact angle and the assembly trajectory at a positive angle equal to 20° with respect to $V_2 = Y$ axis. $V_5 = 1$ shows contact between "*b*" and "*a*" in that direction.

In the relative mobility function M, we have $V_k = 0,1,2,3,4$, representing no relative mobility, the possibility of mobility, presence of an orthogonal assembly trajectory, existence of a circular trajectory, and presence of a non-orthogonal assembly trajectory respectively. Hence, the relative mobility of component "*b*" in relation to component "*a*" is $M(a,b) = (0,4,1{:}20,0,0,0)$ in which $V_1 = V_4 = V_5, V_6 = 0$, indicating that "*b*" cannot be moved along these directions. $V_2 = 4$ signifies that there is a non-orthogonal assembly trajectory using $V_2 = Y$ as reference axis. ($V_3{:}\phi_z$) = (1: −20) indicates that the relative mobility is at −20° (clockwise rotation) with respect to the $V_3 = Z$ axis. The knowledge representation for spaital data models, relating to the example of Fig. 8.11, is shown in Table 8.1.

Assembly Operations

The plan generation formulates the assembly operations, using the knowledge base plus a set of scheduling rules, to achieve the goal. Using learning techniques, the supervision architecture is given capabilities for generating the plan. For each plan level, the main functions consist of dispatching, monitoring, diagnosis, and recovery [21]. The result is a decision tree:

- Dispatching: taking care of global coordination activities to be performed by a high-level cell controller. Approaches to derive the control algorithm are usually based on constant rules used to post constraints. Dispatching is the basic "operation-selection."
- Monitoring: in model-based monitoring, the sensory conditions to test in each situation are well defined in the model. Moreover, discrete monitoring is used to check goal achievement after the execution of operations. Continuous monitoring is used to check sensory conditions during the execution of operations. If an exception is detected, the diagnosis function is called upon.
- Recovery: this function is called to update the global state and the complete operation description is stored in the plan.

Other Automated Assembly Applications

The object-oriented techniques and automated methods eliminate the traditional steps of off-line programming. The assembly architecture, designed through artificial intelligence, solves the use of advanced curve and surface definition features in the relationship between the shapes to be manipulated.

For many years, some predefined tasks in the robotic field have been solved using sensor-based reasoning and/or the trajectory-generation algorithm [9, 12, 33]. These procedures use the constraints of the system to implement operation sequences for achieving specified goals. The basis for establishing a set of design rules for robotic and automated assembly yielded some more generalized design guidelines in 1986 [11]. In 1988, Sanderson et al. [26] introduced a method for the task planning of robotic manipulation in space application.

The major driving applications are part handling and product design. Effective and efficient utilization of assembly techniques requires that the robot task sequence be compatible with the design of components. Then, experimentation was carried out on an automated assembly of mechanical parts in a high mix product environment. The concept of replacing a certain standard simulation technique to provide motion descriptions and task parameters has been carried out by the automated assembly. It may be viewed as the tool of choice for a growing range of applications. The knowledge state of the manipulation motions and operations are defined and validated before accessing the real robot. 3-D modeling has also streamlined the tooling stage. The system is incorporating embedded task planning, with accompanying control synthesis and past architecture, to support goal-directed activities in an uncertain environment. Surface modeling has been defined and produced within tolerances as small as tenths of millimeters in order to be installed on production assemblies and still meet design requirements.

When provided with the assembly tree, the robot planner is capable of considering the objects in a given coordinate frame to deduce which features of a component mate with which others. Fig. 8.12 presents the assembly tree structure, which is handled by a robot planner to perform the specific operations related to Fig. 8.11. Fig. 8.13 shows the capability of the system to handle various levels of component complexities. According to the assembly theory, there are $6! = 720$ possible solutions excluding constraints for the six components.

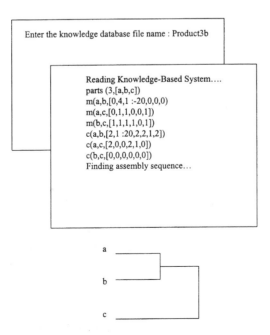

FIGURE 8.12   Robotic Assembly Sequence of the Example in Fig. 8.11.

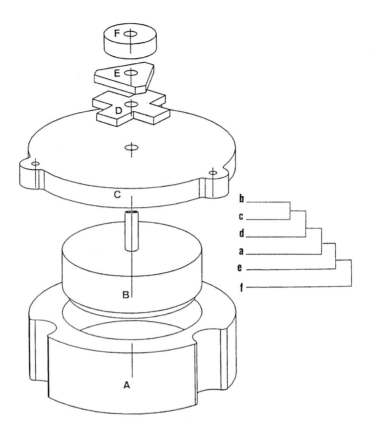

FIGURE 8.13   Equipment Layout with its Robotic Assembly Sequence.

8.5   Conclusion

This chapter discussed the object-oriented techniques and automated methods currently utilized for robotic assembly in manufacturing systems. More precisely, it addressed the problem of knowledge representation in intelligent systems dedicated to assembly planning and robotic assembly. Many real assembly-planning systems are built on the paradigm of artificial intelligence and are implemented as knowledge-based systems. Some of them aim at generating mechanical assembly sequences, which satisfy a set of assembly requirements and may be assembled by humans or robots. All the feasible assembly sequences can be generated as a tree sequence structure and then evaluated, the most appropriate one being chosen to carry out the equipment layout.

For managing geometrical data, topological and abstraction, object-oriented modeling appears to be a suitable approach. According to this course of action, data are stored as examples of abstract data types and all conceptual entities are objects. A class inherits all instances from its superclass. In this context, the question of robotic assembly planning has been discussed as a problem of generating assembly sequences derived from a model-based object recognition. Thus, the output is a list of feasible assembly sequences for use by offline robotic programming to determine the manipulator path generation. A predefined path generation can be very helpful for applications in the fields of telerobotics, machining operations, and automated assembly systems.

Robotic assembly is a difficult process. In fact, numerous transformations must be performed to capture the object features through the set of coordinate frames for each surface. Such a task can become a time-consuming iterative motion. Finally, there is a need to minimize the kinematic computation by providing the robot with knowledge of sequence generation of the components to be assembled.

References

1. Angermuller, G. and Hardeck, W. (1987), CAD Integrated Planning for Flexible Manufacturing Systems with Assembly Tasks, *IEEE CAD Journal,* pp. 1822–1826.
2. Camasinka-Matos, L.M., Seabra Lopes, L., and Barata, J. (1994), Execution Monitoring in Assembly with Learning Capabilities, *IEEE International Conference on Robotics and Automation,* pp. 272–280.
3. Cha, J., Rao, M., Zhou, Z., and Guo, W. (1991), New Progress on Integrated Environment for Intelligent Manufacturing Automation, *Proceedings of the 6th IEEE International Symposium on Intelligent Control,* Arlington, VA.
4. Chakrabasty, S. and Wolter, J. (1994), A Hierarchical Approach to Assembly Planning, *IEEE International Conference on Robotics and Automation,* Vol.1–4, pp. 258–263.
5. Chang, K.H. and Wee, W.G. (1993), A Knowledge-Based Mechanical Assembly Planning, in *Expert Systems in Engineering Applications,* S.G. Tzafestas (Ed.), pp. 291–306.
6. Christiansen, A.D. and Goldberg, K.Y. (1995), Comparing Two Algorithms for Automatic Planning by Robots in Stochastic Environments, *Robotica,* Vol. 13, No. 6, pp. 565–573.
7. Conradson, S., Weinstein, M., Wilker, J.S., and Yencho, S.A. (1987), Automated Material Handling, (Automated Assembly and Product Design as a System), *IEEE Proceedings of the 8th Int. Conf. on Assembly Automation,* pp. 67–68.
8. Ferland, M., O'Shea, J., and Rabemanantsoa, M. (1993), Robotic Interface for CAD/CAM Integration. *Proceedings of Dept. of National Defense on Workshop on Advanced Tech.* in *Knowledge-Based Systems and Robotics,* Ottawa, ON, Canada.
9. Homem de Mello, L.S. and Sanderson, A.C. (1989), A Correct and Complete Algorithm for the Generation of Mechanical Assembly Sequences, *IEEE Journal of Int. Conf. on Robotics and Automation,* CH2750-8, pp. 56–61.
10. Hurt, J. (1989), A Taxonomy of CAD/CAE Systems, *Manufacturing Review,* Vol. 2, No. 3, pp. 170–178.
11. Jackson, A.J. and McMaster, R.S. (1986), Product Design for Robotic and Automated Assembly, *Proc. of IEEE International Conference on Robotics and Automation,* Vol. 2.
12. Jarvis, R.A. (1988), Configuration Space Collision-Free Path Planning for Robotic Manipulators, *Robots in Australia's Future Conference,* ARA, pp. 193–204.
13. Kamrani, A.K., Shashikumar, S., and Patel, S. (1995), An Intelligent Knowledge-Based System for Robotic Cell Design, *Computers & Industrial Engineering,* Vol. 29, pp. 1–4.
14. Kokar, M., Anderson, C., Dean, T., Valavanis, K., and Zadrozny, W. (1990) Knowledge Representations for Learning Control, *Proceedings of the 5th IEEE International Symposium on Intelligent Control,* Philadelphia, PA, USA.
15. Lee, Y.C. and Fu, K.S. (1983), A CSG Based DBMS for CAD/CAM and its Supporting Query Language, *IEEE CAD Journal,* pp. 123–128.
16. Marque-Pacheu, G., Gallausiaux, J.M., and Jormier, G. (1984), Interfacing Prolog and Relational DBMS, in *New Applications of Databases,* E. Gelenbe (Ed.), Academic Press.
17. Massey, L.L., Udoka, S.J., and Ram, B. (1995), Robotic System Design: a Hierarchical Simulation-based Approach, *Computers & Industrial Engineering,* Sept., No. 29, pp.1–4.
18. Nnaji, B.O. (1988), A Framework for CAD-Based Geometric Reasoning for Robot Assembly Language, *Int. Journal Proc. Res.,* Vol. 26, no. 5, pp. 735–764.
19. Pfeiffer, F. and Johanni, R. (1987), A Concept for Manipulator Trajectory Planning, *IEEE Journal of Robotics and Automation,* RA-3, 3, pp. 115–123.
20. Rabemanantsoa, M. and Pierre, S. (1993), A Knowledge-Based Approach for Achieving Assembly Tasks, *Proceedings of the 14th Canadian Congress of Applied Mechanics* CANCAM 93, Kingston, ON, Canada, Vol.1, pp. 51–52.
21. Rabemanantsoa, M. and Pierre, S. (1993), An Integrated Knowledge-Based System for Flexible Assembly Process Manufacturing, *2nd. Int. Conf. on Computer Integrated Manufacturing,* Singapore, pp. 789–798.

22. Rabemanantsoa, M. and Pierre, S. (1993), A Knowledge-Based System for Assembly Process-Planning, *IEEE SESS 93 Int. Conf. on Artificial Intelligence*, Brighton, England, pp. 267–272.

23. Rabemanantsoa, M. and Pierre, S. (1993), A Knowledge-Based Approach for Robot Assembly Planner, *IEEE Proceedings of Canadian Conf. on Electrical and Computer Engineering*, Vancouver, BC, Canada, pp. 829–832.

24. Rabemanantsoa, M. and Pierre, S. (1996), Robotic Assembly for Computer-Integrated Manufacturing, *International Journal of Robotics and Automation*, Vol. 11, No. 3, pp. 132–140.

25. Rabemanantsoa, M. and Pierre, S. (1996), An Artificial Intelligence Approach for Generating Assembly Sequences in CAD/CAM, *Artificial Intelligence in Engineering*, Vol. 10, No. 2, pp. 97–107.

26. Sanderson, A.C., Peshkin, M.A., and Homem de Mello, L.S. (1988), Task Planning for Robotic Manipulation in Space Applications, *IEEE Trans. on Aerospace and Electronic Systems*, Vol. 24, no. 5, pp. 619–628.

27. Shah, J. and Bhatnagar, S. (1989), Group Technology Classification from Feature-Based Geometric Models, *Manufacturing Review*, Vol. 2, pp. 204–213.

28. Shah, J.J. and Rogers, M.T. (1988), Functional Requirements and Conceptual Design of the Feature-Based Modeling System, *CAD Journal*, Vol. 5, pp. 9–15.

29. Swift, K.G. (1987), Knowledge-Based Design for Manufacture, Prentice-Hall, London, England.

30. Tsao, J. and Wolter, J. (1993), Assembly Planning with Intermediate States, *IEEE International Conference on Robotics and Automation*, Vol. 1–3, pp. 71–78.

31. Thompson, D.R. and Ray, A. (1987), A Hierarchically Structured Knowledge-Based System for Welding Automation and Control, *Proceedings IEEE International Symposium on Intelligent Control*, Philadelphia, PA, USA.

32. Valavanis, K.P. and Tzafestas, S.G. (1993), Knowledge-Based (Expert) Systems for Intelligent Control Applications, in *Expert Systems in Engineering Applications*, S.G. Tzafestas (Ed.), pp. 259–268.

33. Woodbury, R.F. and Oppenheim, I.J. (1988), An Approach to Geometric Reasoning in Robotics, *IEEE Trans. on Aerospace and Electronic Systems*, Vol. 24, no. 5, pp. 630–645.

9

CAD-Based Techniques in Task Planning and Programming of Robots in Computer-Integrated Manufacturing

Bijan Shirinzadeh
Monash University

The key to future robotic system utilization in flexible assembly systems (FASs) relies heavily on rapid task planning and programming of robots. Furthermore it has become evident that the complete integration of the design, analysis, and off-line verification of robotic operations in computer-integrated manufacturing (CIM) environment is required. These requirements place an important emphasis on techniques and systems for planning and programming of the assembly tasks within the framework of CIM systems.

This chapter presents techniques for integrated task planning and programming of robotic operations. Such techniques generally employ computer-aided design (CAD) platforms to fully exploit their database and modeling capabilities. Various approaches to off-line programming is presented. The techniques for modeling, task specification, and database structuring are described. Such systems require methods for robot-independent program generation and the subsequent simulation and verification of the application programs including calibration of fixtures and local frames and are also presented here.

9.1   Introduction

Dynamic global competition has compelled many manufacturing industries to be more concerned with productivity, cost reduction, and flexibility due to the reduction of the product life cycles [1, 2]. Recent research efforts have focused on the development of CIM systems to increase productivity and resource management. Robots are chosen, in the manufacturing environment, primarily for their flexibility and rapid response to changeovers. However, in order to take advantage of this flexibility, the production engineer needs to rapidly plan the task and develop application programs for the robotic operations.

There exists a large number of "on-line" and "off-line" robot programming systems and languages [3, 4]. Most industrial robots are still programmed using traditional "on-line" programming systems, commonly referred to as "teach-mode-programming." In this approach, a programmer physically moves the robot by a teach-pendant through the desired locations [5]. The locations are then recorded and can be replayed when needed. "On-line" programming of robots is a time-consuming process and not very efficient in a flexible manufacturing environment. Furthermore, the "on-line" programming technique requires the use of the actual robot which is physically put through the desired sequence of actions [6, 7]. This method is mainly used for large batch manufacturing environments.

A more advanced ap roach allows task specification and programming a robot off-line using task planning and off-line programming techniques [5]. The task planning and off-line programming (OLP) systems have the potential to reduce the programming time [8]. These combine computer modeling and simulation to perform task specifications and generate programs off-line. Such systems do not require direct physical access to the robot or its environment during the planning and programming phase [9]. However, they do require reprogramming of critical locations or frames on the actual manufacturing fixtures. This chapter briefly describes the different approaches to programming systems. Such off-line programming systems provide the basic building blocks for more advanced task planning systems and their integration within the CIM environment. This chapter also presents techniques for integrated task planning and verification of robotic operations. The focus of attention will mainly be directed at robotic assembly operations.

9.2   Strategies for Off-Line Programming (OLP)

Text-Level Programming

The most basic level of off-line programming is the manipulator-level or *text-based programming*. At this level, the focus of attention is the manipulator and other associated devices, such as grippers. The task description consists of how the manipulator is to be moved through the required points [9]. This method provides better control over the robot and it is generally identical to the robot native language [10]. It also provides for a relatively cost effective and reliable programming facility. However, writing text-based programs in languages such as VAL II (Staubli) and ARLA (ABB) is not easy. The user must generally think in 3-D space without any visual aids. In addition, the user must have a background in programming, since most manipulator-level languages are usually extensions of languages such as PASCAL and BASIC.

Graphic-Level Programming

Graphic-level programming is a combination of teach-mode and manipulator-level programming techniques [9]. Graphic-level programming eliminates some of the difficulties in 3-D visualization during off-line programming. In graphic-level programming, the user programs the positions to which the robot is required to move by assigning frames on objects and storing their locations [5]. This level of programming is off-line but robot-dependent, requiring the robot model to be a resident in the system [10]. Most commercial systems use this technique and examples of such systems include IGRIP (Deneb), CimStation (SILMA), and ROBCAD (Technomatix). This technique is sometimes referred to as teach-mode programming on screen [11].

Object-Level Programming

In an object-level program, the user decides the order of assembly, approach, and departure vectors, and the user's description of the task is in terms of the objects that have to be handled [9]. The user must provide a set of commands that describes spatial relationships among objects that have to be handled. For example, a situation where a plate is placed on a block with two holes aligned may be described by:

AGAINST/	top of block, bottom of plate;
COPLANAR/	side of block, side of plate;
ALIGNED/	holeA of block, holeB of plate;

This level of programming is off-line and robot-independent; however, translators are still required to generate manipulator-specific programs. Object-level programming systems, such as RAPT, require the user to provide detailed definition of various features; such as top of block, side of plate, holeA, etc. This specification of features on objects must generally be carried out in conjunction with an advanced modeling system. In addition, this level of programming requires special care in the development of spatial relationships and thorough testing of the program [12]. This is mainly due to the fact that the possibility of ambiguous situations, and thus errors in the program, is extremely high using this approach. Research in this area focuses attention on improving the feature specification and integrated modeling capabilities.

Task-Level Programming

A task-level programming system generates motion and strategy plans based on the specified task or the final goal [9]. A task-level programming, sometimes referred to as objective-level programming, requires a very complete model of the robot and its work cell environment. The system must also generate all possible steps for task planning, grasp planning, path planning, trajectory planning, and post-processing [13]. The system must also be able to deal with uncertainties and sensor actions. The user is not required to define spatial relationships between objects, only high-level task commands. In the proposed objective-level programming language, the user specifies the task the robot must achieve.

The user must also provide information such as the parts to be used, their initial layout, and the final assembly layout. The system is responsible for planning how the assembly is to be done; determining the approach and departure vectors; how and where to grasp the parts; how to use sensors; and how to maneuver parts around obstacles. The final result must be simulated on the graphics screen in order to assist with the final decision (i.e., the engineer must have the final say). A fully functional task-level planning and programming system does not yet exist, although it must be emphasized that attention has been focused on research and development of the modules or sub-systems for such task-level programming systems [14]. Therefore, experimental systems with some of the task-level functionalities are being developed by researchers.

9.3 Strategies for Task Planning and Programming

An idealized task planning and programming system consists of a number of sub-systems or sub-modules. Fig. 9.1 shows a schematic diagram of important sub-modules which have received attention within the framework of off-line task planning and programming systems. These include geometric modeling, task specification, grasp planning, path planning, simulation, and calibration. Individual sub-modules will be described within this section.

Geometric Modeling

Task programming systems require a very complete model of the assembly environment to allow computer-aided planning operations to be performed. Therefore, a geometric modeling platform is an essential component of such planning systems (i.e. planners). In fact, a versatile computer-aided task planning and programming system is generally built on an accurate and user-friendly modeling platform [15].

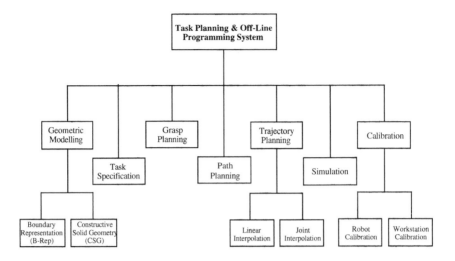

FIGURE 9.1 The hierarchical structure for task planning and off-line programming.

Geometric models must be developed before any task actions can be expanded into sequences of robot operations [16].

Objects in the workspace have to be modeled using solid models rather than wire frames. This approach will assist in retrieving the 3-D information about the objects from the database because this information is required during computer-assisted planning operations. The planning operations may include task specification, grasp planning, path planning, and interference detection of robotic tasks and workpieces [17]. For example, the interference detection may be carried out by testing the intersection of the robot solid model and the union of all other objects. This technique can also be employed for grasp planning which will be described later. Two categories for modeling include constructive solid geometry (CSG) and boundary representation (B-Rep).

Constructive Solid Geometry

Constructive solid geometry (CSG) constructs solid objects from a set of solid geometric shapes [18]. This method uses motional and combinatorial operators to combine a few primitive shapes such as cube, wedge, cone, cylinder, etc. A CSG scheme can be specified by a context-free grammar, as listed below:

<mechanical part>   ∷ = <object>
<object>   ∷ = <primitive> |
                         <object> <motion op> <motion arguments>|
                         <object> <set operator> <object>
<primitive>   ∷ = cube | wedge | cone | cylinder | ...
<motion op>   ∷ = rotate | translate | scale | ...
<set operators>   ∷ = union | intersection | difference | complement | ...

Each primitive of a solid object is described by its volumetric parameters such as length, width, height, etc., and its position relative to a reference frame. These primitives are stored in leaf nodes, as an ordered binary tree and the operators are stored in the non-terminal nodes. During CSG's implementation, the depth of the tree should be monitored very carefully because it can lead to inefficient data retrieval when the tree is too deep. This problem can be overcome by representing each subassembly as a tree in itself and all the subassemblies are combined into the final tree. This method can generally be used for task planning because the final goal can be described as a tree of subassemblies.

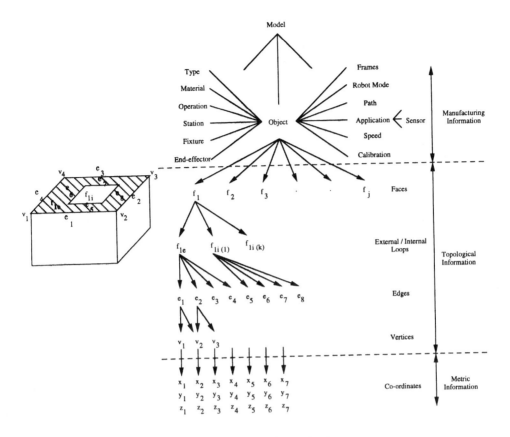

FIGURE 9.2 Structure of boundary representation (B-Rep) for a cuboid.

Boundary Representation (B-Rep)

Boundary representation (B-Rep) technique segments a solid object into its non-overlapping faces or patches, and models these according to their bounding edges and vertices [19]. The resulting data structure constructs a directed graph containing object, face, edge, and vertex nodes. The geometric representation of the object may be described as follows in Fig. 9.2:

- Each face is described by it bounding edges
- Each edge is described by its vertices
- Each vertex is stored as Cartesian coordinates

The external loop of a face (f_{1e}) follows a right hand rule where a face is transversed in counter-clockwise direction. An internal hole in a part may also be considered in B-Rep for possible fixture contact point. The boundary loop (f_{1i}), a hole in a face, is transversed in clockwise direction which is opposite to the external loop [19]. This method produces more efficient sources of geometric data for producing line drawings, graphic interaction (e.g., accessing a face, edge, or vertex), task specification, and graphic simulation as compared to CSG.

Therefore, a model may contain several objects in a scene (i.e., an assembly). Each object is referenced by the topological information and metric information. In addition, strategies are being developed to include information about manufacturing characteristics as a list added to the object characteristics [20]. These may include information such as fixture identification number; manufacturing cell type (e.g., automated, manual); material and manufacturing operation to be performed together with machine type (e.g., robot, milling, etc.); and other programmable aspects (e.g., static frames, dynamic frames, sensor application, calibration, etc.). A schematic diagram of the three types of information list is shown in Fig. 9.2.

Task Specification

Task specification details the task-and-assembly sequence operation for robot programming. The final goal is decomposed into a set of subgoals and is governed by the process sequence, feasibility, and geometric constraints [21]. These constraints imply precedence relationships that will guarantee the correct order for the execution of the operation (i.e., drill object A first before inserting object B into A). Feasibility constraints verifies that all the objects and the extra features, such as "hole in the object," are available or exist. Geometric constraints can be determined by analyzing the geometry of the operation. These constraints are more concerned with the collision problems between a robot and the objects and also between the objects being handled with other objects. Task specification must also work very closely with path planning to ensure a collision-free path.

Many strategies for obtaining the sequence of operations have been developed. These methods are generally based on the three constraints mentioned above. Laperriere et al [22] have developed strategies by using a 'Relation Diagram' method. This method is based on geometric and dimensional information. Rocha and Ramos [21] have developed a strategy called TPMS, where the sequence is derived by using the symbolic plan operation method. Their method accounts for pre-existing manufacturing systems and the robot programming platforms.

Once the sequence has been determined, a computer program which executes that sequence must be generated and stored in the robot controller's memory or operation data files. Each step in the assembly sequence has to be translated into robot executable motions and operations. Every motion and operation are the elements that can be used to change the state of the assembly setup or the environment (i.e., approach object A, close gripper, and screw object A to object B).

The choice of an assembly sequence may depend on additional facts such as time, cost, stability, or ease of assembly. In the past decade, research efforts have focused on the establishment of strategies for automated assembly sequence generation. Such techniques make use of rule-based decision-making approaches to generate precedence for assembly sequence [23]. It must be noted that a complete survey of such research efforts is out of the scope of this chapter. There has also been an increase in research efforts to develop strategies for multiple manipulators [24]. Such studies have developed approaches that regard each manipulator and machine as an agent. The strategy analyzes the behavior and the geometric restriction to sequence of operations for each manipulator and machine.

Grasp Planning

Grasp planning is an important issue that must be considered in order to achieve a truly flexible task planning and off-line programming system. Once the task has been specified, a grasp planner provides all feasible grips of the assembly part and provides an optimal grip [25]. Detailed information, such as description of the part to be grasped, feeder, and assembly task are required in the planning. The planner determines the nonfree regions (Fig. 9.3), based on the gripper size, contact with feeder or assembly, and forbidden part faces such as threaded or fragile faces [26]. These regions are defined as "the regions on a part where the fingertip cannot be placed." The free regions are then used to determine the finger domain, which is an area on a part where a fingertip can be placed.

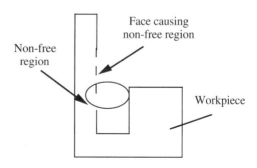

FIGURE 9.3   Non-free region caused by a face on the workpiece.

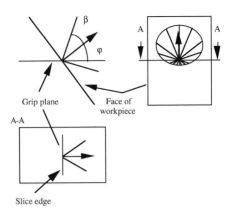

FIGURE 9.4 Schematic illustration of angle β_{gp} of a friction cone in the grip plane.

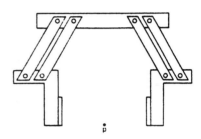

FIGURE 9.5 Schematic diagram of a parallel-jaw gripper.

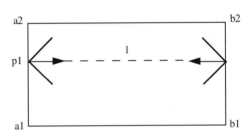

FIGURE 9.6 Grasp domain for a 2-fingered gripper.

A grip plane is a plane where the finger contact points are located. This plane is normally situated on the perpendicular side of the insertion direction, as it will reduce the degrees of freedom [27]. In the finger domain, edges are excluded if grid plane is outside the friction cone belonging to the edge's face, as shown in Fig. 9.4. The calculation of the angle of the friction cone in the grip plane may be performed using the following:

$$\beta_{gp} = \operatorname{atan}((\tan^2\beta - \tan^2\varphi)/(1 + \tan^2\varphi))^{1/2} \tag{9.1}$$

To grasp an object without slipping, it is normally grasped at the centre of mass. In this way, torque on the fingertip surfaces is minimized [27]. However, different types of grippers will have their own effective method of grasping the objects to be manipulated. A parallel-jaw gripper is one of the most commonly used grippers (shown in Fig. 9.5). It uses a parallelogram mechanism and the fingertips extend and retract slightly in order to remain parallel. When the two edges are finally resolved, two grip points can be assigned to the edges by using following equation:

$$P2 = b1 + ((b2 - b1) \cdot (|p1 - a1|)/(|a2 - a1|)) \tag{9.2}$$

These grip points are normally placed on the two parallel surfaces that are not obstructed by other features of the part. The grasping force must be within the friction cones of the grip points as shown in Fig. 9.6 [28]. Thus, an object is normally grasped on the center of mass to minimize the torque on the fingertip surfaces. The length *l* is checked against the maximum gripper opening. If the length *l* is longer than the maximum gripper opening, then the grip points are not feasible and have to be changed. The optimal grasp sites may be analyzed and rated based on the stability, grasp time, and potential mating trajectories [25].

Path Planning

Path planning is an important phase within the task planning process. Research efforts have been mainly focused on establishment of strategies to automatically search all feasible paths and then find the optimal path. The planner automatically excludes those paths that result in collisions between the robot and objects in the work cell, and between objects that are being handled with other objects [29, 30]. There are several advanced strategies for path planning [31]; however, the general path planning can be categorized into three broad techniques: *hypothesize-and-test, penalty function,* and *free space.*

Hypothesize-and-test algorithm hypothesizes a candidate path from start to finish and tests the immediate path for possible collisions. Collision can be detected by examining the intersection between the solid objects. If any collisions are detected, then collision avoidance planning should be executed. This avoidance planner analyzes the obstacles involved in the collision and stores them in the interference data file so that the system is able to avoid the same case. The planner then defines a new immediate configuration by moving the colliding objects. It examines the new path again for possible collision and this is executed recursively until a collision-free path is found.

Penalty function encodes the presence of objects for each manipulator configuration. If there are any collisions, the planner yields an infinite value; however, the value will drop off sharply as the manipulator moves away from the obstacles. Overlaying the workspace with a 3-D grid and calculating the function at each grid point can reduce the computational time.

Free-space finds a path within the known subsets of free space. This method will miss a path if it has not been mapped. The free space is represented by regular geometric shapes with links. The planner only deals with free space rather than the space occupied by obstacles. It detects the free space of the object against the width of the robot plus clearance. If the path is determined to be too narrow, then the path will be eliminated. This procedure is repeatedly executed until an optimum path is found. It must be mentioned that there are also application areas where a desired path is required to be found to follow complex contoured surfaces [32]. Examples of these application areas include deburring, painting, and welding [33–36].

Trajectory Planning

After an appropriate path has been found (i.e., path planning), the task planner has to convert this path description into a time sequence of manipulator configurations. Trajectory planning deals with this problem. There are two types of constraints. These include task related and robot related constraints, shown in Fig. 9.7.

Task constraint is related to the task itself and the time taken for that task to be executed. For instance, a typical task constraint is to maintain the orientation of the gripper unchanged relative to a datum or world coordinate frame during the execution. On the other hand, the time constraint focuses attention

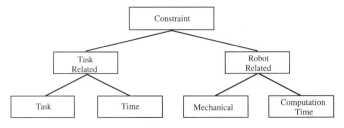

FIGURE 9.7   Constraints for trajectory planning.

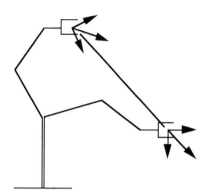

FIGURE 9.8 Example of a robot path undergoing linear interpolation.

on limiting the time period for the action to be performed and also it tries to maximize the equipment use. The robot-related constraint deals with the ability of the system to smooth the chosen path. There are two types of general path-following strategy available: linear interpolation motion and joint interpolated motion. These strategies will be explained, in detail, in the following subsections.

Linear Interpolation Motion

Linear interpolation motion causes the end-effector of the robot to follow the straight line segment drawn between two adjacent points [5]. It continually calculates the inverse transformation in order to obtain joint controller set points. Figure 9.8 illustrates the straight line motion of a revolute robot. The end-effector coordinate frame rotates about an axis k, fixed on the end-effector coordinate frame by an angle θ during the travel of the gripper from the initial to the goal position [37]. The formulation of this method is given below and it is based on the intermediate configuration:

$$P(j) = (P_{goal} - P_{initial}) \cdot (j/n) + P_{initial} \tag{9.3}$$

$$R(j) = R_{initial} \cdot Rot(k, \theta \cdot (j/n)) \tag{9.4}$$

$$T(j) = P(j) \cdot R(j) \tag{9.5}$$

where n represents the number of intermediate configurations plus one, P is the matrix representing the position of the manipulator with respect to the world coordinate frame, R is the matrix representing the rotation of the manipulator with respect to the world coordinate frame, and, T is the matrix describing the transformation of the end-effector coordinate frame with respect to the world coordinate frame.

There are some drawbacks to straight line motion in terms of real time control and this is due to the continual computation of inverse kinematics. Hence, velocity and acceleration controls are needed in order to perform a smooth end-effector movement. However, a predictable trajectory path can be obtained and any collisions during the motion can also be detected.

Joint Interpolation Motion

Joint interpolation strategy calculates the change in the joint variables required to achieve the final location. Figure 9.9 illustrates a joint interpolated motion of a revolute robot. This strategy controls the joint motors so that all the joints reach their final values at the same time [38]. Leu and Mahajan [39] suggested that the joint variables are linear with respect to the path length and the motion should follow the formulation given below:

$$q(j) = [q_{goal} - q_{initial}] \cdot (j/n) \tag{9.6}$$

where j represents the identification number for (i.e., j-th) intermediate configuration increment; q defines the vector representing joint variables; and n is the number of intermediate configurations plus one. It must be emphasized that in the joint interpolation, the end-effector does not produce a predictable

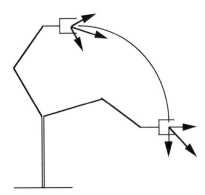

FIGURE 9.9   Example of a robot path undergoing linear interpolation.

trajectory path. This is due to the fact that the joint angles are interpolated between via points; therefore, unexpected collisions between obstacles may occur during the movement.

Calibration

Off-line programming may produce inaccurate robot programs in terms of positioning and orientation [40]. This is due to the differences between the computer models of the manipulator and objects in the working space to that of the actual ones. Offsets of the zero positions of the robot, differences in the link lengths, and misalignment of the joints are some sources of errors [2, 41]. An idealized off-line program-ming system should take into account mechanical deficiencies such as:

- Inaccuracies in the arm element dimensions
- Internal play of joints
- Internal non-linearities of gearing
- Deflection of arm elements and servo-positioning errors

The gap between model and reality can be narrowed by calibrating the robot and obtaining the actual working paths [2]. This can be done by reteaching the robot manually for various working points (e.g., corner of a feeder). A relationship has to be established between the working coordinate system and the measuring coordinate system. A mathematical model for this relationship can be derived based on the nominal coordinate system ($X_{R, i}$), actual coordinate system ($X_{M, i}$), scale (μ), and rotational matrix (D) about x, y, and z axes.

$$X_{R, i} = T + v\, DX_{M, I} \tag{9.7}$$

$$D = D\,(x, \alpha) \cdot D\,(y, \beta) \cdot D\,(z, \gamma) \tag{9.8}$$

where i represents the identification number of position (e.g., i-th position), $X_{R, i}$ represents resultant position vector, $X_{M, i}$ defines the actual measured position vector, and T defines the desired position vector. In general, robot calibration can be classified into two types, static and dynamic [2, 41]. Static calibration identifies those parameters which primarily influence the static positioning of a manipulator. The dynamic calibration identifies those parameters that influence motion characteristics. Various procedures have also been devel-oped to identify calibration parameters. Further, there are several accuracy measurement techniques includ-ing cable measuring system [41], machine vision [42], theodlite system, and laser tracking [43]. Accurate calibration will lead to significant accuracy improvement (generally by a factor of 10).

Post Processing

Once a robot program has been fully developed, it is necessary to translate it into the control language of the target robot [44]. Post processing is divided into three stages: reformatting, translation, and downloading [45]. Reformatting verifies the input file for syntactical errors and adds appropriate motion and function information, determines coordinate frame for command, and establishes sequences for sub-tasks. Translation phase translates the list of information as per rules. This stage deals with the robot's natural language and it will be different for robots that have different format/grammar and thus manipulator-level languages [46]. The program that is generated by the translator must be loaded onto the robot's controller memory. As with any numerically controlled (NC) machines, there are two methods to download the program: downloading the program using a serial interface and downloading the program using a portable medium (e.g., floppy disk) that is compatible with the robot controller.

9.4 CAD-Based Task Planning Implementation

The underlying strategies and individual components for task planning and programming were described in the previous section. This section will focus on procedures and implementation to perform task planning and programming. Although, the procedure has been developed based on a CAD modeling facility, it is similar to a majority of other research-based as well as commercial planning and programming systems [5]. Furthermore, the focus of our attention will be confined to layered and flexible assembly/disassembly operation, as this is one of the most difficult robotic applications.

System Structure

The detailed structure of the system consists of six modules, shown in Fig. 9.10. These include:

- design modeling of assembly parts and work cell
- task specification
- sorting and task decomposition
- verification and simulation
- program transfer to the work cell controller
- work cell controller communication link

Here, we direct our attention to the structure and procedure for a planning and programming system built on a CAD modeling platform (initially on Medusa, currently on AutoCAD). Therefore, assembly parts and fixtures are the components that must be physically manipulated and are required for planning. The assembly parts and other components in the work cell are modeled using the solid modeler in CAD environment. The tasks to be performed are specified during the task planning session. This procedure tags information about the tasks onto the workpiece model. An automated grasp planning may be added at this level. However, for the purpose of the overall system development, the grasp locations are specified manually via frame assignment. Grasp planning is still an important area of research and no commercial systems provide such capabilities.

The planner retrieves the information tagged onto the models and performs sorting and matching of assembly parts for various stages of assembly. This function of the planner, known as access/retrieval, has been the focus of automated object-oriented task decomposition for layered assembly. Thus, the appropriate software modules decompose the higher level task commands and automatically create a "robot-independent" program. It must be emphasized that automated path planning would generally be integrated into the planner at this level of operation.

A trajectory planner solves the inverse kinematic equations that perform linear and joint interpolation. The time and robot related constraints can also be included at this level. This module evaluates the joint angles for linear and joint interpolation and generates a 'robot-dependent' program. The 'robot-dependent' program contains commands and point-to-point motion data files for a specific manipulator. The off-line verification or simulation of the program is the primary aim of any planning and programming system. Therefore, at this level, the user can view the robot movement and assembly operation for final verification and validation of the robot program.

The final process is to download the command and point-to-point motion data files to the work cell controller. The work cell control software retrieves the command and "point-to-point" motion data files, generates commands in a format that is understood by the robot, and sends these commands to the

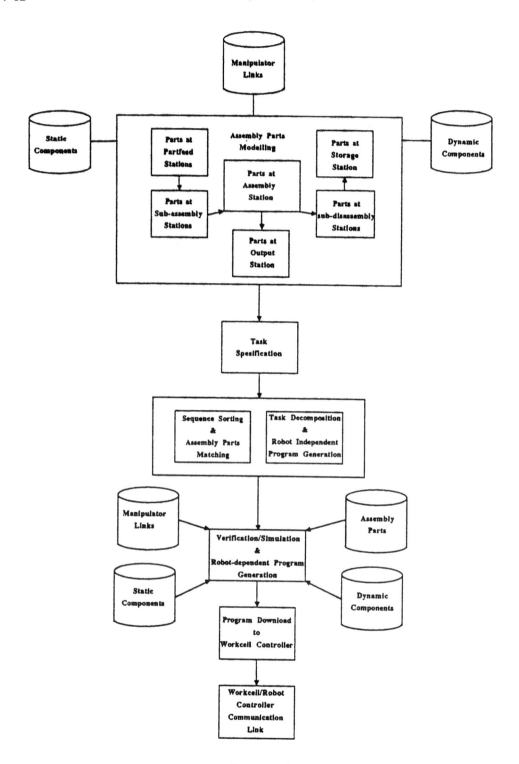

FIGURE 9.10   Structure of the task planning and programming.

robot which in turn performs the assembly operation. The capability to perform workstation calibration is also built to readjust positions in the robot programs.

Design Modeling of Assembly Parts and Work Cell

Modeling of various components of a robot manufacturing cell is divided into four categories: static components, dynamic components, robot manipulator, and assembly parts. The static components of the manufacturing cell are elements which are stationary and may include assembly tables, partfeed machines, fixtures, and part storage. The dynamic components of the cell are elements which may change their locations during the manufacturing operation. The dynamic components are designated with a dynamic reference datum and examples of these components are interchangeable grippers, turntables, and pallets a on conveyor system. The static and dynamic components are modeled as convex polyhedrals and stored in the appropriate format in the database [46]. The models are retrieved and displayed by the verification module during simulation of robotic assembly operations.

The manipulator is modeled as a series of links in the synchronized (i.e., home) position and with a coordinate frame attached to each link [47]. This information is not required for assembly task specification or robot-independent program generation. As with static and dynamic components within the cell, it is only used for the final verification of the assembly task and simulation purposes during the robot-dependent program generation. Assembly parts are modeled with respect to the reference coordinate frame fixed to each station. Thus, the models of the assembly parts are stored within the CAD database just as they would appear at various assembly stages. This procedure will also produce the correct "pose" of the assembly parts referenced by a local coordinate frame fixed to each station throughout the robotic assembly process. The procedure follows a practical structure for assembly/disassembly operations (Fig. 9.11). Figure 9.12 shows the structure for access/retrieval of 3-D models representing various assembly stages.

Therefore, the assembly process is represented in 3-D model drawings and the task planning is performed on the these models. Figure 9.13 shows an example of the assembly part coordinate frames

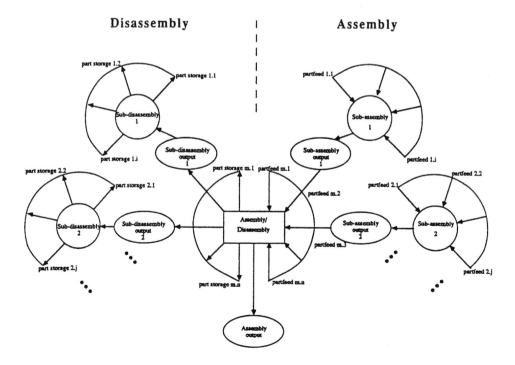

FIGURE 9.11 Representation of assembly/disassembly stages.

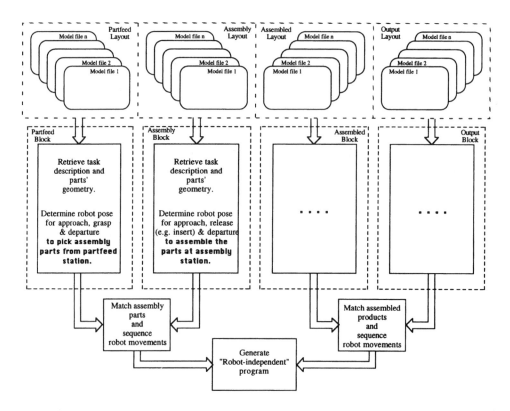

FIGURE 9.12   Structure for access/retrieval and processing of assembly task specifications.

attached to each station. It must be noted that the design modeling of the individual assembly parts and the assembled product is a common practice in industry today. Hence, minimal effort is required in adding appropriate frames and description to these models for planning and programming purposes.

Task Specification

The specification of the task can be carried out using objective commands such as "Pick," "Place," "Insert," etc. In general, frames are assigned (i.e., tagged) to the model to signify the grasp locations. It must be emphasized that task planning has been partially built into the design modeling procedure and it will also be completed by the task decomposition software modules [48]. Other information such as robot speed and robot motion mode (e.g., R: rectangular, J: joint) may also be attached to the model (Fig. 9.14). The layer on which these commands are placed may be turned off, if required.

A macro program has also been developed to allow the specification of complex approach and departure vectors. The program forms construction lines and computes the approach vector, (a_H), orientation vector (o_H), and normal vector (n_H). Hence, the hand orientation is fully defined with respect to a specified frame.

$$T = \begin{bmatrix} n_F & O_F & a_F & p_F \\ \hline O & O & O & 1 \end{bmatrix} \tag{9.9}$$

The approach (a), orientation (o), and normal (n) vectors in their normal form do not have much meaning to the manufacturing engineer and carrying all the elements of the matrix is not very efficient.

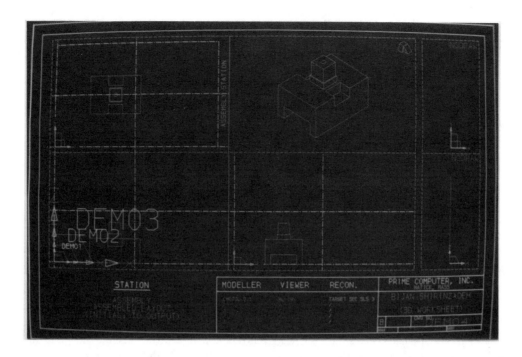

FIGURE 9.13 Design modeling of assembly parts.

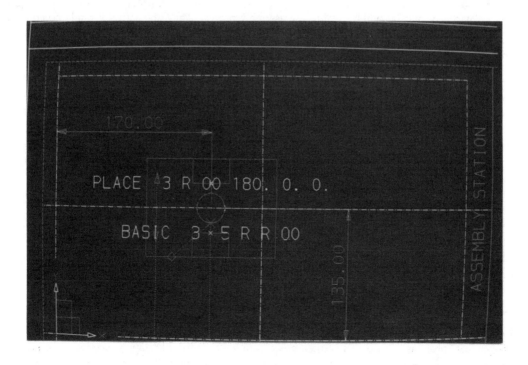

FIGURE 9.14 Example of the task specification on CAD model.

Therefore, once all the components of the vectors in the matrix T_H (i.e., hand coordinate frame) are determined and normalized, the orientation angles, Roll, ϕ (i.e., rotation about the Z axis), Pitch, θ (i.e., rotation about the Y axis), and Yaw, ψ (i.e., rotation about the X axis) are determined using the following relationships:

$$\tan \phi = n_{H_Y}/n_{H_X} \tag{9.10}$$

$$\tan \theta = -n_{H_Z}/(\cos \phi \, n_{H_X} + \sin \phi \, n_{H_Y}) \tag{9.11}$$

$$\tan \psi = (\sin \phi \, a_{H_X} - \cos \phi \, a_{H_Y})/(\sin \phi \, o_{H_X} + \cos \phi \, o_{H_Y}) \tag{9.12}$$

where n_H, a_H, and o_H represent the normal, approach, and orientation vectors in hand coordinate frame, respectively. The above orientation angles (specified in degrees) are reported to the designer and included in the command. Wire frames may be used to create intricate paths when these are required. It must be emphasized that the above procedure is not required for layered assembly (i.e., assembly directions from the above are generated automatically using an offset vector).

Task Decomposition

In commercially available systems such as CATIA, IGRIP, and ROBCAD, the task and individual manipulator motion must be specified sequentially by the user. However, in the approach adopted system a software program was developed to access the CAD data base (i.e., 3-D model data) and automatically retrieve the geometrical data. This software program also retrieves task instructions tagged to the model (i.e., stored in the CAD data base during the task specification phase). The attributes tagged to each planned task are also retrieved. The objects' names are used to match assembly parts in various stages such as "partfeed" and "assembly" within the assembly process. Thus, the software module performs sorting and sequencing of the assembly operations and robot movements and, finally, generating a 'robot-independent' program. Figure 9.12 shows the sequence of retrieval of model data together with the subsequent task and object matching operations. The robot program will then be generated in a format appropriate for subsequent simulation. The geometrical data will also be stored in a structured boundary representation (B-rep) format as discussed previously (Fig. 9.2).

Figure 9.15 shows the structure of the software for the "main block" (e.g., assembly block) of Fig. 9.12, and thus a detailed description of the above software operations. Other blocks utilize a similar structure. As relevant robotic manufacturing information is gathered, these are stored in the data base. The software program initially accesses the model file (which contains the model of the assembly layout), extracts the designer's instructions, and begins to work backward to subassembly and partfeed models (i.e., feeders, pallets, etc.). The name of each object from the assembly layout is then matched against those from the subassembly and partfeed layouts and sequencing of assembly operation is performed, thereby completing the process of programming of the assembly operation. The program then sequences the output operation by retrieving the necessary information from the model files containing the initial location of the assembled product in the assembly station and the final location of the assembled parts in the output station.

Transformation of Robot Locations and Workstation Calibration

The output from the previous software module is a series of tasks, generally referred to as a "robot-independent" program. This contains robot poses fixed to the individual station coordinate frame. The next phase involves transformation of these movements into grasp locations attached to the universe (i.e., world) coordinate frame. This is generally defined as the coordinate frame at the base of the robot (i.e., robot coordinate frame). This software module utilizes transformation vectors to describe the

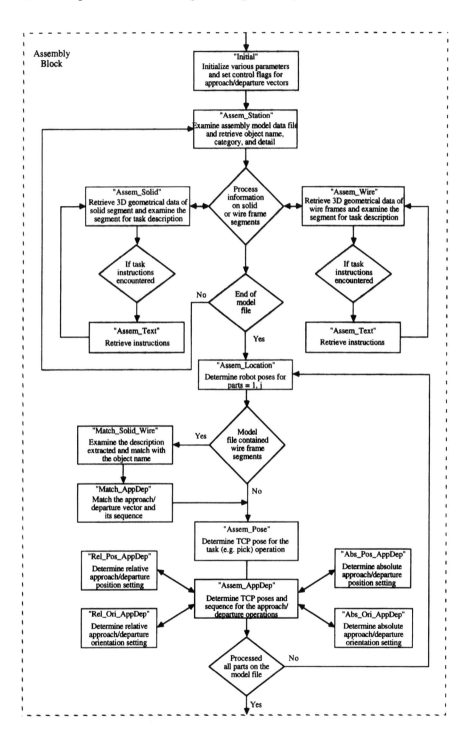

FIGURE 9.15 Detailed flow diagram of the 'Main Block' in the software program.

relationships between station coordinate frames and the universe coordinate frame. These transformations may be modified by simply specifying the translation transform (i.e., vector components *x*, *y*, and *z*) and orientation transform (i.e., roll, pitch, and yaw orientation about *x*, *y*, and *z*). Figure 9.16 shows the schematic arrangement of the coordinate frames.

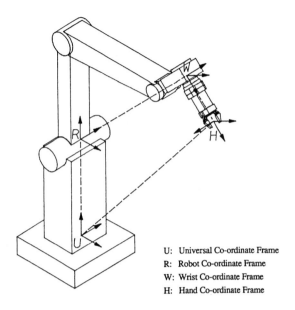

U: Universal Co-ordinate Frame
R: Robot Co-ordinate Frame
W: Wrist Co-ordinate Frame
H: Hand Co-ordinate Frame

U: Universal Co-ordinate Frame
S: Station Co-ordinate Frame
i: Local Co-ordinate Frame
H: Hand Co-ordinate Frame

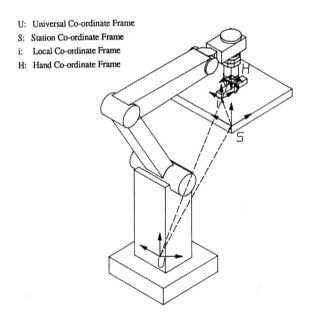

FIGURE 9.16   Schematic representation of assigning coordinate frames.

The following relationship may be used to determine the pose of the manipulator with respect to the universe coordinate frame:

$$^{S}T_U \cdot {}^{U}T_H = {}^{S}T_i \cdot {}^{i}T_{i-1} \cdot \ldots \cdot {}^{1}T_H \tag{9.13}$$

where $^{S}T_U$ is the transformation describing the location of the universe coordinate frame with respect to a particular station coordinate frame, $^{U}T_H$ is the transformation matrix describing the location of the hand with respect to the universe coordinate frame, $^{S}T_i$ is the transform describing the location of the *i-th* local coordinate frame with respect to the station coordinate frame, and $^{i}T_{i-1}$ is the transform describing the (*i*-1)th local coordinate frame with respect to the *i*-th local coordinate frame. Finally, $^{1}T_{H}$ is the transform describing the hand coordinate frame on the assembly part with respect to the 1-th local coordinate frame. Pre-multiplying both sides of the equation by $^{S}T_{U}^{-1}$ will result in the position and orientation of the hand with respect to the universe coordinate frame:

$$^{U}T_{H} = {}^{S}T_{U}^{-1} \cdot {}^{S}T_{i} \cdot {}^{i}T_{i-1} \cdot \ldots \cdot {}^{1}T_{H} \tag{9.14}$$

If there is no local frame and the location of the hand is with respect to the station coordinate frame, then the equation can be simplified to:

$$^{U}T_{H} = {}^{S}T_{U}^{-1} \cdot {}^{S}T_{H} \tag{9.15}$$

where all the transformation matrices are of the form:

$$T = \begin{bmatrix} n_x & o_x & a_x & p_x \\ n_y & o_y & a_y & p_y \\ n_z & o_z & a_z & p_z \\ 0 & 0 & 0 & 1 \end{bmatrix} \tag{9.16}$$

In practice, the transform $^{S}T_{U}^{-1}$ is obtained by jogging the robot (i.e., moving the manipulator arm under manual control) to the coordinate frame of a particular station and performing the status request. The position vector acquired in this manner may be directly used in the above software module. Furthermore, the technique may also be used to calibrate the robot workstation and thus, reduce the robot positioning errors associated with advanced off-line programming techniques. However, it must be emphasized that this approach is concerned with digitizing the workstation and it is not a replacement for robot calibration.

The calibration procedure utilizes a precision-made rod referred to as the "calibration rod." This is attached to the center line of the end-effector. This procedure requires the operator to jog the robot to various control locations ($C_1 - C_4$), align the tip of the calibration rod with the control locations on a particular station, and record their positions [50]. Fig. 9.17 shows a schematic illustration of the procedure and a schematic diagram of the geometrical information. The calibration software determines the reference datum for the workstation (i.e., C_1). It also generates transformation vectors with respect to the robot base coordinate frame U, and determines the calibration factors for the workstation (as shown in Fig. 9.17). It must be noted that the tool center point (TCP) offset for the robot end-effector must be set accordingly.

The industrial robot manipulators possess good repeatability but not absolute accuracy. The approach described employs the manipulator arm to obtain the reference coordinate frame fixed to each station and the programmed movements are calibrated in relation to this frame. This provides a better absolute accuracy over the range of assembly workstation. The technique reduces the inaccuracies in positioning due to long range movement in relation to the robot coordinate frame. The approach has been developed so that it can also use a vision system to perform the calibration [51].

Robot Kinematics and Assembly Process Simulation Module

The simulation module is the final step in order to generate the command and point-to-point motion data files. This is generally referred to as the "robot native program." This module retrieves the hand poses generated by the previous software operations; determines the pose of the wrist with respect to the robot coordinate frame, and solves the robot kinematic equations to determine the robot joint angles for the

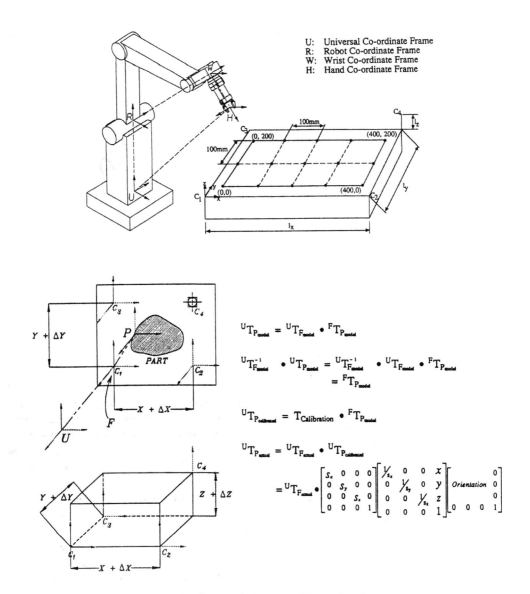

U: Universal Co-ordinate Frame
R: Robot Co-ordinate Frame
W: Wrist Co-ordinate Frame
H: Hand Co-ordinate Frame

$$^U T_{P_{model}} = {}^U T_{F_{model}} \bullet {}^F T_{P_{model}}$$

$$^U T_{F_{model}}^{-1} \bullet {}^U T_{P_{model}} = {}^U T_{F_{model}}^{-1} \bullet {}^U T_{F_{model}} \bullet {}^F T_{P_{model}}$$
$$= {}^F T_{P_{model}}$$

$$^U T_{P_{calibrated}} = T_{Calibration} \bullet {}^F T_{P_{model}}$$

$$^U T_{P_{actual}} = {}^U T_{F_{actual}} \bullet {}^U T_{P_{calibrated}}$$

$$= {}^U T_{F_{actual}} \bullet \begin{bmatrix} S_x & 0 & 0 & 0 \\ 0 & S_y & 0 & 0 \\ 0 & 0 & S_z & 0 \\ 0 & 0 & 0 & 1 \end{bmatrix} \begin{bmatrix} 1/S_x & 0 & 0 & x \\ 0 & 1/S_y & 0 & y \\ 0 & 0 & 1/S_z & z \\ 0 & 0 & 0 & 1 \end{bmatrix} \begin{bmatrix} & & & 0 \\ Orientation & & & 0 \\ & & & 0 \\ 0 & 0 & 0 & 1 \end{bmatrix}$$

FIGURE 9.17   Calibration procedure and geometrical aspects of the workstation.

purpose of simulation. It also determines the location of the robot wrist in quaternion, as required by the robot controller (e.g., ABB IRB 2400), for the purpose of generating the command and motion data file(s).

The simulation can generate motion trajectories according to straight line motion increment or joint interpolated motion increment. As joint angles for a given pose of the wrist are determined, they are checked against the limits of the robot joint angles. If, for a given pose, any of the calculated robot joint angles exceeds the limit, then the module reports this to the user and logs the event. The module retrieves the geometrical data describing robot linkages, stations, feeders, pallets, and assembly parts at various stages of the assembly process and any other equipment and objects that are required to be in the scene. The module determines the location of all the objects in relation to the universe frame using the appropriate transformations. Simulation process is based on the method of coordinate transformation for describing both the manipulator and objects (Fig. 9.18).

The main intention of such graphics systems is to provide the visualization of the entire work cell as well as the animation of the movements of robots and peripherals. In addition, such systems can be

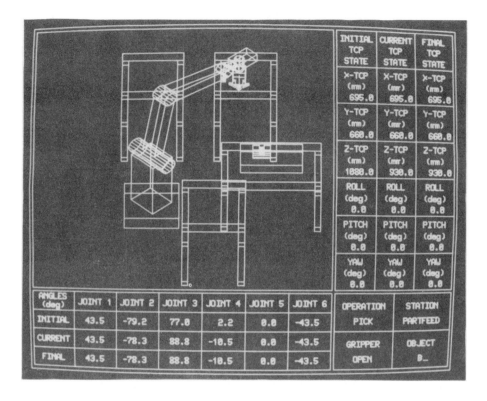

FIGURE 9.18 Photograph showing the simulation of the assembly.

viewed as building blocks toward the integrated CIM solutions that support the manufacturing cycle, from CAD to CAM and distribution [52–54].

9.5 Future Research

The task planning and programming system described is currently being ported to a PC-based CAD package (AutoCAD). The technique is also being examined for implementation on a large modeling and simulation platform such as Deneb's Envision system. In addition, the strategies and formulation are being further developed to include automatic sequence and assembly directions using screw theory. In this approach, disassembly sequence and direction may be generated first and then reversed to determine possible assembly sequence and directions. The planner is being developed further to include a more extensive library of manufacturing operations and to include rules and mathematical formulations for engagement and disengagement of assembly parts.

9.6 Conclusions

Computer-aided design (CAD) has become an integral part of design in manufacturing and many other fields of engineering. It has also become obvious that the cost of programming robots is an important issue in robot-based flexible manufacturing and CIM environment.

This chapter has briefly presented strategies for task planning and programming. This chapter has also described individual modules (i.e., building blocks) for such systems. The structure and development strategies for a CAD-based and off-line task planner has also been described. The development employed a commercial CAD package for the process of design modeling of the assembly parts and task specification.

The approach adopted in development of a dedicated software program to retrieve parts' geometry and instructions which are embedded into the design models was presented.

The strategies to retrieve the appropriate task information from the assembly layout and work backward to automatically generate lower level task commands was also presented. The system is being modified and further developed for operation on PC-based platform and Unix-based Envision (Deneb) modeling and simulation platform.

Acknowledgments

This research is partly supported by an Engineering Research Council (ERC) Small Grant from Monash University, a Monash Research Fund (MRF) grant, and a Harold Armstrong Research Fund. The author wishes to thank his research assistants at Robotics & Mechatronics Research Laboratory, Department of Mechanical Engineering, Monash University.

References

1. H. K. Rampersad, State of the art in robotic assembly. Int. J. Industrial Robot, Vol. 22, No. 2, pp. 10–13, 1995.
2. K. Schroer, Robot calibration—closing the gap between model and reality. Int. J. Industrial Robot, Vol. 21, No. 6, pp. 3–5, 1994.
3. J. C. Latombe, Une analyse structuree d'outils de programmation pour la robotique industrielle. Proc. Int Seminar on Programming Methods and Languages for Industrial Robots. INRIA, Rocquencourt, France, June 1979.
4. S. Bonner, K. G. Shin, A comparative study of robot languages. Computer Dec, pp. 82–96, 1982.
5. D. M. Lee, W. H. ElMaraghy, ROBOSIM: a CAD-based off-line programming and analysis system for robotic manipulators. Computer-Aided Engineering Journal, Vol. 7, No. 5, pp. 141–148, October 1990.
6. A. R. Thangaraj, M. Doelfs, Reduce downtime with off-line programming, Robotics Today, Vol. 4, No. 2, pp. 1–3, 1991.
7. Y. Regev, The evolution of off-line programming. Industrial Robot, Vol. 22, No. 3, pp. 3, 1995.
8. W. A. Gruver, B. I. Soroka, J. J. Craig, T. L. Turner, Evaluation of commercially available robot programming languages. Proc. 13th Int. Symposium on Industrial Robots, Chicago, 1983.
9. J. J. Craig, Issues in the design of off-line programming systems. Fourth Int. Symposium on Robotics Research, University of California, Santa Cruz, pp. 379–389, 1987.
10. M. Kortus, T. Ward, M. H. Wu, An alternative approach to off-line programming. Industrial Robot, Vol. 20, No. 4, pp. 17–20, 1995.
11. P. Sorenti, GRASP for simulation and off-line programming of robots in industrial applications. Proc. of Conference on Welding Engineering Software, Essen, DVS N0 156, pp. 55–58, Sept 1993.
12. R. J. Popplestone, A. P. Ambler, T. M. Bellos, An interpreter for a language for describing assemblies. Artificial Intelligence, Vol. 14, pp. 79–107, Aug. 1980.
13. L. I. Liberman, M. A. Wesley, AUTOPASS: an automatic programming system for computer controlled mechanical assembly. IBM Journal of Research and Development, pp. 321–333, July 1977.
14. R. Mattikalli, D. Baraff, P. Khosla, Finding all stable orientations of assemblies with friction. IEEE Trans. Robotics and Automation, Vol. 12, No. 2, pp. 290–301, 1996.
15. J. J. Craig, Introduction to Robotics, 2nd, Addison-Wesley Publishing, 1989.
16. B. Shirinzadeh, A simplified approach to robot programming from CAD system. Fourth Int. Conference on Manufacturing Engineering, pp. 141–145, Brisbane, Australia, 1988.
17. T. N. Wong, S. C. Hsu, An off-line programming system with graphics simulation. Int. J. Advanced Manufacturing Technology, Vol. 6, pp. 205–223, 1991.
18. J. Shah, P. Sreevalsan, A. Mathew, Survey of CAD/feature-based process planning and NC programming techniques. Computer-Aided Engineering Journal, Vol. 8, No. 1, pp. 25–33, 1991.

19. B. Shirinzadeh, A CAD-based design and analysis system for reconfigurable fixtures in robotic assembly. Computing & Control Engineering Journal, Vol. 5, No. 1, pp. 41–46, 1994.

20. B. Shirinzadeh, Strategies for planning and implementation of flexible fixturing systems in a computer integrated manufacturing environment. J. Computers in Industry, Vol. 30, pp. 175–183, 1996.

21. J. Rocha, C. Ramos, Task planning for flexible and agile manufacturing systems, McGraw-Hill Inc, 1987.

22. L. Laperriere, H. A. ElMaraghy, Automatic generation of robotic assembly sequences. Int. J. Advanced Manufacturing Technology, Vol. 6, pp. 299–316, 1991.

23. V. N. Rajan, S. Y. Nof, Minimal precedence constraints for integrated assembly and execution planning. IEEE Trans. Robotics and Automation, Vol. 12, No. 2, pp. 175–186, 1996.

24. H. Chu, H. A. ElMaraghy, Integration of task planning and motion control in a multi-robot assembly workcell. Robotics & Computer-Integrated Manufacturing, Vol. 10, No. 3, pp. 235–255, 1993.

25. H. L. Welch, R. B. Kelley, The analysis of potential mating trajectories and grasp sites. Int. J. of Advanced Manufacturing Technology, Vol. 8, pp. 320–328, 1993.

26. Z. Shiller, S. Dubowsky, Robot path planning with obstacles, actuator, gripper, and payload constraints. Int. J. Robotics Research, Vol. 8, No. 6, pp. 3–18, 1989.

27. S. Ahmad, J. T. Feddema, Static grip selection for robot-based automated assembly systems. J. Robotic Systems, Vol. 4, No. 6, pp. 687–717, 1987.

28. Y. L. Xiong, D. J. Sanger, D. R. Kerr, Geometric modeling of bounded and frictional grasps, Robotica, Vol. 11, pp. 185–192, 1993.

29. S. Cameron, Obstacle avoidance and path planning. Industrial Robot, Vol. 21, No. 5, pp. 9–14, 1994.

30. T. D. Luk, Planning collision-free paths in Cartesian space. Robotics Today, Vol. 5, No. 3, pp. 1–3, 1992.

31. T. Lozano-Perez, Automatic planning of manipulator transfer movements. IEEE Trans. Systems, Man, Cybernetics SMC-11, 10, pp. 681–689, 1981.

32. Y. Itoh, M. Idesawa, T. Soma, A study on robot path planning from a solid model. J. Robotics Systems, Vol. 3, No. 2, pp. 191–203, 1986.

33. S. Stifter, Collision detection in the robot simulation system SMART. Int. J. Advanced Manufacturing Technology, Vol. 7, pp. 277–284, 1992.

34. M. J. Tsai, S. Lin, M. C. Chen, Mathematical model for robotic arc-welding off-line programming system. Int. J. Computer Integrated Manufacturing, Vol. 5, No. 4 & 5, pp. 300–309, 1992.

35. R. O. Buchal, D. B. Cherchas, F. Sassani, J. P. Duncan, Simulated off-line programming of welding robots. Int. J. Robotics Research, Vol. 8, No. 3, pp. 31–43, 1989.

36. S. H. Suh, J. J. Lee, Y. J. Choi, S. K. Lee, Prototype integrated robotic painting system: software and hardware development. Journal of Manufacturing Systems, Vol. 12, No. 6, pp. 463–472, 1993.

37. H. Wapenhans, J. Holzl, J. Steinle, F. Pfeiffer, Optimal trajectory planning with application to industrial robot. Int. J. Advanced Manufacturing Technology, Vol. 9, pp. 49–55, 1994.

38. K. S. Moon, K. Kim, F. Azadivar, Optimum continuous path operating conditions for maximum productivity of robotic manufacturing systems. J. Robotics & Computer-Integrated Manufacturing, Vol. 8, No. 4, pp. 193–199, 1991.

39. M. C. Leu, R. Mahajan, Computer graphic simulation of robot kinematics and dynamics. Proc. 8th International Conference on Robots, pp. 1–20, June, 1984.

40. G. Wiitenberg, Developments in off-line programming: an overview. Industrial Robot, Vol. 22, No. 3, pp. 21–23, 1995.

41. J. F. Quinet, Calibration for offline programming purpose and its expectations. J. Industrial Robot, Vol. 22, No. 2, pp. 10–13, 1995.

42. B. Shirinzadeh, Y. Shen, Three dimensional calibration of robotic manufacturing cell using machine vision techniques. Proc. of Pacific Conference on Manufacturing, pp. 360–365, Seoul, Korea, 1996.

43. M. Vincze, J. P. Prenninger, H. Gander, A laser tracking system to measure position and orientation or robot end-effectors under motion. Int. J. Robotics Research. Vol. 13, No. 4, pp. 305–314, 1994.

44. K. V. Kamisetty, Development of a CAD/CAM robotic translator for programming the IBM 7535 SCARA robot off-line. J. Computers in Industry, Vol. 20, pp. 219–228, 1992.

45. K. V. Steiner, M. Keefe, Interactive graphics simulation with multi-level collision algorithm. Journal of Manufacturing Systems, Vol. 11, No. 6, pp. 462–469, 1992.

46. B. Shirinzadeh, H. Tie, Object-oriented task planning and programming system for layered assembly and disassembly operations. Proc. of the ARA/IFR International Conference, Robots for Competitive Industries, pp. 354–360, Brisbane, 1993.

47. P. Fanghella, C. Galletti, E. Giannotti, Computer-aided modeling and simulation of mechanisms and manipulators. J. Computer-Aided Design, Vol. 21, No. 9, pp. 577–583, 1989.

48. B. Shirinzadeh, H. Tie, G. Lin, Computer integrated task planning and programming of robotic assembly operations. Proc. Second International Conference, CIM, Vol. 2, pp. 599–605, Singapore, 1993.

49. P. K. Venuvinod, Automated analysis of 3-D polyhedral assemblies: assembly directions and sequences. Journal of Manufacturing systems, Vol. 12, No. 3, pp. 246–252, 1993.

50. B. Shirinzadeh, H. Tie, Experimental investigation on the performance of a reconfigurable fixturing system. International Journal of Advanced Manufacturing Technology, Vol. 10, pp. 330–341, 1995.

51. B. Shirinzadeh, C. Paragreen, W. Lee, Calibration of robotic cell for CAD-based planning using machine vision techniques. Proc. of Twelfth International Conference on Robotics and Factories of the Future, pp. 132–137, London, 1996.

52. T. S. Kang, B. O. Nnaji, Feature representation and classification for automatic process planning systems. Journal of Manufacturing Systems, Vol. 12, No. 2, pp. 133–145, 1993.

53. S. Chakrabarty, J. Wolter, A structure-oriented approach to assembly sequence planning. IEEE Trans. Robotics and Automation, Vol. 13, No. 1, pp. 14–29, 1997.

54. C. P. Tung, A. Kak, Integrating sensing, task planning, and execution for robotic assembly. IEEE Trans. Robotics and Automation, Vol. 12, No. 2, pp. 187–201, 1996.

55. G. Kim, S. Lee, G. Bekey, Interleaving assembly planning and design. IEEE Trans. Robotics and Automation, Vol. 12, No. 2, pp. 246–251, 1996.

10

Physical Model Technique for Design of Robotic Manipulators in Manufacturing Systems

Feng Gao
Hebei University of Technology

10.1 Introduction

The mechanical design of industrial robots requires the application of engineering expertise in a variety of areas. Important disciplines include machine design, structure design, and mechanical, control, and electrical engineering. Traditionally, mechanism design has been based largely on use of specifications including number of axes, workspace volume, payload capacity, and end-effector speed. Robots have not been designed to perform specific tasks but to meet general performance criteria.

In the design of industrial robots, the design of the robotic mechanisms is one of the most important activities because these mechanisms determine the performance characteristics of the robotic machines.

Although some methods and criteria have been developed for robotic mechanism design and many types of robotic mechanisms have been proposed, there has not been a unified method for the design.

One reason for this is that there has not been a method to visualize the performance characteristics of robots and many criteria just show the local performance characteristics of robots.

In this chapter, a novel method of the physical model technique for the design of robotic mechanisms is introduced. We establish several kinds of physical models of solution space for serial and parallel robots [1-9]; propose many important global performance criteria for evaluation of robotic mechanisms; and utilize the physical model technique and performance criteria to construct the performance atlases of robotic mechanisms which are used to analyze the relationships between the performance criteria and link lengths of robotic mechanisms. The performance atlases are a very efficient and useful tool for design of robotic mechanisms.

10.2 Technique of Physical Models of Solution Space for Robotic Manipulators

As you know, link lengths of robotic mechanisms may vary between zero and infinity. This means that links can be very long or short and can be measured by different units or unit systems (such as meter, millimeter, foot, and inch). It is very difficult to investigate the relationship between the performance criteria and link lengths of all the robots. Therefore, it is convenient to eliminate the physical sizes of the robotic mechanisms to investigate the robotic design method.

Model of Solution Space for Robots with Three Moving Links

Human arms, legs, and fingers, and limbs of animals and insects can be viewed as mechanisms with three moving links. Fingers of dextrous hands [10–22], arms of industrial robots [23–27, 32–35], and legs of walking machines [28–30] often consist of three moving links. Therefore, an understanding of the relationships between the criteria and the dimensions of three-moving-link mechanisms is of great importance for design of fingers, arms, and legs of robotic machines. The robotic mechanisms with three moving links can be used for 3-, 4-,…, 7-DOF serial robots, as shown in Fig. 10.1. For example, if joints A, B, and C are revolute pairs, it can be viewed as a 3-DOF robotic mechanism. When joints A and C are spherical pairs and joint B is a revolute pair, they comprise a 7-DOF redundant spatial robotic mechanism. Therefore, to investigate the robotic mechanisms with three moving links has generality.

Because the link lengths may have a wide range of possible values, it is convenient to avoid explicit use of the physical sizes of the mechanisms during analysis and design. We shall define normalized parameters of the mechanisms and construct a physical model of the solution space as follows.

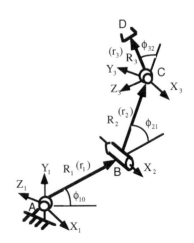

FIGURE 10.1 A serial robotic mechanism with three moving links. (a) 3-D model, (b) 2-D model with 3 coordinates.

To facilitate the analysis, we normalize the parameters of a serial mechanism as shown in Fig. 10.1 If the link lengths of the mechanism are R_i $(i = 1, 2, \text{ and } 3)$, we define

$$r_i = R_i / L; \quad i = 1, 2, \text{ and } 3 \tag{10.1}$$

$$L = (R_1 + R_2 + R_3) / 3 \tag{10.2}$$

where, r_i is the nondimensional parameter of the robotic mechanisms with three moving links. From Eqs. (10.1) and (10.2), we obtain:

$$r_1 + r_2 + r_3 = 3 \tag{10.3}$$

$$0 \le r_i \le 3, \quad i = 1, 2, \text{ and } 3. \tag{10.4}$$

Let r_1, r_2, and r_3 be orthogonal coordinate axes. Using Eqs. (10.3) and (10.4), we generate a physical model of the solution space, as shown in Fig. 10.2, consisting of an equilateral triangle ABC for which any possible combination of the link lengths is represented by r_1, r_2, and r_3. The resulting model graphically provides a means to represent all possible mechanisms with three moving links. In this model, points A, B, and C have coordinates $(3, 0, 0)$, $(0, 3, 0)$, and $(0, 0, 3)$, respectively; then edges BC, AB, and AC satisfy conditions $r_1 = 0$, $r_2 = 0$, and $r_3 = 0$, respectively. Within triangle ABC, we inscribe another equilateral triangle EFG, where edges EG, FG, and EF satisfy $r_1 = r_2 + r_3$, $r_2 = r_1 + r_3$, and, $r_3 = r_1 + r_2$, respectively (as shown in Fig. 10.2(b)). Triangle EFG divides triangle ABC into regions I, II, III, and IV. Table 10.1 describes the geometric characteristics of the four classes of mechanisms.

By using the solution space model, we can investigate relationships between performance criteria and link lengths of the robotic mechanisms.

TABLE 10.1 Dimensional Properties of Mechanisms with Three Moving Links

Region	Dimensional Characteristics	Distribution
I	$r_1 > r_2 + r_3$	$\triangle$ AEG
II	$r_2 > r_1 + r_3$	$\triangle$ FGC
III	$r_3 > r_1 + r_2$	$\triangle$ BEF
IV	$r_1 \le r_2 + r_3 \cap r_2 \le r_1 + r_3 \cap r_3 \le r_1 + r_2$	$\triangle$ EFG

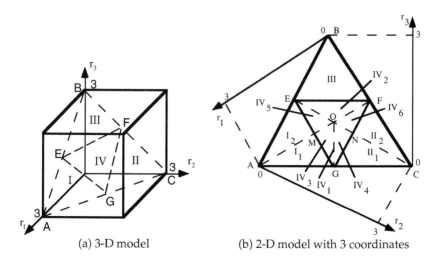

(a) 3-D model   (b) 2-D model with 3 coordinates

FIGURE 10.2 Model of solution space for mechanisms with three moving links.

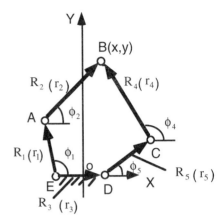

FIGURE 10.3 A typical 2-DOF PPM. (a) 3-D model,
(b) 2-D model with 3 coordinates.

Model of Solution Space for 2-DOF Parallel Planar Manipulators

Since parallel robots have advantages, compared with serial robots (i.e., their stiffness, speed, payload, and precision are higher), much research has been devoted to them.

2-DOF parallel-planar manipulators (PPMs) are an important class of robotic manipulators that can follow an arbitrary planar curve. Because of their usefulness in applications, these mechanisms have attracted the attention of researchers who have investigated their workspace, mobility, and methods for analysis and design [36–42].

Although much research has been devoted to 2-DOF PPMs, there has not been a study of the relationships between the criteria and link lengths of 2-DOF PPMs. The reason for this is that there has not been an effective method to solve the problem.

Figure 10.3 shows a typical 2-DOF PPM. Since any of the actual link lengths of the manipulator lies in the range zero to infinity, we have to eliminate the physical size of the manipulator from the discussion. Let:

$$r_i = R_i / L \quad (i = 1, 2, 3, ..., 5) \tag{10.5}$$

where R_i is the actual length of link i and r_i is the nondimensional relative length of link i. And

$$L = \frac{1}{4} \sum_{i=1}^{5} R_i \tag{10.6}$$

where L is the average length of all links for the manipulator. Gosselin [31] has shown that the parallel manipulator should be symmetric, so that we have the following results:

$$R_5 = R_1 \quad \text{and} \quad R_4 = R_2; \ r_5 = r_1 \quad \text{and} \quad r_4 = r_2 \tag{10.7}$$

These conditions show us that the two input links should have the same length and the two coupling bars also have the same length. Therefore, the sum of the five link lengths is

$$2r_1 + 2r_2 = r_3 = 4 \tag{10.8}$$

If the manipulator can be assembled, the range of the relative link lengths should be 0 to 2, so

$$0 < r_i < 2 \quad (i = 1, 2, 4, 5) \quad \text{and} \quad 0 \leq r_3 < 2 \tag{10.9}$$

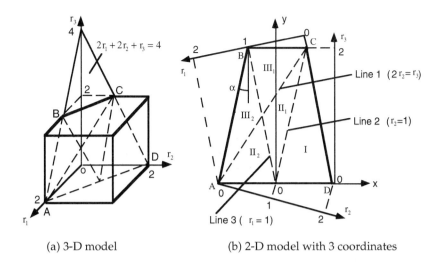

(a) 3-D model (b) 2-D model with 3 coordinates

FIGURE 10.4 Model of the solution space for 2-DOF PPMs.

Using these equations and inequalities, the model of the solution space for 2-DOF PPMs can be constructed as shown in Fig. 10.4(a), which is an isosceles trapezoid ABCD. Within the model, all 2-DOF PPMs can exist and the relationships between the different criteria and dimensions of the manipulators can be investigated.

For convenience, the model of the solution space, which is an isosceles trapezoid ABCD in the $r_1 r_2 r_3$ coordinate system as shown in Fig. 10.4(a) can be changed into the isosceles trapezoid ABCD in the xy coordinate system as shown in Fig. 10.4(b), by using the following equations:

$$\begin{pmatrix} x \\ y \end{pmatrix} = \begin{pmatrix} 1/\cos\alpha \\ 0 \end{pmatrix} - \begin{bmatrix} 1/\cos\alpha & 1/(4\ \cos\alpha) \\ 0 & -1 \end{bmatrix} \begin{pmatrix} r_1 \\ r_3 \end{pmatrix} \tag{10.10}$$

where

$$\alpha = \sin^{-1} 1/4$$

When the input links r_1 and r_5 are located at the extreme positions and the angular velocities $\dot\phi_2$ and $\dot\phi_4$ are not equal to zero, the angular velocities $\dot\phi_1$ and $\dot\phi_5$ must be equal to zero, that is

$$\dot\phi_1 = 0 \quad \text{and} \quad \dot\phi_5 = 0 \tag{10.11}$$

From Eqs. 10.7, 10.8, and 10.11,

$$\begin{pmatrix} \dot\phi_1 \\ \dot\phi_5 \end{pmatrix} = \frac{1}{r_1 r_5 \sin(\phi_1 - \phi_5)} \begin{bmatrix} r_2 r_5 \sin(\phi_5 - \phi_2) & r_4 r_5 \sin(\phi_4 - \phi_5) \\ r_1 r_2 \sin(\phi_1 - \phi_2) & r_1 r_4 \sin(\phi_4 - \phi_1) \end{bmatrix} \begin{pmatrix} \dot\phi_2 \\ \dot\phi_4 \end{pmatrix}$$

we can derive three conditions:

Line 1: $\quad 2r_2 = r_3$ $\qquad\qquad\qquad\qquad\qquad$ (10.12)

Line 2: $\quad 2r_2 = 2r_1 + r_3 \quad \text{or} \quad r_2 = 1$ $\qquad$ (10.13)

Line 3: $\quad 2r_1 = 2r_2 + r_3 \quad \text{or} \quad r_1 = 1$ $\qquad$ (10.14)

Using Eqs. (10.12), (10.13), and (10.14), three lines can be drawn on the model of solution space, which divide the model into five regions I, II_1, II_2, III_1, and III_2, as shown in Fig. 10.4(b). By analyzing the characteristics of the mechanisms in the five regions, respectively, the mechanisms can be classified as four types, as shown in Table 10.2.

In Table 10.2, we define

UDCM = Unrestrained double crank 2-DOF PPMs;
RDCM = Restrained double crank 2-DOF PPMs;
DRM = Double rocker 2-DOF PPMs;
CPM = Change point 2-DOF PPMs.

Model of Solution Space for 3-DOF Parallel Planar Manipulators

Since 3-DOF parallel planar manipulators (PPMs) can follow both an arbitrary planar curve and an orientation, they are an important class of robotic manipulators. Figure 10.5 shows a typical 3-DOF PPM. Since any of the actual link lengths of the manipulator lies in the range zero to infinity, we have to eliminate the physical size of the manipulator from the discussion.

Nondimensional Parameters of 3-DOF PPMs

Let

$$r_i = R_i / L \quad (i = 1, 2, 3, \ldots, 12) \qquad (10.15)$$

TABLE 10.2 Complete Classification of 2-DOF PPMs

Type No.	Region	Symbol of Type	Dimensional Characteristics
1	I	UDCM	$r_2 > 1$
2	II_1	RDCM	$r_2 < 1 \cap r_1 < 1 \cap 2r_2 > r_3$
	II_2	RDCM	$2r_2 > r_3 \cap r_1 > 1$
3	III_1	DRM	$2r_2 < r_3 \cap r_1 < 1$
	III_2	DRM	$2r_2 < r_3 \cap r_1 > 1$
4	Line 1	CPM	$2r_2 = r_3$
	Line 2	CPM	$2r_2 = 2r_1 + r_3 \quad$ or $\quad r_2 = 1$
	Line 3	CPM	$2r_1 = 2r_2 + r_3 \quad$ or $\quad r_1 = 1$

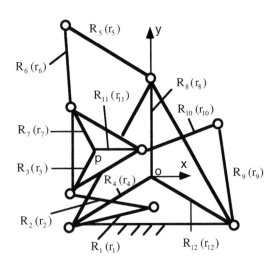

FIGURE 10.5 A typical 3-DOF PPM.

where R_i is the actual length of link i and r_i is the nondimensional relative length of link i. And

$$L = \frac{1}{12} \sum_{i=1}^{12} R_i \qquad (10.16)$$

where L is the average length of all links for the manipulator. Gosselin [31] has shown that the parallel manipulator should be symmetric, so that we have the following results:

$$
\begin{aligned}
R_9 &= R_5 = R_1 \\
R_{10} &= R_6 = R_2 \\
R_{11} &= R_7 = R_3 \\
R_{12} &= R_8 = R_4
\end{aligned}
\quad \text{and} \quad
\begin{aligned}
r_9 &= r_5 = r_1 \\
r_{10} &= r_6 = r_2 \\
r_{11} &= r_7 = r_3 \\
r_{12} &= r_8 = r_4
\end{aligned}
\qquad (10.17)
$$

Therefore, only four parameters $(r_1, r_2, r_3, \text{ and } r_4)$ are needed to consider. From Eqs. (10.15), (10.16), and (10.17), we see that the sum of the 12 nondimensional relative link lengths is

$$r_1 + r_2 + r_3 + r_4 = 4 \qquad (10.18)$$

Model of Solution Space for 3-DOF PPMs

If the manipulator can be assembled, the range of the nondimensional relative link lengths should be 0 to 2, so

$$0 < r_i < 2 \quad (i = 1, 2, \ldots, 12) \qquad (10.19)$$

Using these equations and inequalities, the model of the solution space for 3-DOF PPMs can be constructed, which is an irregular octahedron ABCDEF as shown in Fig. 10.6. Within the model, all 3-DOF PPMs can exist and the relationships between the different criteria and dimensions of the manipulators can be investigated.

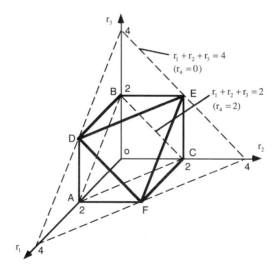

FIGURE 10.6 The physical model of the solution space for 3-DOF PPMs. (a) $r_4 = 0$, (b) $0 < r_4 < 1$, (c) $r_4 = 1$, (d) $1 < r_4 < 2$, (e) $r_4 = 2$.

Planar Closed Configurations with Coordinates $r_1r_2r_3$

When r_4 takes several values (such as 0, 0.2, 0.4, 0.6, 0.8, 1.0,..., 1.8), a set of plane equations can be obtained by means of Eq. 10.18. Therefore, r_4 is an auxiliary coordinate axis in the model of the solution space. When taking $r_4 = 0, 0 < r_4 < 1, r_4 < 1, 1 < r_4 < 2$, and $r_4 = 2$, respectively, five types of planar closed configurations with the three coordinates $r_1r_2r_3$ are obtained (as shown in Fig. 10.7). When $r_4 = 2$, the mechanisms have no motion.

As shown in Fig. 10.8, a planar closed configuration has three coordinates $r_1r_2r_3$. Since only two of them are independent in a plane, we can use two orthogonal coordinates x y to express $r_1r_2r_3$ by using the following equations. When r_1, r_2, and r_3 are given, we can calculate coordinates x and y from Eq. (10.20). When x and y are given, we can calculate r_1, r_3, and r_3 from Eq. (10.21).

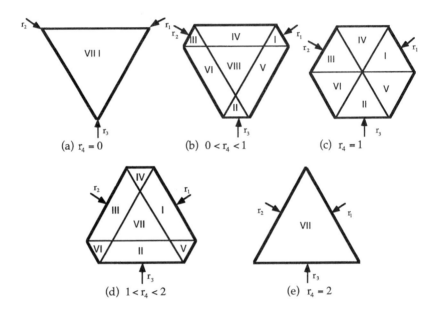

FIGURE 10.7   Five types of planar closed configurations.

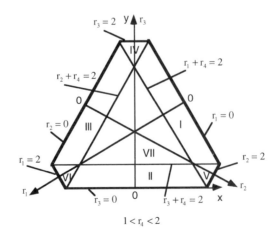

FIGURE 10.8   The coordinate transformation.

The Eqs. (10.20) and (10.21) are useful for constructing the performance atlases within the planar closed configurations.

$$\begin{pmatrix} x \\ y \end{pmatrix} = \begin{bmatrix} -\sqrt{3}/3 & \sqrt{3}/3 & 0 \\ 0 & 0 & 1 \end{bmatrix} \begin{pmatrix} r_1 \\ r_2 \\ r_3 \end{pmatrix} \tag{10.20}$$

$$\begin{pmatrix} r_1 \\ r_2 \\ r_3 \end{pmatrix} = \begin{bmatrix} -\sqrt{3}/2 & -1/2 & 2 - r_4/2 \\ \sqrt{3}/2 & -1/2 & 2 - r_4/2 \\ 0 & 1 & 0 \end{bmatrix} \begin{pmatrix} x \\ y \\ 1 \end{pmatrix} \tag{10.21}$$

Classification of all 3-DOF PPMs

Among the four links, r_1, r_2, r_3, and r_4 in a 3-DOF PPM, when the sum of any two link lengths is equal to the sum of the other two link lengths, it is designated as a change point 3-DOF PPM. Physically, it means that all four links (r_1, r_2, r_3, and r_4) can be collinear. According to the definition of change point 3-DOF PPM, we yield the conditions that express three kinds of change point 3-DOF PPMs as follows:

$$r_1 + r_4 = r_2 + r_3 = 2 \tag{10.22}$$

$$r_2 + r_4 = r_1 + r_3 = 2 \tag{10.23}$$

$$r_3 + r_4 = r_1 + r_2 = 2 \tag{10.24}$$

The Eqs. (10.22), (10.23), and (10.24) can define three planes, respectively. By using the three planes, the model of the solution space as shown in Fig. 10.6 can be divided into eight parts. Within the planar closed configurations, Eqs. (10.22), (10.23), and (10.24) express three lines (as shown in Figs. 10.7 and 10.8), which partition the planar closed configurations into eight regions I, II, III, IV, V, VI, VII, and VIII. Each region defines a type of 3-DOF PPMs. According to analysis of the dimensional characteristics of the 3-DOF PPMs within the regions I~VIII, the inequalities Eqs. (10.25)–(10.32) can be derived.

$$r_1 + r_2 < r_3 + r_4 \cap r_1 + r_3 < r_2 + r_4 \cap r_1 + r_4 < r_2 + r_3 \tag{10.25}$$

$$r_2 + r_1 < r_3 + r_4 \cap r_2 + r_3 < r_1 + r_4 \cap r_2 + r_4 < r_1 + r_3 \tag{10.26}$$

$$r_3 + r_1 < r_2 + r_4 \cap r_3 + r_2 < r_1 + r_4 \cap r_3 + r_4 < r_1 + r_2 \tag{10.27}$$

$$r_3 + r_1 > r_2 + r_4 \cap r_3 + r_2 < r_1 + r_4 \cap r_3 + r_4 > r_1 + r_2 \tag{10.28}$$

$$r_2 + r_1 > r_3 + r_4 \cap r_2 + r_3 < r_1 + r_4 \cap r_2 + r_4 > r_1 + r_3 \tag{10.29}$$

$$r_1 + r_2 > r_3 + r_4 \cap r_1 + r_3 < r_2 + r_4 \cap r_1 + r_4 > r_2 + r_3 \tag{10.30}$$

$$r_4 + r_1 > r_2 + r_3 \cap r_4 + r_2 < r_1 + r_3 \cap r_4 + r_3 > r_1 + r_2 \tag{10.31}$$

$$r_4 + r_1 < r_2 + r_3 \cap r_4 + r_2 < r_1 + r_3 \cap r_4 + r_3 < r_1 + r_2 \tag{10.32}$$

Table 10.3 shows the eleven types of 3-DOF PPMs, where 3-crank T-i (i = 1, 2, 3, and 4) means the 3-crank-type-i 3-DOF PPMs; 3-rocker T-i (i = 1, 2, 3, and 4) expresses the 3-rocker-type-i 3-DOF PPMs; and Change point T-i (i = 1, 2, and 3) denotes the change-point-type-i 3-DOF PPMs. Therefore, there are four types of 3-crank 3-DOF PPMs, four types of 3-rocker 3-DOF PPMs and three types of change point 3-DOF PPMs (as shown in Table 10.3).

Model of Solution Space for 3-DOF Delta Parallel Robots

Delta parallel robots (DPR) are a specific class of parallel manipulators that can position a platform in a region of 3-D space so that the platform remains parallel to a specified reference plane. DPRs have applications in the manipulation of lightweight objects for the electronic, food, and pharmaceutical industries. Although researchers have investigated the forward and inverse kinematics, inverse dynamics, sizes, singularities, and working volume of DPRs [44–50], insufficient attention has been given to the analysis of relationships involving the link lengths of DPRs.

We shall consider a DPR, as shown in Fig. 10.9, where R_i is the length of the ith link; R_1, R_5, R_9, are the lengths of the input links; R_2, R_6, R_{10} are the   lengths of the parallelogram supporting rods; R_3, R_7, R_{11} are the lengths of the output links; and R_4, R_8, R_{12} are the lengths of the fixed link.

TABLE 10.3 Complete Classification of 3-DOF PPMs

No.	Region	Proposed Name	Characteristics of link lengths	Dimensional Characteristics
1	I	3-crank T-1	r_1 =the shortest	Inequality (10.25)
2	II	3-crank T-2	r_2 =the shortest	Inequality (10.26)
3	III	3-crank T-3	r_3 =the shortest	Inequality (10.27)
4	IV	3-rocker T-4	r_3 =the longest	Inequality (10.28)
5	V	3-rocker T-1	r_2 =the longest	Inequality (10.29)
6	VI	3-rocker T-2	r_1 =the longest	Inequality (10.30)
7	VII	3-rocker T-3	r_4 =the longest	Inequality (10.31)
8	VIII	3-crank T-4	r_4 =the shortest	Inequality (10.32)
9	Line 1	Change point T-1	$r_1 + r_4 = r_2 + r_3$	Equality (10.22)
10	Line 2	Change point T-2	$r_2 + r_4 = r_1 + r_3$	Equality (10.23)
11	Line 3	Change point T-3	$r_3 = r_4 + r_1 + r_2$	Equality (10.24)

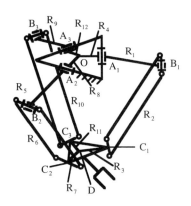

FIGURE 10.9   A typical 3-DOF DPR.

Nondimensional Parameters of DPRs

We define

$$L = \frac{1}{12} \sum_{i=1}^{12} R_i \tag{10.33}$$

By dividing the link length R_i $(i = 1, 2,..., 12)$ by L, we obtain twelve nondimensional parameters

$$r_i = R_i / L \quad (i = 1, 2,..., 12) \tag{10.34}$$

Therefore,

$$\sum_{i=1}^{12} r_i = 12 \tag{10.35}$$

Gosselin [31] proposed that parallel robotic manipulators should be symmetric because the tasks to be performed by the manipulators are unknown and unpredictable. By symmetry, one obtains

$$
\begin{aligned}
R_1 &= R_5 = R_9 \\
R_2 &= R_6 = R_{10} \\
R_3 &= R_7 = R_{11} \\
R_4 &= R_8 = R_{12}
\end{aligned}
\quad \text{and} \quad
\begin{aligned}
r_1 &= r_5 = r_9 \\
r_2 &= r_6 = r_{10} \\
r_3 &= r_7 = r_{11} \\
r_4 &= r_8 = r_{12}
\end{aligned}
\tag{10.36}
$$

The symmetry assumption will be used throughout this paper. From Eqs. (10.35) and (10.36), the relative link lengths of the DPR satisfy the relationship

$$r_1 + r_2 + r_3 + r_4 = 4 \tag{10.37}$$

It is well known that a DPR cannot be synthesized if one of the link lengths exceeds the following maximum values:

$$r_1 = 4, \quad r_2 = 4, \quad r_3 = 2, \quad \text{and} \quad r_4 = 2 \tag{10.38}$$

If one of all the link lengths r_i $(i = 1,2,3, \text{ and } 4)$ satisfies Eqs. (10.37) and (10.38), the DPR can be synthesized, but it cannot move. Eq. (10.38) is the zero-mobility condition for DPRs, which defines four zero-mobility planes.

Model of Solution Space for DPRs

Let r_1, r_2, and r_3 be orthogonal coordinate axes. Using Eqs. (10.37) and (10.38) and the conditions

$$0 < r_1 < 4, \quad 0 < r_2 < 4, \quad 0 \leq r_3 < 2, \quad \text{and} \quad 0 \leq r_4 < 2, \tag{10.39}$$

we can establish the model of the solution space as shown in Fig. 10.10. The model is an irregular hexahedron ABCDEFG as shown in Fig. 10.10(b), for which any possible combination of the link lengths is represented by values of the links r_1, r_2, r_3, and r_4. The resulting model provides a means to reduce the 12-D infinite space to the 3-D finite one.

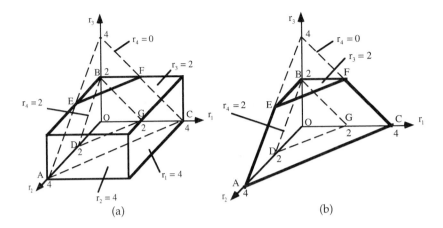

FIGURE 10.10   The model of the solution space for 3-DOF DPRs. (a) $r_1 + r_4 = r_2 + r_3$, (b) $r_2 + r_4 = r_1 + r_3$.

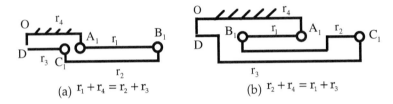

FIGURE 10.11   Two types of the branches of change point DPR. (a) $r_4 = 0$, (b) $0 < r_4 < 1$, (c) $r_4 = 1$, (d) $1 < r_4 < 2$, (e) $r_4 = 2$.

Planar Closed Configurations

Consider the link r_i ($i = 1, 2, 3,$ and 4) in a DPR, where the sum of any two link lengths, except r_3 and r_4, is equal to the sum of the other two link lengths. For this situation we call the DPR a change point one. Physically, this means that all four links may be collinear (as shown in Fig. 10.11). Therefore, two change point planes are defined by:

$$r_1 + r_4 = r_2 + r_3 \tag{10.40}$$

$$r_2 + r_4 = r_1 + r_3. \tag{10.41}$$

By giving several values for r_4, a set of plane equations can be obtained from Eq. (10.37). Therefore, r_4 is an auxiliary coordinate axis in the model of the solution space. When r_4 is given, Eq. (10.37) can be expressed as

$$r_1 + r_2 + r_3 = 4 - r_4 \tag{10.42}$$

which is an equation of a plane. When $r_4 = 0$; $r_4 < 1$; $r_4 = 1$; $r_4 > 1$; and $r_4 = 2$, respectively, a set of planes is obtained. These planes are called "planar closed configurations" with coordinates $r_1\, r_2\, r_3$, which are isosceles trapezoids, as shown in Fig. 10.12.

Using Eqs. (10.40) and (10.41), lines 1 and 2 can be drawn on the five types of planar closed configurations as shown in Fig. 10.12. These two lines are change point lines on which all the change point DPRs exist.

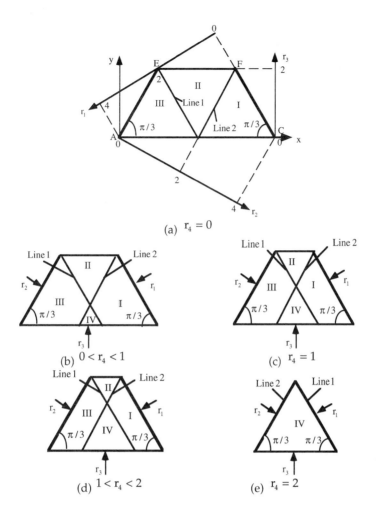

FIGURE 10.12   Five types of planar closed configurations.

Coordinate Transformation

As shown in Fig. 10.12(a), the planar closed configurations have coordinates $r_1\ r_2\ r_3$. Only two of these are independent. For convenience, we should utilize two orthogonal coordinates $x\ y$ to express $r_1\ r_2\ r_3$. Thus, by use of

$$\begin{pmatrix} x \\ y \end{pmatrix} = \begin{bmatrix} 2\sqrt{3}/3 & \sqrt{3}/3 \\ 0 & 1 \end{bmatrix} \begin{pmatrix} r_2 \\ r_3 \end{pmatrix}, \tag{10.43}$$

we can transform coordinates $r_1\ r_2\ r_3$ into $x\ y$. Eq. (10.43) is very useful for construction of the performance atlases. In addition, if the values of x and y are known, the values of r_1, r_2, and r_3 can be calculated by

$$\begin{pmatrix} r_1 \\ r_2 \\ r_3 \end{pmatrix} = \begin{bmatrix} -\sqrt{3}/2 & -1/2 & 4 - r_4 \\ \sqrt{3}/2 & -1/2 & 0 \\ 0 & 1 & 0 \end{bmatrix} \begin{pmatrix} x \\ y \\ 1 \end{pmatrix} \tag{10.44}$$

Classification of DPRs

By using Eqs. (10.40) and (10.41), lines 1 and 2 can be drawn on the planar closed configurations (as shown in Fig. 10.12). These lines separate planar closed configurations into four regions I, II, III, and IV. Each expresses a class of DPRs. For convenience of classification, we define a virtual planar slider-crank mechanism $OA_1B_1C_1D$, which is a branch of the Delta mechanism as shown in Fig. 10.13.

Because Delta mechanisms treated in this section are symmetric, we only need to consider the four links r_i ($i = 1, 2, 3,$ and 4) which describe a virtual planar slider-crank mechanism. In Fig. 10.13, point D is the center of the platform when the three input links have the same angle with respect to their start positions, i.e., when the three input links r_1, r_5 and r_9 are located at the same plane with the frame. We assume point D is a virtual slider that can slide only along the line OD, because only when the center of the platform (point D) moves along the line OD do the three input links have the same motion and it is possible to investigate the mobility of Delta mechanisms. From the virtual planar slider-crank mechanism, we derive the velocity equation

$$\dot{\phi}_1 = J\dot{\phi}_2$$

We know that when input link r_1 is located at the extreme position and the angular velocity $\dot{\phi}_2$ is non-zero, the angular velocity $\dot{\phi}_1$ must be zero. Thus, we obtain the results:

$$\dot{\phi}_1 = 0 \quad \text{and} \quad J = 0$$

From this, we derive the conditions which can be used to determine the extreme positions of link r_1, as shown in Fig. 10.14, that is,

$$\phi_2 = \frac{\pi}{2} \text{ or } -\frac{\pi}{2}$$

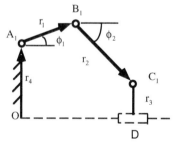

FIGURE 10.13 Virtual planar slider-crank mechanism for DPRs. (a) $\phi_2 = -\frac{\pi}{2}$, (b) $\phi_2 = \frac{\pi}{2}$.

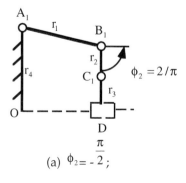

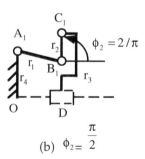

(a) $\phi_2 = -\frac{\pi}{2}$;   (b) $\phi_2 = \frac{\pi}{2}$

FIGURE 10.14 Virtual mechanisms with extreme positions of input link r_1. (a) Perspective view, (b) top view.

or these equations.

$$\phi_{1\,max} = \sin^{-1}[(r_2 + r_3 - r_4)/r_1] \qquad (10.45)$$

$$\phi_{1\,MIN} = \sin^{-1}[(r_3 - r_2 - r_4)/r_1] \qquad (10.46)$$

That is, if the virtual slider-crank mechanism satisfies one of the Eqs. (10.45) and (10.46), the revolving input link does not exist. On the contrary, if it does not satisfy both of them at the same time, the conditions for existence of the revolving input link r_1 are obtained:

$$|(r_2 + r_3 - r_4)/r_1| \geq 1 \qquad (10.47)$$

$$|(r_3 - r_2 - r_4)/r_1| \geq 1 \qquad (10.48)$$

From Eqs. (10.47) and (10.48), we can further derive the conditions for existence of the revolving input link of the virtual slider-crank mechanism as

$$r_1 + r_4 \leq r_2 + r_3 \qquad (10.49)$$

$$r_1 + r_3 \leq r_2 + r_4 \qquad (10.50)$$

Using these conditions, we can classify the Delta mechanisms. Table 10.4 describes the resulting classification. We achieve six classes of Delta parallel robots, that is, one type of 3-crank Delta mechanisms, three types of 3-rocker Delta mechanisms, and two types of change point Delta mechanisms. Table 10.4 also shows the dimensional characteristics of these classes.

Model of Solution Space for F /T Sensors Based on Stewart Platform

Force/torque (F/T) sensors can be used for monitoring forces of variable directions and intensity (such as wind-tunnel testing, adaptive control of machines and thrust stand testing of rocket engines [51]), measuring inertia force (computer input device: Smartpen [52]) or contact force to feed it back to the command signal and estimating the location of the contacts between robot and environment through force measurements so that they have been applied to manufacturing, robotics, military, electronic, and computer industries, and so on. Therefore, a multiaxis F/T sensor is a critical component of an automated system for extending the capability of manipulation and assembly, especially with contact tasks that require mechanical operations involving interaction with the environment or objects.

Although many kinds of F/T sensors have been developed, Stewart-platform transducer F/T sensor is a novel and specific one. Some research has been devoted to this kind of sensor. Kerr [53] presented Stewart-platform transducer force sensor, and Nguyen [54] and Ferraresi [55] analyzed Stewart-platform

TABLE 10.4 Classification of DPRs

No.	Region	Name	Dimensional Characteristics
1	I	3-Crank DPR	$r_1 + r_4 < r_2 + r_3 \cap r_1 + r_3 < r_2 + r_4$
2	II	3-Rocker 1 DPR	$r_1 + r_4 < r_2 + r_3 \cap r_1 + r_3 > r_2 + r_4$
3	III	3-Rocker 2 DPR	$r_1 + r_4 > r_2 + r_3 \cap r_1 + r_3 > r_2 + r_4$
4	IV	3-Rocker 3 DPR	$r_1 + r_4 > r_2 + r_3 \cap r_1 + r_3 < r_2 + r_4$
5	line 1	Change Point 1DPR	$r_1 + r_4 = r_2 + r_3$
6	line 2	Change Point 2DPR	$r_1 + r_3 = r_2 + r_4$

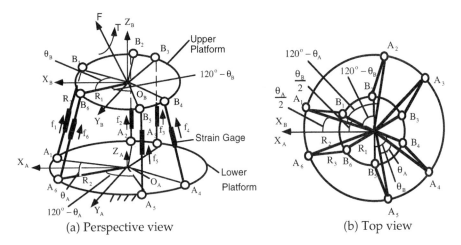

(a) Perspective view   (b) Top view

FIGURE 10.15   The mechanism of the sensor based on Stewart platform.

force sensor. Because the structure of Stewart-platform force sensor is a kind of parallel mechanism, the technique for analysis and design of parallel mechanisms can be used for the F/T sensor.

In this section we propose a model of the solution space for the structures of the F/T sensors based on the Stewart platform, which is the foundation for investigation of the design method of the sensor.

Consider the F/T sensor mechanism based on Stewart platform as shown in Fig. 10.15. Because the link lengths may have a wide range of possible values, it is convenient to avoid explicit use of the physical sizes of the mechanisms during analysis and design. We shall define normalized parameters of the sensor mechanisms and construct the model of the solution space.

Since the parallel mechanisms should be symmetric [31], there are four parameters in the sensor mechanism as shown in Fig. 10.15, that is, R_1, R_2, R_3, and θ_{AB}, where

$$\theta_{AB} = |\theta_A - \theta_B| \quad (0° < \theta_{AB} < 120°)$$
$$\theta_A = \angle A_2A_3 = \angle A_4A_5 = \angle A_6A_1$$
$$\theta_B = \angle B_2B_3 = \angle B_4B_5 = \angle B_6B_1$$
$$180° - \theta_A = \angle A_1A_2 = \angle A_3A_4 = \angle A_5A_6$$
$$180° - \theta_B = \angle B_1B_2 = \angle B_3B_4 = \angle B_5B_6$$

Let

$$L = (R_1 + R_2 + R_3)/3 \tag{10.51}$$

$$r_1 = R_1/L, \quad r_2 = R_2/L, \quad \text{and} \quad r_3 = R_3/L \tag{10.52}$$

where R_i is the length of link i, r_i is the normalized, nondimensional length of link i, and L is the average link length of the mechanism. Therefore, the sum of the normalized link lengths is

$$r_1 + r_2 + r_3 = 3 \tag{10.53}$$

If the mechanism can be assembled, the normalized link lengths satisfy

$$0 < r_1 < 1.5, \quad 0 < r_2 < 1.5, \quad \mu < r_3 < 3 \tag{10.54}$$

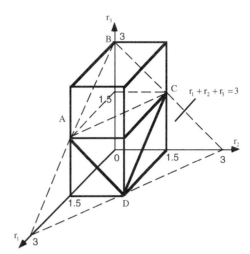

FIGURE 10.16 A physical model of the solution space for the sensor mechanisms.

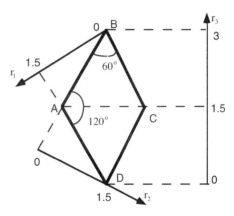

FIGURE 10.17 Planar closed configuration of the solution space.

where

$$\mu = \sqrt{r_1^2 + r_2^2 - 2r_1r_2\cos\left(\frac{\theta_{AB}}{2}\right)} \quad \text{and} \quad 0 < \mu \tag{10.55}$$

From the conditions (10.53) and (10.54), a physical model of the solution space for the sensor mechanisms can be constructed as shown in Fig. 10.16, which is the rhombus ABCD. By placing the model on a plane, the planar closed configuration of the solution space with three coordinates can be obtained as shown in Fig. 10.17. Within the model ABCD, we can investigate relationships between the criteria and parameters of the sensor mechanisms.

10.3 Performance Evaluation Criteria for Design of Robotic Mechanisms

The design of robotic mechanisms can be simplified by the criteria that permit the analysis of tradeoffs. In recent years, researchers have presented some performance criteria based on workspace geometry [56–64], isotropy [65–68], dexterity [69–73], global conditioning index [43], and singularity [74–76]. By use of the criteria, optimization has been used as a design tool [31, 77–81].

In this section, we list the major performance criteria that provide a foundation for using the technique of the solution space models to design robotic mechanisms.

Three Kinds of Workspaces

We consider a robot consisting of a manipulator and an end-effector. The robot structure may be serial or parallel. The workspace of a robot defines the useful positions and orientations of the end-effector. In this section, we consider the following three types of workspaces:

- Reachable Workspace: the set of points attainable by a point attached to the end-effector.
- Dextrous Workspace: the set of points of the workspace in which an end-effector can have arbitrary orientations.
- Global Workspace: the set of points that the end-effector can reach when all orientations of the end-effector are given [31].

Three Kinds of Singularities

To improve robot performance, singular configurations must be identified at the design stage because singularities lead to an instantaneous change in the degrees of freedom of the mechanism. They also result in a loss of the controllability and degradation of the natural stiffness that may lead to high joint forces or torques. There are three kinds of singularities [75]. The first type occurs when

$$\det(J) = 0 \qquad (10.56)$$

where [J] is the Jacobian matrix of the robotic mechanism. The corresponding configurations are located at the boundary of the workspace of the manipulator or on internal boundaries between regions of the workspace in which the number of solutions of the inverse kinematic problem differs.

A second type of singularity occurs when

$$\det(J) \to \infty. \qquad (10.57)$$

This type of singularity results in an unwanted degree of freedom of the manipulator. Such a configuration must be avoided because the manipulator is not controllable. This type of singularity, however, cannot occur in serial manipulators.

A third type of singularity only occurs for parallel manipulators,

$$\det(J) \to \frac{0}{0}. \qquad (10.58)$$

Global Conditioning Index

The global conditioning index is an important measure for control of the manipulator. It is defined as the inverse of the condition number of the Jacobian matrix [J] integrated over the reachable workspace and divided by the volume of the workspace. In particular, a large value of the index ensures that the manipulator can be precisely controlled. To evaluate the global behavior of a manipulator, Gosselin [43] defined the global conditioning index,

$$\eta = A/B, \qquad (10.59)$$

where

$$A = \int_w \left(\frac{1}{\|J\| \|J^{-1}\|} \right) dw$$
$$\qquad (10.60)$$
$$B = \int_w dw$$

Here, B is the volume of the reachable workspace w, and

$$\|J\| = \sqrt{tr\left(J\frac{1}{n}[I]J^T\right)} \tag{10.61}$$

where n is the dimension of the square matrix $[J]$, and $[I]$ is the identity matrix having the same dimension as $[J]$.

Global Velocity Index

The global velocity index is a measure of robotic speed. We define the global maximum and minimum velocity indices to be the extreme values of the end-effector velocities integrated over the reachable workspace and divided by the volume of the workspace.

The linear velocity V and angular velocity ω of the end-effector are related to the input velocities $\dot{\phi}$ by

$$\binom{V}{\omega} = [J](\dot{\phi}) = \begin{bmatrix} J_v \\ J_\omega \end{bmatrix}(\dot{\phi}) \tag{10.62}$$

From Eq. (62), we obtain

$$\|V\|^2 = (\dot{\phi})^T[J_v]^T[J_v](\dot{\phi}) \tag{10.63}$$

$$\|\omega\|^2 = (\dot{\phi})^T[J_\omega]^T[J_\omega](\dot{\phi}) \tag{10.64}$$

Let

$$\|\dot{\phi}\|^2 = (\dot{\phi})^T(\dot{\phi}) = 1, \tag{10.65}$$

$$L_v = (\dot{\phi})^T[J_v]^T[J_v](\dot{\phi}) - \lambda_v[(\dot{\phi})^T(\dot{\phi}) - 1] \tag{10.66}$$

and

$$L_v = (\dot{\phi})^T[J_\omega]^T[J_\omega](\dot{\phi}) - \lambda_\omega[(\dot{\phi})^T(\dot{\phi}) - 1] \tag{10.67}$$

where λ_v and λ_ω are Lagrange multipliers. From Eqs. (10.66) and (10.67), necessary conditions for extreme values of the linear and angular velocities of the end-effector are

$$\frac{\partial L_v}{\partial \lambda_v} = 0 : (\dot{\phi})^T(\dot{\phi}) - 1 = 0, \quad \frac{\partial L_v}{\partial(\dot{\phi})} = 0 : [J_v]^T[J_v](\dot{\phi}) - \lambda_v(\dot{\phi}) = 0 \tag{10.68}$$

$$\frac{\partial L_\omega}{\partial \lambda_\omega} = 0 : (\dot{\phi})^T(\dot{\phi}) - 1 = 0, \quad \frac{\partial L_\omega}{\partial(\dot{\phi})} = 0 : [J_\omega]^T[J_\omega](\dot{\phi}) - \lambda_\omega(\dot{\phi}) = 0 \tag{10.69}$$

From Eqs. (10.68) and (10.69), we see that λ_v and λ_ω are eigenvalues of $[J_v]^T[J_v]$ and $[J_\omega]^T[J_\omega]$, respectively, so that extreme values of the velocities are

$$\|V_{max}\| = \sqrt{\lambda_{V\,max}} \quad \text{and} \quad \|V_{min}\| = \sqrt{\lambda_{V\,min}} \tag{10.70}$$

$$\|\omega_{max}\| = \sqrt{\lambda_{\omega\,max}} \quad \text{and} \quad \|\omega_{min}\| = \sqrt{\lambda_{\omega\,min}} \tag{10.71}$$

Because [J] depends on the configuration of the manipulator, extreme values of the end-effector velocity cannot characterize the robotic performance. We define global velocity index, a criterion involving the end-effector velocity, as follows:

$$\gamma_{V\,max} = \frac{C_{V\,max}}{B} \quad \text{and} \quad \gamma_{V\,min} = \frac{C_{v\,min}}{B} \tag{10.72}$$

$$\gamma_{\omega\,max} = \frac{C_{\omega\,max}}{B} \quad \text{and} \quad \gamma_{\omega\,min} = \frac{C_{\omega\,min}}{B} \tag{10.73}$$

where $\gamma_{V\,max}$ and $\gamma_{V\,min}$ are the global maximum and minimum linear velocity indices, respectively; $\gamma_{\omega\,max}$ and $\gamma_{\omega\,min}$ are the global maximum and minimum angular velocity indices, respectively; B is the volume of the reachable workspace; and

$$C_{v\,max} = \int_{w}(V_{max})dw \quad \text{and} \quad C_{V\,min} = \int_{w}(V_{min})dw,$$

$$C_{\omega\,max} = \int_{w}(\omega_{max})dw \quad \text{and} \quad C_{\omega\,min} = \int_{w}(\omega_{min})dw.$$

Global Payload Index

The global payload index is a measure of the capability of payload that can be handled by the robotic mechanisms. We define the global maximum and minimum payload indices to be the extreme values of the robotic payload integrated over the reachable workspace of the robot and divided by the volume of the workspace.

The external force F and torque T applied at the end-effector are related to the input force or torque τ by

$$(\tau) = [J]^{T}\binom{F}{T} \tag{10.74}$$

If $\det(J^{T}) \neq 0$, we obtain

$$\binom{F}{T} = [J^{T}]^{-1}(\tau) = \begin{bmatrix} J_{F} \\ J_{T} \end{bmatrix}(\tau) \tag{10.75}$$

Using the same method as in the previous section,

$$\|F_{max}\| = \sqrt{\lambda_{F\,max}} \quad \text{and} \quad \|F_{min}\| = \sqrt{\lambda_{F\,min}} \tag{10.76}$$

$$\|T_{max}\| = \sqrt{\lambda_{T\,max}} \quad \text{and} \quad \|T_{min}\| = \sqrt{\lambda_{T\,min}} \tag{10.77}$$

where $\lambda_{F\,max}$ and $\lambda_{F\,min}$ are the maximum and minimum eigenvalues of the matrix $[J_{F}]^{T}[J_{F}]$, respectively; and $\lambda_{T\,max}$ and $\lambda_{T\,min}$ are the maximum and minimum eigenvalues of the matrix $[J_{T}]^{T}[J_{T}]$, respectively.

When the input torque is a unit vector, the global payload indices are

$$\gamma_{F\,max} = \frac{C_{F\,max}}{B} \quad \text{and} \quad \gamma_{F\,min} = \frac{C_{F\,min}}{B} \tag{10.78}$$

$$\gamma_{T\,max} = \frac{C_{T\,max}}{B} \quad \text{and} \quad \gamma_{T\,min} = \frac{C_{T\,min}}{B}o \tag{10.79}$$

where $\gamma_{F\,max}$ and $\gamma_{F\,min}$ are the global maximum and minimum force payload indices, respectively; $\gamma_{T\,max}$ and $\gamma_{T\,min}$ are the global maximum and minimum torque payload indices, respectively; and

$$C_{F\,max} = \int_w \|F_{max}\| \, dw \quad \text{and} \quad C_{F\,min} = \int_w \|F_{min}\| \, dw \tag{10.80}$$

$$C_{T\,max} = \int_w \|T_{max}\| \, dw \quad \text{and} \quad C_{T\,min} = \int_w \|T_{min}\| \, dw \tag{10.81}$$

Global Deformation Index

The global deformation index is a measure of the stiffness of the robotic end-effector. We define the global maximum and minimum deformation indices to be the extreme values of the end-effector deformation integrated over the reachable workspace of the robot and divided by the volume of the workspace.

The input forces or torques are related to the deformation by

$$(\tau) = [K](\Delta q) \tag{10.82}$$

where

$$[K] = [I](k_1 \quad k_2 \dots k_m)^T$$
$$(\Delta q) = (\Delta q_1 \quad \Delta q_2 \dots \Delta q_m)^T \tag{10.83}$$

In Eq. (10.83), Δq_i is the deformation of joint i of the input link i; k_i is the stiffness of actuator i at joint i ($i = 1, 2, ..., m$); and $[I]$ is an $m \times m$ identity matrix. We assume

$$k_i = k, \tag{10.84}$$

which means that all actuators have the same stiffness. From Eqs. (10.74) and (10.82), the deformation (D) of the end-effector is

$$(D) = [J](\Delta q) = [J][K]^{-1}[J]^T(F \quad T)^T = [C](F \quad T)^T \tag{10.85}$$

From Eqs. (10.84) and (10.85), we obtain a representation for the compliance matrix of the manipulator as

$$[C] = [J][K]^{-1}[J]^T = \frac{1}{k}[J][J]^T \tag{10.86}$$

$[C]^{-1}$ is the stiffness matrix. Because we only investigate the relationship between the criteria and link lengths of robots, let k be equal to 1. From Eqs. (10.85) and (10.86), we obtain

$$(D) = \begin{pmatrix} D_P \\ D_O \end{pmatrix} = [C]\begin{pmatrix} F \\ T \end{pmatrix} = \begin{bmatrix} C_P \\ C_O \end{bmatrix}\begin{pmatrix} F \\ T \end{pmatrix} \tag{10.87}$$

where (D_P) and (D_O) express the position deformation and orientation deformation of the end-effector, respectively. Thus, we have

$$(D_P) = [C_P](F\ T)^T \tag{10.88}$$

$$(D_O) = [C_O](F\ T)^T \tag{10.89}$$

Using the method of Section 3.4, we derive

$$\|D_{P\max}\| = \sqrt{\lambda_{D_P\,\max}} \quad \text{and} \quad \|D_{P\min}\| = \sqrt{\lambda_{D_P\,\min}} \tag{10.90}$$

$$\|D_{O\max}\| = \sqrt{\lambda_{D_O\,\max}} \quad \text{and} \quad \|D_{O\min}\| = \sqrt{\lambda_{D_O\,\min}} \tag{10.91}$$

where $\lambda_{D_P\,\max}$ and $\lambda_{D_P\,\min}$ are maximum and minimum eigenvalues of $[C_P]^T[C_P]$, respectively, and $\lambda_{D_O\,\max}$ and $\lambda_{D_O\,\min}$ are maximum and minimum eigenvalues of $[C_O]^T[C_O]$.

The global positional and orientational deformation indices can be represented by

$$\gamma_{D_P\,\max} = \frac{A_{P\,\max}}{B} \quad \text{and} \quad \gamma_{D_P\,\min} = \frac{A_{P\,\min}}{B} \tag{10.92}$$

$$\gamma_{D_O\,\max} = \frac{A_{O\,\max}}{B} \quad \text{and} \quad \gamma_{D_O\,\min} = \frac{A_{O\,\min}}{B} \tag{10.93}$$

where $\gamma_{D_P\,\max}$ and $\gamma_{D_P\,\min}$ are the global maximum and minimum position deformation indices, respectively; $\gamma_{D_O\,\max}$ and $\gamma_{D_O\,\min}$ are the global maximum and minimum orientation deformation indices, respectively; and

$$A_{P\,\max} = \int_w \|D_{P\,\max}\| dw \quad \text{and} \quad A_{P\,\min} = \int_w \|D_{P\,\min}\| dw\,,$$

$$A_{O\,\max} = \int_w \|D_{O\,\max}\| dw \quad \text{and} \quad A_{O\,\min} = \int_w \|D_{O\,\min}\| dw\,.$$

Global Error Index

The global error index is a measure of accuracy of the robotic end-effector. We define the global maximum and minimum error indices to be the extreme values of the end-effector error integrated over the reachable workspace of the robot and divided by the volume of the workspace.

Errors in the end-effector motion can be divided into two parts,

$$(\Delta E) = (\Delta E_\theta) + (\Delta E_D) \tag{10.94}$$

where (ΔE_θ) is the error concerning the input motion tolerance $\Delta\theta$; and (ΔE_D) is the error relative to the tolerance of the dimensions of the robotic mechanism. Therefore, we obtain the position error of the end-effector as

$$(\Delta E_P) = (\Delta E_{\theta P}) + (\Delta E_{DP}) \tag{10.95}$$

and the orientation error of the end-effector as

$$(\Delta E_O) = (\Delta E_{\theta O}) + (\Delta E_{DO}). \tag{10.96}$$

Since

$$\begin{pmatrix} \Delta E_{\theta P} \\ \Delta E_{\theta O} \end{pmatrix} = [J](\Delta \theta) = \begin{bmatrix} J_V \\ J_\omega \end{bmatrix} (\Delta \theta) \tag{10.97}$$

Comparing Eq. (10.97) and (10.62), we obtain

$$\gamma_{E_{\theta P} \, max} = \gamma_{V \, max} \quad \text{and} \quad \gamma_{E_{\theta P} \, min} = \gamma_{V \, min} \tag{10.98}$$

$$\gamma_{E_{\theta O} \, max} = \gamma_{\omega \, max} \quad \text{and} \quad \gamma_{E_{\theta O} \, min} = \gamma_{\omega \, min} \tag{10.99}$$

where $\gamma_{E_{DP} \, max}$ and $\gamma_{E_{DP} \, min}$ are the global maximum and minimum position error indices with respect to the input motion tolerance $\Delta \theta$, respectively; $\gamma_{E_{\theta O} \, max}$ and $\gamma_{E_{\theta O} \, min}$ are the global maximum and minimum orientation error indices with respect to the input motion tolerance $\Delta \theta$, respectively.

Next, relative to the dimensional tolerance of the mechanism, we consider errors

$$\begin{pmatrix} \Delta E_{DP} \\ \Delta E_{DO} \end{pmatrix} = [J_D](\Delta D) = \begin{bmatrix} J_{DP} \\ J_{DO} \end{bmatrix} (\Delta D) \tag{10.100}$$

where (ΔD) is the dimensional tolerance of the robotic mechanism, and $[J_D]$ is the error transformation matrix with respect to the dimensional error.

Using the method in the section entitled "Global Velocity Index," we derive

$$\left\| \Delta E_{DP_{max}} \right\| = \sqrt{\lambda_{E_{DP} \, max}} \quad \text{and} \quad \left\| \Delta E_{DP_{min}} \right\| = \sqrt{\lambda_{E_{DP} \, min}} \tag{10.101}$$

$$\left\| \Delta E_{DO_{max}} \right\| = \sqrt{\lambda_{E_{DO} \, max}} \quad \text{and} \quad \left\| \Delta E_{DO_{min}} \right\| = \sqrt{\lambda_{E_{DO} \, min}} \tag{10.102}$$

where $\lambda_{E_{DP} \, max}$ and $\lambda_{E_{DP} \, min}$ are maximum and minimum eigenvalues of matrix $[J_{DP}]^T[J_{DP}]$, respectively; $\lambda_{E_{DO} \, max}$ and $\lambda_{E_{DO} \, min}$ are maximum and minimum eigenvalues of $[J_{DO}]^T[J_{DO}]$. Therefore, the global position and orientation error indices relative to the dimensional tolerances can be obtained by

$$\gamma_{E_{DP} \, max} = \frac{A_{E_{DP} \, max}}{B} \quad \text{and} \quad \gamma_{E_{DP} \, min} = \frac{A_{E_{DP} \, min}}{B} \tag{10.103}$$

$$\gamma_{E_{DO} \, max} = \frac{A_{E_{DO} \, max}}{B} \quad \text{and} \quad \gamma_{E_{DO} \, min} = \frac{A_{E_{DO} \, min}}{B} \tag{10.104}$$

where $\gamma_{E_{DP} \, max}$ and $\gamma_{E_{DP} \, min}$ are the global maximum and minimum position error indices relative to the tolerance of the dimensions of the mechanism, respectively; $\gamma_{E_{DO} \, max}$ and $\gamma_{E_{DO} \, min}$ are the global maximum and minimum orientation error indices relative to the dimensional tolerance of the mechanism, respectively; and

$$A_{E_{DP} \, max} = \int_w \left\| \Delta E_{DP \, max} \right\| dw \quad \text{and} \quad A_{E_{DP} \, min} = \int_w \left\| \Delta E_{DP \, min} \right\| dw,$$

$$A_{E_{DO} \, max} = \int_w \left\| \Delta E_{DO \, max} \right\| dw \quad \text{and} \quad A_{E_{DO} \, min} = \int_w \left\| \Delta E_{DO \, min} \right\| dw,$$

The criteria described in this section involve kinematic and dynamic properties and characterize robot performance. These results provide a basis for analyzing and designing robotic mechanisms. Because the solution space model applies to both serial and parallel robotic mechanisms, the criteria have wide application to many different types of robotic mechanisms.

10.4   Performance Atlases for Design of Serial Robots with Three Moving Links

Human arms, legs, fingers, and limbs of animals and insects can be viewed as mechanisms with three moving links. Fingers of dextrous hands, arms of industrial robots, and legs of walking machines often consist of three moving links. Therefore, an understanding of the relationships between the criteria and the dimensions of three-moving-link mechanisms is of great importance for design of fingers, arms, and legs of robotics machines.

In this section, we use the technique of physical models of solution space proposed in Section 10.2 and the evaluation criteria presented in Section 10.3 to investigate the relationships between the criteria and the dimensions of mechanisms with three moving links. In addition, the distribution of the 63 commercially available industrial and research robots within the model of solution space is discussed.

Robotic Principle Motion

Although a biological system can provide compact actuation of fingers, arms, and legs with multiple degrees of freedom, planar motion provides the most significant contribution to kinematic and dynamic performance. We define the principle motions produced by robotic mechanisms with three moving links as that produced by joints sharing a common and parallel axis. Since industrial robots often are built in this manner, this type of motion is very important for design.

By using the theory of the principle motions, many robotic mechanisms with three moving links as shown in Fig. 10.3 can be simplified as 3-DOF planar mechanisms. For this reason, the treatment described in this chapter is restricted to mechanisms that produce principle motions, that is, 3-DOF planar serial mechanisms as shown in Fig. 10.18.

Atlases of Workspace Criteria

The workspace of a manipulator determines attainable positions and orientations of its end-effector. We shall consider the reachable workspace, dextrous workspace, and global workspace. The atlases of reachable, dextrous, and global workspaces can be represented on the solution space as shown in Figs. 10.19, 10.21, and 10.22.

Reachable Workspace Atlas

The reachable workspace of a 3-DOF planar serial mechanism, as shown in Fig. 10.18, consists of all points traversed by the end effector. Two types of reachable workspaces are the loop and the plate as

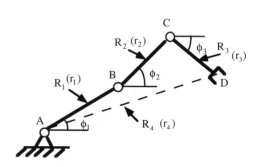

FIGURE 10.18   A 3-DOF planar mechanism.

TABLE 10.5 Two Types of Reachable Workspace

Region	Shape	r_{MAX}	r_{MIN}	Area
I		$r_1 + r_2 + r_3$	$r_2 - r_1 - r_3$	$4\pi(3 - r_1)r_1$
II		$r_1 + r_2 + r_3$	$r_2 - r_1 - r_3$	$4\pi(3 - r_2)r_2$
III		$r_1 + r_2 + r_3$	$r_3 - r_1 - r_2$	$4\pi(3 - r_3)r_3$
IV		$r_1 + r_2 + r_3$	0	9π

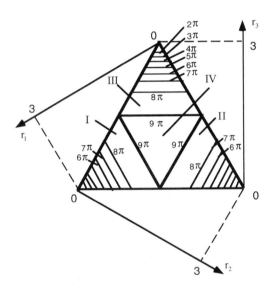

FIGURE 10.19 Reachable workspace atlas. (a) Single loop, (b) plate, (c) double loops.

shown in Table 10.5. Using this information, one can calculate the areas of the reachable workspace of all possible 3-DOF planar serial mechanisms and plot the contours of equal area on the model of the solution space. Figure 10.19 shows the reachable workspace atlas of 3-DOF planar serial mechanisms, which provides a convenient graphical method to analyze the relationship between the reachable workspace and the link lengths. By inspection of Fig. 10.19, we conclude the following:

- In region IV, the reachable workspace has maximum area equal to 9π.
- In regions I, II, and III, the area of the reachable workspace is inversely proportional to $r_p, p = 1, 2, 3$, respectively.
- The reachable workspace is symmetric about axes $r_1 = r_2$, $r_1 = r_3$, and $r_2 = r_3$, respectively.

Dextrous Workspace Atlas

The dextrous workspace of a robot is an important criterion for performance evaluation. Since the robot end-effector may have to achieve a given pose from a particular direction, it is of practical interest to determine these directions.

For the convenience of analysis, we assume a 3-DOF robot to be a virtual four-bar linkage ABCDA, shown in Fig. 10.18. AD is named the virtual frame (r_4); AB is called the virtual input link (r_1); BC is the virtual couple link (r_2); and CD expresses the virtual output link (r_3). For the virtual output link r_3 of the "virtual four-bar linkage" to be a crank that is capable of continuous rotation through 360°, the following Grashoff's conditions must be satisfied:

$$r_1 + r_3 \le r_2 + r_4 \cap r_2 + r_3 \le r_1 + r_4 \cap r_4 + r_3 \le r_1 + r_2 \quad (r_3 \text{ is the shortest}) \quad (10.105)$$

or

$$r_1 + r_4 \le r_2 + r_3 \cap r_2 + r_4 \le r_1 + r_3 \cap r_3 + r_4 \le r_1 + r_2 \quad (r_4 \text{ is the shortest}) \quad (10.106)$$

where, if r_3 is the shortest of the four links, we utilize inequalities (10.105); if r_4 is the shortest, we utilize inequalities (10.106). From Grashoff's theory, we obtain the following criterion:

If the output link r_3 is a crank, i.e., the conditions (10.105) or (10.106) are satisfied, the dextrous workspace of the 3-DOF robot exists in this position, and conversely.

By means of the criterion and the solution space, we obtain the following classes of dextrous workspace: (1) null, (2) single-loop, (3) plate, and (4) double-loop. Three of these classes are shown in Fig. 10.20.

Figure 10.21 shows the relationship between the areas of the dextrous workspaces and the link lengths of all 3-DOF planar serial mechanisms. From Figs. 10.2(b) and 10.21, we conclude the following:

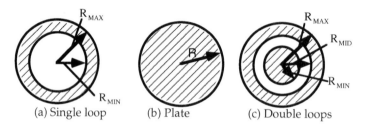

(a) Single loop   (b) Plate   (c) Double loops

FIGURE 10.20   Classification of dextrous workspace.

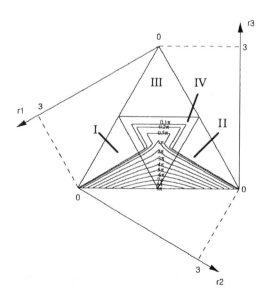

FIGURE 10.21   Dextrous workspace atlas.

- When r_3 is small and $r_1 = r_2$, the area of the dextrous workspace is large and approximately equal to 9π.
- Within regions I_2, II_2, and III as shown in Fig. 10.2(b), the dextrous workspaces of the robots do not exist, which are null.
- Within regions I_1 and II_1, the dextrous workspaces of the robots are a single-loop.
- Within regions IV_1, IV_2, and IV_6, the dextrous workspaces of the robots is a plate.
- When $r_1 = r_2 \cap r_3 < r_1$ (region IV_1), the robots have large dextrous workspaces. This result has important consequences for design of robots with dextrous manipulation.
- Within regions IV_3 and IV_4, the dextrous workspaces of the robots are a double-loop.

Global Workspace Atlas

Using the results of Gosselin and Angeles [31], we derive the global workspace areas of the 3-DOF planar serial manipulator as follows:

$$A_{GW} = \int_0^{2\pi} \pi[(r_1 + r_2)^2 - (r_1 - r_2)^2]d\phi_3 = 8\pi^2 r_1 r_2 \qquad (10.107)$$

The global workspace implies that when all orientations ϕ_3 of the output link r_3 are given, the end-effector can reach points in the space spanned by coordinates x, y, ϕ_3.

By use of Eq. (10.107), contours in the global workspace of all 3-DOF planar serial manipulators can be plotted on the model of the solution space. Fig. 10.22 illustrates the relationship between the volume of the global workspace and the link lengths. We conclude the following:

- When r_3 is small and $r_1 = r_2$, the global workspace has maximum volume and is approximately equal to $18\pi^2$.
- If r_3 is specified and $r_1 = r_2$, the volume of the global workspace is large.
- The volume of the global workspace is inversely proportional to r_3.

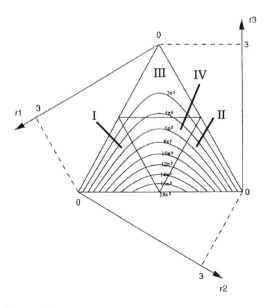

FIGURE 10.22 Global workspace atlas.

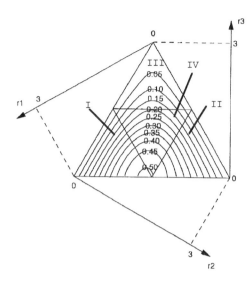

FIGURE 10.23   Atlas of global conditioning index.

Atlas of Global Conditioning Index

By using Eq. (10.59), contours in the global conditioning index of all 3-DOF planar serial manipulators can be plotted on the model of the solution space.

In particular, a large value of the index ensures that the manipulator can be precisely controlled. Fig. 10.23 describes the relationship between the global conditioning index and the link lengths of all 3-DOF planar robotic mechanisms.

From Fig. 10.23, we see that

- If r_3 is small, $r_1 \cong r_2$ and $r_1 < r_2$, the global conditioning index of the robots is large.
- The global conditioning index of mechanisms located on lines parallel to axis r_3 is inversely proportional to r_3.

Atlases of Global Velocity Index

Using Eqs. (72) and (73), the atlases of the global maximum and minimum linear velocity indices are plotted on the solution space model of robotic mechanisms with three moving links. Figs. 10.24 and 10.25 illustrate relationships between the global maximum and minimum linear velocity indices and the link lengths of 3-DOF planar robotic mechanisms, respectively. From Figs. (10.24) and (10.25), we conclude the following:

- If $r_1 = r_2 + r_3$ and $r_1 \cong r_2$, the global maximum linear velocity index of the robots is small as shown in Fig. 10.24.
- In region IV, we obtain a large value of the global minimum linear velocity index. This result is very important for design of robots in which the end-effector should move quickly as shown in Fig. 10.25.
- In the case of serial robots, the global maximum and minimum angular velocity indices are equal and constant.

Atlases of Global Payload Index

By means of Eqs. (10.78) and (10.79), the atlases of global maximum and minimum force payload indices are constructed on the solution space model. Figs. 10.26 and 10.27 are contours of global maximum and minimum force payload indices for 3-DOF planar robotic mechanisms, respectively. And Fig. 10.28

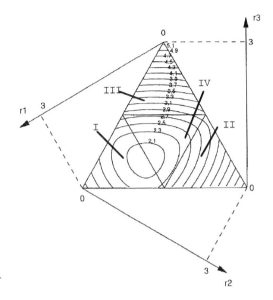

FIGURE 10.24   Atlas of global maximum linear velocity index.

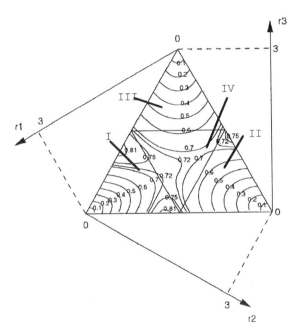

FIGURE 10.25   Atlas of global minimum linear velocity index.

describes the global maximum torque payload index. From Figs. 10.26, 10.27, and 10.28, we obtain the following:

- If r_3 is small and $r_1 = r_2$, the global maximum force payload index is small; if r_3 is given and $r_1 = r_2$, the global maximum force payload index is large, and values of the global maximum force payload index of mechanisms located on lines parallel to the axis r_3 is inversely proportional to r_3 and symmetric about the axis $r_1 = r_2$ as shown in Fig. 10.26.
- If r_3 is given and $r_1 = r_2$, the global minimum force payload index is small; the global minimum force payload index of mechanisms located on lines parallel to the axis r_3 is proportional to r_3 and symmetric about the axis $r_1 = r_2$ as shown in Fig. 10.27.

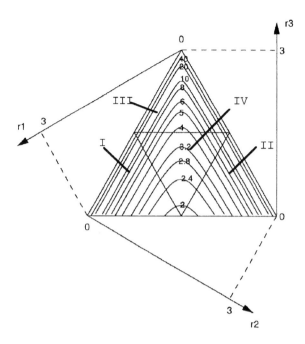

FIGURE 10.26   Atlas of global maximum force payload index.

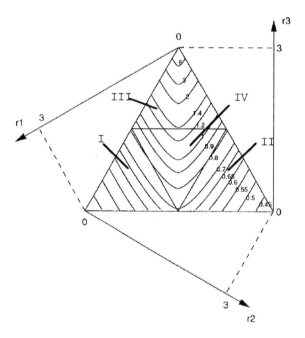

FIGURE 10.27   Atlas of global minimum force payload index.

- If r_3 is small and $r_1 = r_2$, the global maximum torque payload index is small; if r_3 is given, and $r_1 = r_2$, the global maximum torque payload index is large; and the global maximum torque payload index of the mechanisms located on lines parallel to the axis r_3 is proportional to r_3 and symmetric about the axis $r_1 = r_2$ as shown in Fig. 10.28.

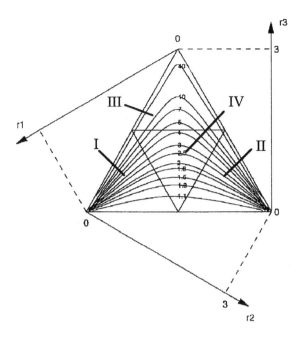

FIGURE 10.28   Atlas of global maximum torque payload index.

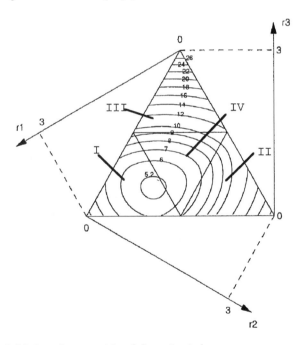

FIGURE 10.29   Atlas of global maximum position deformation index.

Atlases of Global Deformation Index

The global deformation index is a measure of stiffness of the end-effector. The global maximum and minimum deformation indices can be calculated by Eqs. (10.92) and (10.93).

Figures 10.29 and 10.30 represent contours of the global maximum and minimum position deformation indices for 3-DOF planar robotic mechanisms, respectively. Figure 10.31 represents contours

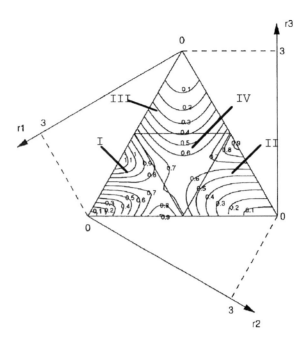

FIGURE 10.30   Atlas of global minimum position deformation index.

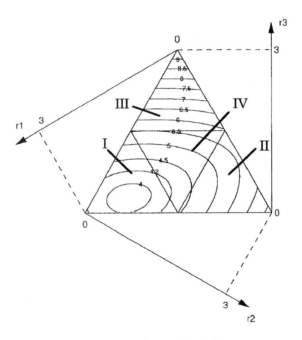

FIGURE 10.31   Atlas of global maximum orientation deformation index.

of the global maximum orientation deformation index. From Figs. 10.29, 10.30, and 10.31, we see that

- If $r_1 = r_2 + r_3$ and $r_2 \cong r_3$, the mechanisms have a small global maximum position deformation index as shown in Fig. 10.29. This result is important for robot design because we would like to reduce position deformation.
- In region IV, the global minimum position deformation index is large as shown in Fig. 10.30.

- If $r_1 > r_2 > r_3$ in region I, the global maximum orientation deformation index is small as shown in Fig. 10.31.

We have developed performance atlases for the design of serial robotic mechanisms with three moving links by plotting performance criteria on the model of the solution space. The atlases describe the relationships between the performance criteria and the link lengths of all the robotic mechanisms and can be used to design the robotic mechanisms. The technique described in this section can be utilized to analyze and design other types of robotic mechanisms.

Analysis of Commercially Available Robots

The parameters of 63 commercially available industrial and research robots are listed in Table 10.6. Figure 10.32 shows how these robots are distributed in the solution space, where each point is one of 63 industrial robot's geometric representation by link lengths. From Fig. 10.32, we see that all of the robots are located at the area ($r_1 \cong r_2$, $r_3 < r_1$ and $r_3 < r_2$) in the solution space model of three-moving-link robotic mechanisms.

The performance criteria proposed in this chapter involve kinematic and dynamic properties and characterize robot performance, which provide a basis for analyzing and designing optimum robotic mechanisms. The physical model technique, solution space model is a useful tool for investigation of the relationships between the performance criteria and link lengths of both serial and parallel robotic mechanisms. From the performance criterion atlases and an analysis of the results, we reach the following conclusions. To design optimum robotic mechanisms, we should select the robotic mechanisms with large reachable, dextrous, and global workspaces, large global conditioning index, large global minimum velocity index, large global minimum payload index, small global maximum deformation index, and small global maximum error index. For the robotic mechanisms with three moving links, the best region for selecting the dimensions of three-moving-link robots is $r_1 \cong r_2$, $r_3 < r_1$ and $r_3 < r_2$. In this region, the robotic mechanisms have optimum performance characteristics and most commercially available robots occur in the region ($r_1 \cong r_2$, $r_3 < r_1$ and $r_3 < r_2$) on the solution space model. This conclusion is very important for designing serial robotic mechanisms for use in manufacturing tasks of machine tending, assembly, and welding. The physical model technique is a useful tool for analysis and design of parallel robotic mechanisms [1,2,3,6,7,8].

10.5 Atlases for Design of F/T Sensor Based on Stewart Platform

In the design of F/T sensors, the structural design is particularly important since the sensors detect forces and moments through the measurement of strains at specific points of their bodies. Much research has been devoted to the sensor structure design. Watson [83] proposed the six-axis force sensor with three vertical deformation components. Stanford Institute [84] investigated the tube design six-axis force sensor. Schott [85] designed the double ring-shaped six-axis force sensor. Brussel [86] and Kroll [87] studied the six-axis force sensor with four vertical deformation components. Shimano [88] was the first researcher who presented and designed the cellar six-axis force sensor. Uchiyama [89] and Bayo [90] studied the systematic design procedure of the Maltese cross-bar-type force sensor. Little [91] designed the force sensor having three beams. Bicchi [82] investigated the miniaturized cylindrical force sensor for mounting on the fingertips of the dextrous hand. Diddens et al. [52] designed the ring-shaped three-axis micro force sensor for mounting in the Smartpen. Kaneko [92] proposed the twin-head type six-axis force sensor. Kerr [53] presented the Stewart-platform transducer force sensor. Nguyen [54] and Ferraresi [55] analyzed the Stewart-platform force sensor. Because the structure of the Stewart-platform force sensor is a kind of parallel mechanism, the technique for analysis and design of parallel mechanisms can be used for the design of this type of sensor.

Because it is very important to investigate the performance criteria for evaluation of the force sensors, many researchers have paid attention to this problem. Uchiyama et al. proposed an index for the evaluation of a structural isotropy of the force sensor body [93] and studied a systematic design procedure

TABLE 10.6 Dimensions of Commercial Robots

No.	Name	Original Dimensions (in mm unless specified)			Scaled Dimensions		
		R_1	R_2	R_3	r_1	r_2	r_3
1	Pentel Puha-2	315	315	0	1.5	1.5	0.0
2	Pentel Puha-1	160	100	0	1.8462	1.1538	0.0
3	Sankyo Skilam-1	400	250	0	1.8462	1.1538	0.0
4	Sankyo Skilam-2	200	160	0	1.6666	1.3333	0.0
5	Argonne Nat. Lab. E-2	18.875''	40''	5.75''	0.8762	1.8568	0.2669
6	Alpha II	177.8	177.8	129.5	1.0995	1.0995	0.8010
7	Rhino XR-3	228.6	228.6	9.5	1.4695	1.4695	0.0610
8	Intelledex 660T	304.8	304.8	228.6	1.0909	1.0909	0.8182
9	Milacron T3-756	44''	55''	1''	1.3200	1.6500	0.0300
10	Nachi-8601	1135	1500	135	1.2292	1.6246	0.1462
11	Seiko 600-5	310	310	50	1.3881	1.3881	0.2238
12	Pentel-3 GL-50	250	250	0	1.5000	1.5000	0.0
13	ABB IRB L6/2	22.5''	26''	0	1.3918	1.6082	0.0
14	Binks 88-800	39.6''	50''	0	1.3259	1.6741	0.0
15	Daewoo, NOVA-10	650	850	100	1.2187	1.5938	0.1875
16	Waseda Univ. -1	300	250	270	1.0976	9.1466	0.9878
17	Waseda Univ. -2	295	250	150	1.2734	1.0791	0.6475
18	Waseda Univ. -3	305	230	175	1.2887	0.9718	0.7395
19	Kayaba Co. Ltd.	580	410	235	1.4204	1.0041	0.5755
20	MELARM Lab.	630	550	0	1.6017	1.3983	0.0
21	Sankyo SR8437	400	400	0	1.5000	1.5000	0.0
22	Sankyo SR8438	300	250	0	1.6364	1.3636	0.0
23	Fanuc S-6	600	559.02	100	1.4297	1.3320	0.2383
24	Fanuc S-700	700	816.24	200	1.2236	1.4268	0.3496
25	Fanuc LR Mate-100	250	220	80	1.3636	1.2000	0.4364
26	Fanuc M-400	1150	780	0	1.7875	1.2124	0.0
27	Fanuc S-12	800	604.67	100	1.5950	1.2056	0.1994
28	Fanuc S-800	720	939.63	200	1.1615	1.5158	0.3227
29	Fanuc S-900	1050	1265.90	225	1.2397	1.4946	0.2657
30	Fanuc S-10	700	610	115	1.4737	1.2842	0.2421
31	Fanuc ARC Mate-100	600	559.02	100	1.4297	1.3320	0.2383
32	Fanuc ARC Mate-120	800	604.67	100	1.5950	1.2056	0.1994
33	Fanuc P-100	860	1209.34	0	1.2468	1.7532	0.0
34	Fanuc P-150	1000	1264.36	0	1.3249	1.6751	0.0
35	Fanuc S-500	900	1607.02	180	1.0048	1.7942	0.2010
36	Fanuc S-420i	950	1321.97	200	1.1529	1.6044	0.2427
37	Mitsubishi RV-M1	250	160	72	1.5560	0.9959	0.4481
38	Mitsubishi RV-M2	250	220	65	1.4019	1.2336	0.3645
39	Motoman-K10S	615	770	100	1.2424	1.5556	0.2020
40	Panasonic KS-V20	250	200	120	1.3158	1.0526	0.6316
41	Panasonic HR-50	275	275	0	1.5000	1.5000	0.0
42	Panasonic HR-150	425	425	0	1.5000	1.5000	0.0
43	Sony SRX-4 CH-LA	400	250	0	1.8462	1.1538	0.0
44	Sony SRX-4 CH-LZ	350	250	0	1.7500	1.2500	0.0
45	Stäubli RX-90	450	450	0	1.5000	1.5000	0.0
46	Stäubli RX-90L	450	650	0	1.2273	1.7727	0.0
47	Stäubli RX-130	625	625	0	1.5000	1.5000	0.0
48	Stäubli RX-130L	625	925	0	1.2097	1.7903	0.0
49	Stäubli RX-170	850	750	0	1.5938	1.4062	0.0
50	Stäubli RX-170L	850	1050	0	1.3421	1.6579	0.0
51	Seiko TT8030	400	400	40	1.4286	1.4286	0.1428
52	Seiko TT8550	225	225	0	1.5000	1.5000	0.0
53	Seiko TT8800	450	350	0	1.6875	1.3125	0.0
54	Seiko TT8010	300	200	40	1.6667	1.1111	0.2222
55	Seiko TT4000SC	356	305	0	1.6157	1.3843	0.0

TABLE 10.6 Dimensions of Commercial Robots (continued)

No.	Name	Original Dimensions (in mm unless specified)			Scaled Dimensions		
		R_1	R_2	R_3	r_1	r_2	r_3
56	Seiko TT8010C	300	200	40	1.6667	1.1111	0.2222
57	Adept-1	16.73''	14.76''	0	1.5938	1.4062	0.0
58	Adept-3	559	508	0	1.5717	1.4283	0.0
59	Adept-1850	1000	850	70	1.5625	1.3281	0.1094
60	BUAA C1 Hand	46	20	13	1.7468	0.7595	0.4937
61	IRC XDH-9S/9A Hand	50	58	0	1.3889	1.6111	0.0
62	Tokushima Univ. Hand	48	30	0	1.8462	1.1538	0.0
63	Utah DH master	39.08	43.53	11.18	1.2500	1.3924	0.0

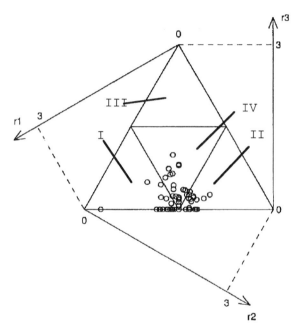

FIGURE 10.32 The distribution of commercial robots within the solution space. (a) $\theta_{AB} = 10°$, (b) $\theta_{AB} = 20°$, (c) $\theta_{AB} = 30°$, (d) $\theta_{AB} = 40°$.

to minimize a performance index for the force sensors [89]. Bayo et al [90] investigated the criteria of the condition number, stiffness, and strain gauge sensitivity of the sensors. Diddens et al [52] used a three-dimensional finite-element model to optimize the strain-gauge positions of the sensor.

Although extensive research has been directed toward the analysis and design of the structures of force sensors, there has not been an effective way to relate performance criteria and the parameters of the sensor structures. In this section, a geometric model of the solution space for the structures of the F/T sensors based on the Stewart platform is proposed and used to investigate relationships between the criteria and parameters of the structures of the F/T sensors.

Atlases of Condition Number of Jacobian Matrix

The condition number of Jacobian matrix is an important criterion for evaluation of the sensor mechanisms based on Stewart platform as shown in Fig. 10.15, since it is the measure of isotropy of a sensor.

Using the analysis approach of the parallel mechanisms, we can derive Jacobian matrix $[J]$ for the force sensor mechanisms. From Fig. 10.15, one derives

$$\sum_{i=1}^{6} f_i \$_i = (F) + \in (T) \tag{10.108}$$

where f_i is the force acted on the link i, and $\$_i$ expresses the unit screw on the link i, i.e.,

$$\$_i = (s_i) + S_{oi}, \quad (s_i)^T(s_i) = 1, \quad (s_i)^T(s_{oi}) = 0 \tag{10.109}$$

$$(s_i) = (B_i - A_i)/|B_i - A_i| \tag{10.110}$$

$$(s_{oi}) = (A_i) \times (s_i) = [(A_i) \times (B_i)]/|B_i - A_i| \tag{10.111}$$

where (A_i) and (B_i) are the coordinates of the points A_i and B_i in the coordinate system $o_A - x_A y_A z_A$. From Eq. (10.108), we can obtain

$$\begin{pmatrix} F \\ T \end{pmatrix} = [G](f) \tag{10.112}$$

where

$$(f) = (f_1 \quad f_2 \quad Pf_3 \quad f_4 \quad f_5 \quad f_6)^T$$

and

$$[G] = \begin{bmatrix} s_1 & s_2 & s_3 & s_4 & s_5 & s_6 \\ s_{o1} & s_{o2} & s_{o3} & s_{o4} & s_{o5} & s_{o6} \end{bmatrix} \tag{10.113}$$

By comparing Eq. (10.74) with Eq. (10.112), we see that Jacobian matrix $[J]$ of the mechanism has the following relationship:

$$[J] = \left[[G]^{-1} \right]^T \tag{10.114}$$

So the condition number of Jacobian matrix $[J]$ can be calculated by the following equations:

$$\text{Cond} = \|J\| \|J^{-1}\|, \tag{10.115}$$

where

$$\|J\| = \sqrt{tr\left(J \frac{1}{n} [I] J^T \right)} \tag{10.116}$$

Because

$$1 \leq \text{Cond} < \infty,$$

let

$$E = \frac{1}{\text{Cond}} \tag{10.117}$$

where E is the inverse of the condition number of Jacobian matrix.

By using the Eq. (10.117), the atlases of condition number can be plotted within the model of the solution space shown in Fig. 10.17. Figure 10.32 only shows four of the atlases of the condition number that can describe the relationship between the condition number and the four parameters of the sensor mechanisms.

From the atlases of the condition number as shown in Fig. 10.32, we obtain the following important result: if r_3 is small, $r_1 \cong r_2$, and both r_1 and r_2 are large; that is, in the region nearby the vertex D on the model of the solution space, the mechanisms have large values of the condition number and approximately equal to one, which means that the mechanisms in that region are isotropy. This result is very important and useful for design of the sensor mechanisms, because a large value of the condition number ensures that the force sensor can have high accuracy.

Atlases of Force and Torque Sensitivity

As shown in Fig. 10.15, from Eqs. (10.112) and (10.114), we know that the external force F and torque T applied at the upper platform center are related to the forces f_i $(i = 1,2,\ldots,6)$ by

$$\binom{F}{T} = [J^T]^{-1}(f) = \begin{bmatrix} J_F \\ J_T \end{bmatrix}(f) \tag{10.118}$$

So

$$\| F \|^2 = (f)^T [J_F]^T [J_F](f) \tag{10.119}$$

$$\| T \|^2 = (f)^T [J_T]^T [J_T](f) \tag{10.120}$$

Let

$$\| f \|^2 = (f)^T (f) = 1 \tag{10.121}$$

$$L_F = (f)^T [J_F]^T [J_F](f) - \lambda_F[(f)^T(f) - 1] \tag{10.122}$$

$$L_T = (f)^T [J_T]^T [J_T](f) - \lambda_T[(f)^T(f) - 1] \tag{10.123}$$

where λ_F and λ_T are scalar Lagrange multipliers. Using the same method as that in the section entitled "Global Velocity Index," we determine that when the force (f) (see Eq. (10.121)) is a unit vector, the maximum and minimum values of the external force F and torque T applied at the upper platform center as shown in Fig. 10.15 are

$$\| F_{\max} \| = \sqrt{\lambda_{F \max}} \quad \text{and} \quad \| F_{\min} \| = \sqrt{\lambda_{F \min}} \tag{10.124}$$

$$\| T_{\max} \| = \sqrt{\lambda_{T \max}} \quad \text{and} \quad \| T_{\min} \| = \sqrt{\lambda_{T \min}} \tag{10.125}$$

where $\| F_{\max} \|$ and $\| F_{\min} \|$ are the maximum and minimum values of the external forces, respectively; $\| T_{\max} \|$ and $\| T_{\min} \|$ express the maximum and minimum values of the external torques, respectively; and λ_F and λ_T are eigenvalues of $[J_F]^T[J_F]$ and $[J_T]^T[J_T]$, respectively.

Because we hope that the small force and torque applied at the upper platform center could make the strain gages fixed on the six links have big strain, which means that the sensor has high force and torque sensitivity, and the maximum values of the external force $\| F_{\max} \|$ and torque $\| T_{\max} \|$ should be as small as possible, we only need to consider $\| F_{\max} \|$ and $\| T_{\max} \|$.

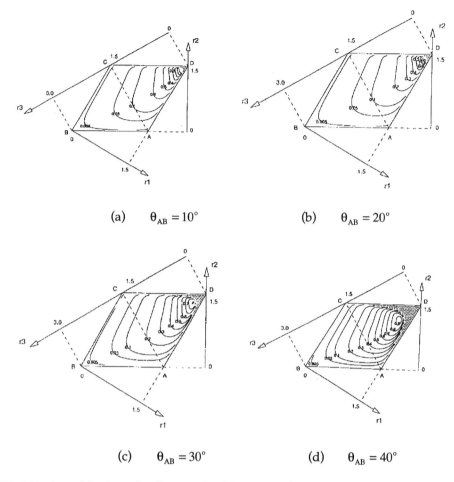

FIGURE 10.33   Four of the atlases of condition number. (a) $\theta_{AB} = 20°$, (b) $\theta_{AB} = 30°$, (c) $\theta_{AB} = 40°$, (d) $\theta_{AB} = 50°$.

By using the Eqs. (10.124) and (10.125), the atlases of the indices $\|F_{\max}\|$ and $\|T_{\max}\|$ can be plotted within the model of the solution space.

Figure 10.33 shows four of the atlases of the index $\|F_{\max}\|$, which can describe the relationship between the index $\|F_{\max}\|$ and the four parameters of the sensor mechanisms. From the atlases of the index $\|F_{\max}\|$, we obtain the following important results: if r_3 is small, ($r_1 \cong r_2$), and both r_1 and r_2 are large, that is, in the region near by the vertex D on the model of the solution space, the sensor mechanisms have small values of the index $\|F_{\max}\|$, which means that the force sensitivity of the sensors in this region is high.

Figure 10.34 shows four of the atlases of the index $\|T_{\max}\|$, which can describe the relationship between the index $\|T_{\max}\|$ and the four parameters of the sensor mechanisms. From the atlases of the index $\|T_{\max}\|$, we see that when the sensor mechanisms are located at the region near by the edges AB, AD, and CD, the sensor mechanisms have small value of the index $\|T_{\max}\|$, which means that the sensors in the region have high torque sensitivity.

Atlases of Sensor Stiffness

Since the deformations of the upper platform of the sensor mechanism can describe the stiffness of the sensor, we need to investigate the deformations. The force (f) is related to the deformation by, as shown in Fig. 10.15,

$$(f) = [K](\Delta q) \qquad (10.126)$$

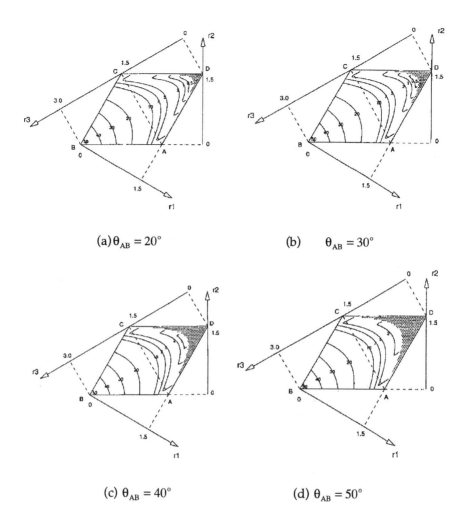

(a) $\theta_{AB} = 20°$ (b) $\theta_{AB} = 30°$

(c) $\theta_{AB} = 40°$ (d) $\theta_{AB} = 50°$

FIGURE 10.34 Four of the atlases of the index $\|F_{max}\|$. (a) $\theta_{AB} = 10°$, (b) $\theta_{AB} = 30°$, (c) $\theta_{AB} = 40°$, (d) $\theta_{AB} = 50°$.

where

$$[K] = [I](k_1 \quad k_2 \quad \cdots \quad k_6)^T \tag{10.127}$$

$$(\Delta q) = (\Delta q_1 \quad \Delta q_2 \quad \cdots \quad \Delta q_6)^T \tag{10.128}$$

Here, Δq_i is the deformation of link i, which is connected with both upper platform and lower platform; k_i is the stiffness of link i ($i = 1,2,\ldots, 6$); and $[I]$ is a 6×6 identity matrix. We assume

$$k_i = k, \tag{10.129}$$

which implies that all six links have the same stiffness. Because of attention paid to investigation of the relationship between the stiffness and the four parameters of the sensor mechanisms, we suppose $k = 1$. From Eqs. (10.85) to (10.89), we obtain

$$(D) = \begin{pmatrix} D_P \\ D_O \end{pmatrix} = [C]\begin{pmatrix} F \\ T \end{pmatrix} = \begin{bmatrix} C_P \\ C_O \end{bmatrix}\begin{pmatrix} F \\ T \end{pmatrix} \tag{10.130}$$

where (D_P) and (D_O) express the positional deformation and orientational deformation of the upper platform, respectively.

Let

$$\left\| (F\ T)^T \right\|^2 = (F\ T)^T (F\ T) = 1 \tag{10.131}$$

$$L_P = (F\ T)[C_P]^T[C_P](F\ T)^T - \lambda_P[(F\ T)^T(F\ T) - 1] \tag{10.132}$$

$$L_O = (F\ T)[C_O]^T[C_O](F\ T)^T - \lambda_O[(F\ T)^T(F\ T) - 1] \tag{10.133}$$

where λ_P and λ_O are scalar Lagrange multipliers. As previously, from Eqs. (10.132) and (10.133), necessary conditions for extreme values of the positional and orientational deformations of the upper platform are

$$\left\| D_{P\ \max} \right\| = \sqrt{\lambda_{D_p\ \max}}, \quad \text{and} \quad \left\| D_{P\ \min} \right\| = \sqrt{\lambda_{D_p\ \min}}, \tag{10.134}$$

$$\left\| D_{O\ \max} \right\| = \sqrt{\lambda_{D_O\ \max}}, \quad \text{and} \quad \left\| D_{O\ \min} \right\| = \sqrt{\lambda_{D_O\ \min}}, \tag{10.135}$$

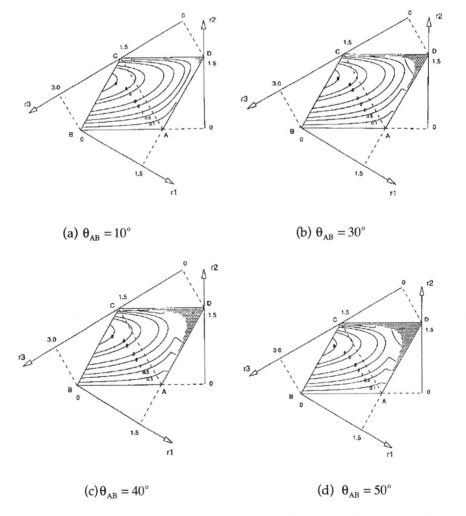

(a) $\theta_{AB} = 10°$ (b) $\theta_{AB} = 30°$

(c) $\theta_{AB} = 40°$ (d) $\theta_{AB} = 50°$

FIGURE 10.35 Four of the atlases of the index $\|T_{\max}\|$. (a) $\theta_{AB} = 30°$, (b) $\theta_{AB} = 40°$, (c) $\theta_{AB} = 50°$, (d) $\theta_{AB} = 60°$.

where, $\lambda_{D_p\,max}$ and $\lambda_{D_p\,min}$ are maximum and minimum eigenvalues of $[C_P]^T[C_P]$, respectively; $\lambda_{D_O\,max}$ and $\lambda_{D_O\,min}$ are maximum and minimum eigenvalues of $[C_O]^T[C_O]$, respectively; $\|D_{P\,max}\|$ and $\|D_{P\,min}\|$ are the maximum and minimum values of the positional deformations of the upper platform, respectively; and $\|D_{O\,max}\|$ and $\|D_{O\,min}\|$ express the maximum and minimum values of the orientational deformations of the upper platform, respectively.

Because we hope that when the force F and torque T are applied at the upper platform center, the positional and orientational deformations of the upper platform are small at the same time, which means that the sensor has high stiffness and the maximum values of the positional and orientational deformations should be as small as possible, we only need to consider $\|D_{P\,max}\|$ and $\|D_{O\,max}\|$.

By using the Eqs. (10.134) and (10.135), the atlases of the indices $\|D_{P\,max}\|$ and $\|D_{O\,max}\|$ can be plotted within the model of the solution space shown in Fig. 17.

Figure 10.35 and 10.36 show four of the atlases of the indices $\|D_{P\,max}\|$ and $\|D_{O\,max}\|$, respectively, which can describe the relationship between the indices $\|D_{P\,max}\|$ and $\|D_{O\,max}\|$ and the four parameters of the sensor mechanisms. From the atlases of the index $\|D_{P\,max}\|$, we see that if r_3 is small, $r_1 \cong r_2$, and both r_1 and r_2 are large, that is, in the region near by the vertex D on the model of the solution space, the sensor mechanisms have small values of the index $\|D_{P\,max}\|$ and $\|D_{O\,max}\|$ at the sametime,

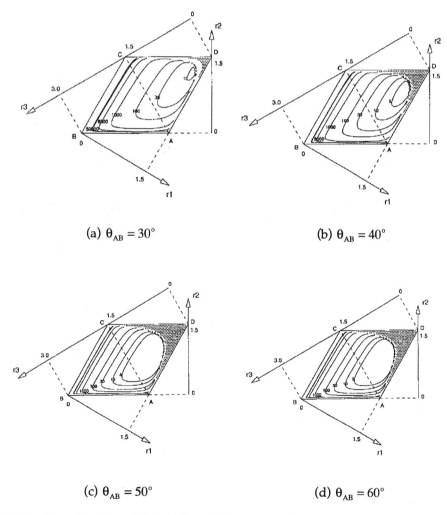

(a) $\theta_{AB} = 30°$ (b) $\theta_{AB} = 40°$

(c) $\theta_{AB} = 50°$ (d) $\theta_{AB} = 60°$

FIGURE 10.36 Four of the atlases of the index $\|D_{P\,max}\|$. (a) $\theta_{AB} = 30°$, (b) $\theta_{AB} = 40°$, (c) $\theta_{AB} = 50°$, (d) $\theta_{AB} = 60°$.

which means that the positional and orientational stiffness of the sensors with the mechanisms are high in this region.

Sensor Design

By using the performance atlases as shown in Figs. 10.31 through 10.36, the four parameters R_1, R_2, R_3, and θ_{AB} of the sensor mechanisms can be easily selected to optimize sensor performance. From the performance atlases, we see that when r_3 is small, $r_1 \cong r_2$, and both r_1 and r_2 are large, that is, in the region near by the vertex D on the model of the solution space, the sensor mechanisms have optimum performances.

Figure 10.37 is the CAD layout of sensor mechanism. Because we hope to make the sensor as small as possible, the elastic joints are utilized to replace the spherical joints as shown in Fig. 10.37, which makes it possible to manufacture the small size sensors.

To use the model of the solution space for design of the F/T sensor mechanisms based on Stewart platform is a novel and useful method for investigation of the optimal sensor design. The three kinds of atlases of the criteria, including condition number, force and torque sensitivity, and stiffness of the sensors clearly show relationships between the performance criteria and parameters of all the sensor mechanisms. By using the performance atlases, an optimal design for the sensor mechanism can be achieved. Because

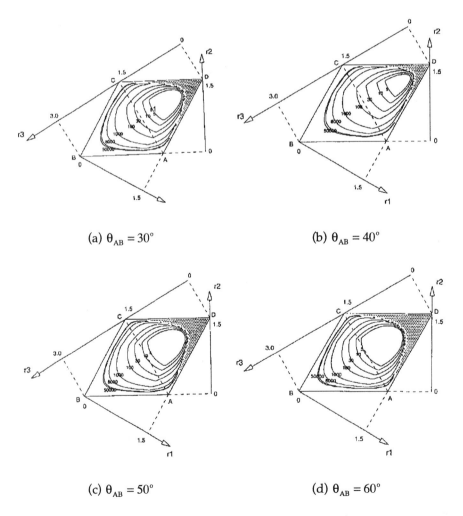

(a) $\theta_{AB} = 30°$ (b) $\theta_{AB} = 40°$

(c) $\theta_{AB} = 50°$ (d) $\theta_{AB} = 60°$

FIGURE 10.37   Four of the atlases of the index $\|D_{O\,max}\|$.

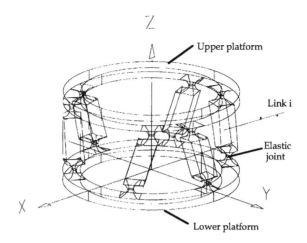

Z

Upper platform

Link i

Elastic joint

X

Y

Lower platform

FIGURE 10.38 CAD layout of sensor house.

the elastic joints were proposed for replacing the spherical joints, the Stewart-based sensor can be designed as small as possible.

10.6 Conclusions

The physical model technique is a simple, useful, and efficient tool for design of robotic manipulators. By using the technique and the performance criteria, the relationships between the criteria and dimensions of robotic mechanisms can be obtained as the performance atlases, which are easily utilized for optimal robotic design.

References

1. F. Gao, X. Q. Zhang, Y. S. Zhao, and W. B. Zu, "Distribution of some properties in a physical model of the solution space of 2-DOF parallel planar manipulators," Mechanism and Machine Theory, Vol. 30, No. 6, 1995, pp. 811–817.
2. F. Gao, X. Q. Zhang, Y. S. Zhao, and H. R. Wang, "A physical model of the solution space and the atlases of the reachable workspaces for 2-DOF parallel planar manipulators," Mechanism and Machine Theory, Vol. 31, No. 2, 1996, pp. 173–184.
3. F. Gao, Y. S. Zhao, and Z. H. Zhang, "Physical model of the solution space of 3-DOF parallel planar manipulators," Mechanism and Machine Theory, Vol. 31, No. 2, 1996, pp. 161–171.
4. F. Gao, W. A. Gruver, and Y. Zhang, "Performance charts for design of robotic mechanisms with three moving links," IEEE SMC96, Oct. 14–17, 1996, Beijing, China.
5. F. Gao and W. A. Gruver, "Performance Evaluation Criteria for Analysis and Design of Robotic Mechanisms," IEEE CIRA'97, July 1997, Monterey, California.
6. F. Gao and W. A. Gruver, "The global conditioning index in the solution space of two degrees of freedom planar parallel manipulators," Proc. of the 1995 IEEE International Conference on SMC, October 1995, Vancouver, Canada, Vol. 5, pp. 4055–4059.
7. F. Gao, X. J. Liu, and W. A. Gruver, "Performance Evaluation of Two Degree of Freedom Planar Parallel Robots," accepted for publication in Mechanism and Machine Theory, 1997.
8. F. Gao, W. A. Gruver, et al., "A geometric model for the analysis and design of Delta parallel robots," ASME Design Engineering Technical Conferences and Computers in Engineering Conference, August 18–22, 1996, Irvine, California.
9. F. Gao and W. A. Gruver, "Criteria based analysis and design of three degree of freedom planar robotic manipulators," IEEE ICRA'97, April 20–25, 1997, Albuquerque, New Mexico.

10. F. Skinner, "Designing a multiple pretension manipulator," Mechanical Engineering, September 1975, pp 30–37.

11. F. R. E. Crossley and F. G. Umholtz, "Design for a three-fingered hand," Mechanism and Machine Theory, Vol. 12, 1977, pp 85–93.

12. A. Rovetta, P. Vincentini, and I. Franchetti, "On development and realization of a multipurpose grasping system," Proc. of the 11th International Symposium on Industrial Robots, Tokyo, 1981, pp 273–280.

13. T. Okada, "Computer control of multi-jointed finger system for precise object-handling," IEEE Transactions on Systems, Man and Cybernetics, Vol. SMC-12 (3), 1982, pp 289–299.

14. J. K. Salisbury, "Kinematic and force analysis of articulated hands," Ph.D. Thesis, Stanford University, 1982, Report No. STAN-CS-82-921.

15. S. C. Jacobsen, J. E. Wood, D. F. Knutti, and K. B. Biggers, "The Utah/MIT dextrous hand: work in progress," The International Journal of Robotics Research, 1984, Vol. 3, No. 4, pp 21–50.

16. P. Dario and G. Buttazzo, "An anthropomorphic robot finger for investigating artificial tactile perception," The International Journal of Robotics Research, 1987, Vol. 6, No. 3, pp 25–48.

17. B. A. Grupen, T. C. Henderson, and I. D. McCammon, "A survey of general-purpose manipulation," The International Journal of Robotics Research, 1989, Vol. 8, No. 1, pp 38–62.

18. S. A. Stansfield, "Robotic grasping of unknown objects: a knowledge-based approach," The International Journal of Robotics Research, 1991, Vol. 10, No. 4, pp 314–326.

19. K. H. Hunt, A. E. Samuel, and P. R. McAree, "Special configurations of multi-finger multi-freedom grippers–a kinematic study," The International Journal of Robotics Research, 1991, Vol. 10. No. 2, pp 123–134.

20. R. N. Rohling and J. M. Hollerbach, "Modeling and parameter estimation of the human index finger," Proc. of the IEEE International Conference on Robotics and Automation, 1994, pp 223–230.

21. Y. Yang, Y. Zhang, and Q. X. Zhang, "A performance evaluation of HB-2 dextrous robotic hand," Proc. of the IEEE International Conference on Systems, Man and Cybernetics, 1995, Vancouver, Vol. 1, pp 922–927.

22. P. R. McAree, A. E. Samuel, K. H. Hunt, and C. G. Gibson, "A dexterity measure for the kinematic control of a multifinger, multifreedom robot hand," The International Journal of Robotics Research, 1991, Vol. 10, No. 5, pp 439–453.

23. T. Yoshikawa, "Manipulability of robotic mechanism," The International Journal of Robotics Research, 1985, Vol. 4, No. 2, pp 3–9.

24. T. W. Nye, D. J. LeBlanc, and R. J. Cipra, "Design and modeling of a computer-controlled planar manipulator," The International Journal of Robotics Research, 1987, Vol. 6, No. 1, pp 85–95.

25. C. J. J. Paredis and P. K. Khosla, "Kinematic design of serial link manipulators from task specifications," The International Journal of Robotics Research, 1993, Vol. 12, No. 3, pp 274–287.

26. D. Tesar and M. S. Butler, "A generalized modular architecture for robot structures," Manufacturing Review, 1989, Vol. 2, No. 2, pp 91–118.

27. D. Tesar and M. Sklar, "Dynamic analysis of hybrid serial manipulator systems parallel modules," ASME Trans. Journal of Mechanisms, Transmission, and Automation in Design, 1988, Vol. 104, pp 218–228.

28. P. Gorce, O. Vanel, and C. Ribreau, "Equilibrium study of 'human' robot," Proc. of the IEEE International Conference on Systems, Man and Cybernetics, 1995, Vancouver, Vol. 2, pp 1309–1314.

29. T. A. McMahon, "Mechanics of locomotion," The International Journal of Robotics Research, 1984, Vol. 3, No. 2, pp 4–28.

30. S. Hirose and O. Kunieda, "Generalized standard foot trajectory for a quadruped walking vehicle," The International Journal of Robotics Research, 1991, Vol. 10, No. 1, pp 3–12.

31. C. Gosselin and J. Angeles, "The optimum design of planar three-degree-of-freedom parallel manipulator," ASME Journal of Mechanisms, Transmissions, and Automation in Design, Vol. 110, No. 1, 1988, pp. 35–41.

32. R. J. Schilling, Fundamentals of Robotics Analysis and Control, Prentice-Hall, Inc., 1990.

33. C. Y. Ho and J. Sriwattanathamma, Robot Kinematics: Symbolic Automation and Numerical Synthesis, Ablex Publishing Corporation, 1989.
34. R. E. Parkin, Applied Robotic Analysis, Prentice-Hall, Inc., 1991.
35. H. Asada and Y. T. Kamal, Direct-Drive Robots Theory and Practice, MIT Press, 1987.
36. C. M. Gosselin and M. Guillot, "The synthesis of manipulators with prescribed workspaces," Trans. of ASME, J. of Mechanical Design, Vol. 113, 1991, pp. 451–455.
37. D. McCloy, "Some comparisons of serial driven and parallel driven manipulators," Robotica, Vol. 8, 1990, pp. 355–362.
38. D. McCloy, "Planar linkages for parallel-driven manipulators," The 4th Conference of the Irish Manufacturing Committee, Limerick, Ireland, 1987.
39. A. Bajpai and B. Roth, "Workspace and mobility of a closed loop manipulator," International J. Robotics Research, No. 2, 1986, pp. 131–142.
40. H. Asada and K. Youcef-Toumi, "Analysis and design of a direct drive arm with a five-bar link parallel driven mechanism," Proc. of the American Control Conference, San Diego, 1984.
41. R. Stoughton and T. Kokkinia, "Some properties of a new kinematic structure for robot manipulators," ASME Design Automation Conference, DET-Vol. 10-2, 1987, pp. 73–79.
42. V. Kumar, "Characterization of workspaces of parallel manipulators," ASME J. Mechanical Design, Vol. 114, 1992, pp. 368–375.
43. C. Gosselin and J. Angeles, "A global performance index for the kinematic optimization of robotic manipulators," Trans. ASME, J. of Mechanical Design, Vol. 113, 1991, pp. 220–226.
44. F. Sternheim, "Computation of the direct and inverse geometric models of the DELTA4 parallel robot," Robotersysteme, Vol. 3, 1987, pp. 199–203.
45. F. Sternheim, "Tridimensional computer simulation of parallel robot. Results for the DELTA4 machine," Proc. of the 18th International Symposium on Industrial Robots, Lausanne, 1988.
46. R. Clavel, "DELTA, a fast robot with parallel geometry," Proc. of the 18th Int. Symposium on industrial Robots, IFS Publications, 1988, pp. 91–100.
47. K. Miller and R. Clavel, "The Lagrange-based model of Delta-4 robot dynamics," Robotersysteme, Vol. 8, 1992, pp. 49–54.
48. F. Pierrot, A Fournier, and P. Dauchez, "Toward a fully parallel 6-DOF robot for high-speed applications," International Journal of Robotics and Automation, Vol. 7, No. 1, 1992, pp. 15–22.
49. F. Pierrot, C. Reynaud, and A. Fournier, "DELTA: a simple and efficient parallel robot," Robotica, Vol. 8, 1990, pp. 105–109.
50. F. Pierrot and A. Fournier, "Fast models for the DELTA parallel robot," Proc. of I.F.I.P., Rome, Italy, 1990, pp. 123–130.
51. E. O. Doebelin, "Measurement systems applications and design," McGraw Hill, New York, 1985.
52. D. Diddens, D. Reynaerts, and H. V. Brussel, "Design of a ring-shaped three-axis micro force/torque sensor," Sensors and Actuators A, 46–47, 1995, pp. 225–232.
53. D. R. Kerr, "Analysis, properties and design of a Stewart-platform transducer," J. Mech. Transm. Autom. design, Vol. 111, 1989, pp. 25–28
54. C. C. Nguyen, S. S. Antrazi, Z.-L. Zhou, and C. E. Campbell, Jr., "Analysis and experimentation of a Stewart platform-based force/torque sensor," International Journal of Robotics and Automation, Vol. 7, No. 3, 19, pp. 133–140.
55. C. Ferraresi, S. Pastorelli, M. Sorli, and N. Zhmud, "Static and dynamic behavior of a high stiffness Stewart platform-based force/torque sensor," Journal of Robotic Systems, Vol. 12, No. 12, 1995, pp. 883–893.
56. K. C. Gupta and B. Roth, "Design considerations for manipulator workspace," Trans. ASME, Journal of Mechanical Design, Vol. 104, 1982, pp. 704–711.
57. J. K. Davidson, "A synthesis procedure for design of 3-R planar robotic workcells in which large rotations are required at the workpiece," Journal of Mechanical Design, Vol. 114, 1992, pp. 547–558.
58. C. D. Lin and F. Freudenstein, "Optimization of the workspace of a three-link turning-pair connected robot arm," The International Journal of Robotics Research, Vol. 5, No. 2, 1986, pp. 104–111.

59. C. M. Gosselin and M. Jean, "Determination of the workspace of planar parallel manipulators with joint limits," Robotics and Autonomous Systems, Vol. 17, 1996, pp. 129–138.
60. V. Kumar, "Characterization of workspaces of parallel manipulators," Trans. ASME, Journal of Mechanical Design, Vol. 114, 1992, pp. 368–375.
61. A. Bajpai and B. Roth, "Workspace and mobility of a closed-loop manipulator," The International Journal of Robotics Research, Vol. 5, No. 2, 1986, pp. 131–142.
62. M. Trabia and J. K. Davidson, "Design conditions for the orientation and attitude of a robot tool carried by a 3-R spherical wrist," Trans. ASME, Journal of Mechanisms Transmissions, and Automation in Design, Vol. 111, 1989, pp. 176–186.
63. G. R. Pennock and D. J. Kassner, "The workspace of a general geometry planar three-degree-of-freedom platform-type manipulator," Trans. ASME, Journal of Mechanical Design, Vol. 115, 1993, pp. 269–276.
64. C. Gosselin, "Determination of the workspace of 6-DOF parallel manipulators," Trans. ASME, Journal of Mechanical Design, Vol. 112, 1990, pp. 331–336.
65. J. Angeles and C. S. Lopez-Cajun, "Kinematic isotropy and the conditioning index of serial robotic manipulators," The International Journal of Robotics Research, Vol. 11, No. 6, 1992, pp. 560–571.
66. J. Angeles, "The Design of Isotropic manipulator architectures in the presence of redundancies," The International Journal of Robotics Research, Vol. 11, No. 3, 1992, pp. 196–201.
67. C. A. Klein, "Spatial robotic isotropy," The International Journal of Robotics Research, Vol. 10, No. 4, 1991, pp. 426–437.
68. M. Kircanski, "Kinematic isotropy and optimal kinematic design of planar manipulators and a 3-DOF spatial manipulator," The International Journal of Robotics Research, Vol. 15, No. 1, 1996, pp. 61–77.
69. C. M. Gosselin, "The optimum design of robotic manipulators using dexterity indices," Robotics and Autonomous System, Vol. 9, 1992, pp. 213–226.
70. C. A. Klein and B. E. Blaho, "Dexterity measures for the design and control of kinematically redundant manipulators," The International Journal of Robotics Research, Vol. 6, No. 2, 1987, pp. 72–83.
71. F. C. Park and R. W. Brockett, "Kinematic dexterity of robotic mechanisms," The International Journal of Robotics Research, Vol. 13, No. 1, 1994, pp. 1–15.
72. P. R. McAree, A. E. Samuel, K. H. Hunt, and C. G. Gibson, "A dexterity measure for the kinematic control of a multifinger, multifreedom robot hand," The International Journal of Robotics Research, Vol. 10, No. 5, 1991, pp. 439–453.
73. T. Yoshikawa, "Manipulability of robotic mechanisms," The International Journal of Robotics Research, Vol. 4, No. 2, 1985, pp. 3–9.
74. J. P. Merlet, "Singular configurations of parallel manipulators and Grassmann geometry," The International Journal of Robotics Research, Vol. 8, No. 5, 1989, pp. 45–56.
75. J. Sefrioui and C. M. Gosselin, "Singularity analysis and representation of planar parallel manipulators," Robotics and Autonomous Systems, Vol. 10, 1992, pp. 209–224.
76. D. K. Pai, "Genericity and singularities of robot manipulators," IEEE Transactions on Robotics and Automation, Vol. 8, No. 5, 1992, pp. 545–595.
77. C. J. J. Paredis and P. K. Khosla, "Kinematic design of serial link manipulators from task specifications," The International Journal of Robotics Research, Vol. 12, No. 3, 1993, pp. 274–287.
78. K. V. D. Doel and D. K. Pai, "Performance measures for robot manipulators: a unified approach," The International Journal of Robotics Research, Vol. 15, No. 1, 1996, pp. 92–111.
79. C. Gosselin and J. Angeles, "The optimum kinematic design of a spherical three-degree-of-freedom parallel manipulator," Trans. ASME, Journal of Mechanisms Transmissions, and Automation in Design, Vol. 111, 1989, pp. 202–207.
80 C. Gosselin and J. Angeles, "Kinematic inversion of parallel manipulators in the presence of incompletely specified tasks," Trans. ASME, Journal of Mechanical Design, Vol. 112, 1990, pp. 494–500.

81. C. Gosselin and E. Lavoie, "On the kinematic design of spherical three-degree-of-freedom parallel manipulators," The International Journal of Robotics Research, Vol. 12, No. 4, 1993, pp. 394–402.

82. A. Bicchi, "A criterion for optimal design of multi-axis force sensors," Robotics and Autonomous Systems, Vol. 10, No. 4, 1992, pp. 269–286.

83. P. C. Watson and S. H. Drake, "Pedestal and wrist force sensors for automatic assembly," Proc. the 5th Int. Symp. on Industrial Robots, 1975, pp. 501–511.

84. "Robot Technology," Kogon Page Ltd., London, 1983.

85. J. Schott, "Tactile sensor with decentralized signal conditioning," The 9th IMEKO World Congress, Beilin, 1982.

86. H. V. Brussel, et al., "Force sensing for advanced robot control," Proc. of the 5th Int. Cof. on Robot Vision and Sensory Controls, 1980.

87. E. Kroll, et al., "Decoupling load components and improving robot interfacing with an easy-to-use 6-axis wrist force sensor," Theory of Machines and Mechanisms, Proc. of the 7th World Congress, 1986.

88. B. Shimano, et al., "On force sensing information and its use in controlling manipulators," Proc. of the 8th Int. Symp. on Industrial Robots, 1979.

89. M. Uchiyama, E. Bayo, and E. Palma-Villalon, "A systematic design procedure to minimize a performance index for robot force sensors," Trans. ASME, Journal of Dynamic Systems, Measurement, and Control, Vol. 113, 1991, pp. 388–394.

90. E. Bayo and J. R. Stubbe, "Six-axis force sensor evaluation and a new type of optimal frame truss design for robotic applications," Journal of Robotic Systems, Vol. 6, No. 2, 1989, pp. 191–208.

91. R. Little, "Force/Torque sensing in robotic manufacturing," Sensors, The Journal of Machine Perception, Vol. 9, No. 11, 1992.

92. M. Kaneko, "Twin-head six-axis force sensors," IEEE Transactions on Robotics and Automation, Vol. 12, No. 1, 1996, pp. 146–154.

93. M. Uchiyama, Y. Nakamura, and K. Hakomori, "Evaluation of robot force sensor structure using singular value decomposition," Journal of the Robotics Society of Japan, Vol. 5, No. 1, 1987, pp. 4–10.

Index